Springer Series in Optical Sciences

Volume 233

Springer Series in Optical Sciences is led by Editor-in-Chief William T. Rhodes, Florida Atlantic University, USA and provides an expanding selection of research monographs in all major areas of optics:

- lasers and quantum optics
- ultrafast phenomena
- optical spectroscopy techniques
- optoelectronics
- information optics
- applied laser technology
- industrial applications and
- other topics of contemporary interest.

With this broad coverage of topics the series is useful to research scientists and engineers who need up-to-date reference books.

More information about this series at http://www.springer.com/series/624

Andreas Heinrich
Editor

3D Printing of Optical Components

Springer

Editor
Andreas Heinrich
Center for Optical Technologies
Aalen University
Aalen, Germany

ISSN 0342-4111 ISSN 1556-1534 (electronic)
Springer Series in Optical Sciences
ISBN 978-3-030-58959-2 ISBN 978-3-030-58960-8 (eBook)
https://doi.org/10.1007/978-3-030-58960-8

This Springer imprint is published by the registered company Springer Nature Switzerland AG
The registered company address is: Gewerbestrasse 11, 6330 Cham, Switzerland

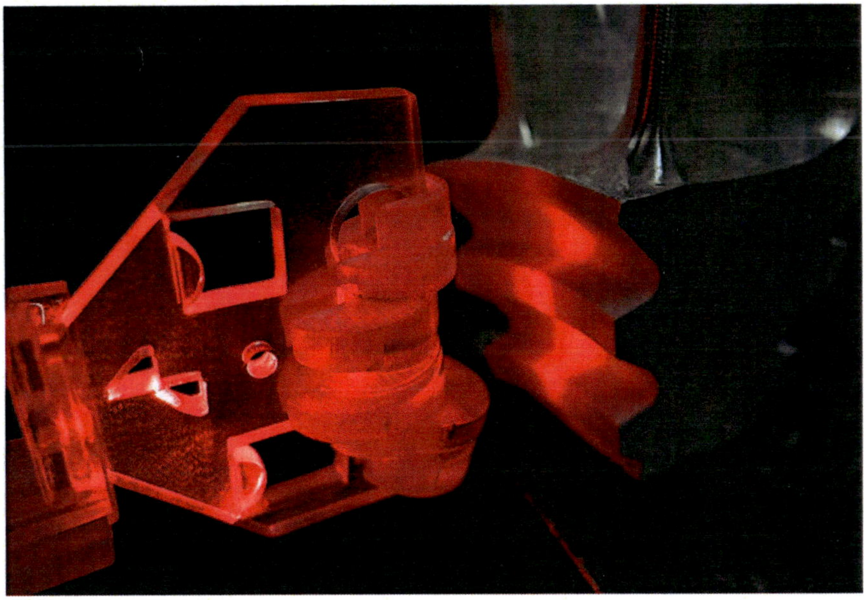

Preface

In recent years, the development of additive manufacturing methods has progressed rapidly. Much of the earlier work focused on realizing mechanical components. But additive manufacturing technology also holds great potential in the field of optics because it offers new degrees of design freedom, which allows completely new approaches to be explored.

High-precision two-photon polymerization is one example of how optical components can be manufactured additively. Among other things, this technique enables the production of microlenses with complex shapes. These microlenses are characterized by high optical quality and do not require post-processing after being manufactured. By contrast, realizing larger optical components or elements with very different material properties has proven to be a challenge. Conventional 3D-printing systems are used as an alternative. These systems can be used both to produce transmissive optics from glass or plastic and to realize reflective metallic objects. Such printing systems can be used to develop macroscopic optical elements; however, they typically achieve lower optical quality than two-photon polymerization.

The objective of this book is to present the current possibilities and characteristic properties of the additive manufacturing of optical components, as well as the current challenges and future prospects of this field. Additive manufacturing is shown to enable completely new solutions in optics, solutions that can be expected to become even more diverse in future.

Chapter 1 of this book introduces additive manufacturing, with a particular focus on conventional 3D-printing processes. The key concepts, typical materials, and various methods of additive manufacturing are described, and their potential applications are discussed.

The additive manufacturing of reflectors using the SLM process is presented in Chap. 2. For each application, the lighting requirements are presented and used to deduce the design parameters of the product. The light distribution produced by a macroreflector is simulated and validated as an example. The entire additive manufacturing process chain is also examined.

Chapter 3 examines the potential of 3D-printed polymer optics. The key focus of the chapter is a discussion of several completely different examples of additively manufactured optics to illustrate the potential and limitations of additive manufacturing in this area. The 3D printing of macroscopic optical elements such as light guides, liquid lenses, luminescent optics, random lasers, and mirror elements is discussed, as well as the inkjet printing of microscopic lenses. This chapter also examines the development of new additive manufacturing technologies, such as robot-based printing and the detuning of inkjet-printed lenses within an electric field.

Chapter 4 discusses the additive manufacturing of glass. Glass has shaped the optics and photonics like no other material. Once silicate glasses became available for 3D printing, two main approaches emerged: direct 3D printing of low-melting glasses at high temperatures and indirect glass printing of glass precursors using technologies borrowed from polymer 3D printing. This chapter discusses how precursors can be printed at room temperature then converted into transparent glass by a heat treatment process.

The high-precision 3D-printing technique of 3D lithography by two-photon or multi-photon absorption is discussed in Chap. 5. This area has developed significantly over the past two decades and opens up new possibilities in a wide variety of photonic applications. Chapter 5 describes the principles of this process, as well as the materials that can be used for it. It discusses how this method is not only able to realize refractive and diffractive optics, but also meta-optics extending from the sub-micrometer range to the millimeter range. This is demonstrated with printed optics intended for direct use as well as master models for replication.

Chapter 6 presents direct femtosecond laser writing for the manufacturing of micro-optical components and systems. This chapter primarily focuses on the design of such components. A selection of imaging and lighting optics are presented and discussed to demonstrate the potential of this manufacturing technology.

The quality of additively manufactured optics depends on the properties of the materials that are used. Accordingly, the final Chap. 7 discusses hybrid polymers. Hybrid polymers are a class of optical materials that combine the properties of inorganic glass and organic polymers. The properties of such polymers can be specifically adapted, which is desperately needed when printing micro-optical elements. This chapter therefore considers the chemical concepts of hybrid polymers, as well as their synthesis and processing. Applications of hybrid polymers to produce micro-optical and photonic elements using established water scale processes, inkjet methods, and direct laser writing with two-photon polymerization are also discussed.

The editor would like to thank all contributors to this book for their remarkable chapters. Special thanks to Dr. Sam Harrison, Editor at Springer, for his assistance with the creation of this book, and Mr. Murugesan Tamilsevan, Project Coordinator at Springer, for his skillful management of the production process.

Aalen, Germany Andreas Heinrich

Contents

Chapter 1
Introduction to Additive Manufacturing

Miranda Fateri and Andreas Gebhardt

Abstract Additive manufacturing (AM) works by creating objects layer by layer in a manner similar to a 2D printer with the "printed" layers stacked on top of each other. The layer-wise manufacturing nature of AM enables fabrication of freeform geometries which cannot be fabricated using conventional manufacturing methods as a one part. Depending on how each layer is created and bonded to the adjacent layers, different AM methods have been developed. In this chapter, the basic terms, common materials, and different methods of AM are described, and their potential applications are discussed.

Keywords Additive manufacturing · 3D printing · Digital manufacturing · Rapid prototyping · Rapid manufacturing · Rapid tooling · Freeform fabrication Materials · Manufacturing · Industry 4.0

1.1 Characteristics of Additive Manufacturing Processes

Additive manufacturing (AM)[1] is generally used (even if there are local designations in some countries) for directly making (manufacturing, production) of three-dimensional (3D) objects from digital data. The fabricated objects using AM represent prototypes or (final) products or parts of it. Commonly the term "three-dimensional printing" or "3D printing" is used, even as standardized term. In addition, there are many proprietary and industry-driven terms in use, some of which are used even generically.

AM complements today's dominating subtractive and formative manufacturing methods and hence is called the third pillar of the entire manufacturing technology.

[1] According to ISO/ASTM52900.

M. Fateri (✉)
Hochschule Aalen, Aalen, Germany
e-mail: Miranda.Fateri@hs-aalen.de

A. Gebhardt
FH Aachen, University of Applied Sciences, Aachen, Germany

A. Heinrich (ed.), *3D Printing of Optical Components*, Springer Series in Optical Sciences 233, https://doi.org/10.1007/978-3-030-58960-8_1

Subtractive methods use semifinished products such as prefabricated sheets, straps, or tubes. They show defined properties that do not depend on the impact direction, which is called isotropic behavior. The desired accuracy of the final parts is achieved by removing material, e.g., turning, milling, or similar processes, using machine tools and their embedded control. Subtractive manufacturing is suitable for all kinds of production such as series production from high-volume manufacturing down to "one-off fabrication."

Formative methods apply heat to initially transform the raw material into a pasty or fused state. Subsequently, solid molds are used to obtain the desired shape (e.g., casting). Formative methods usually require higher amount of liquid material than the final part represents. The costly tool and the reusable material are the reasons why formative methods are preferably used for high-volume production.

Additive methods generate the material as incremental volume elements and arrange it according to the desired 3D contour. Subsequently, additively fabricated parts are made by staggering incremental volume elements. This is done side by side (following the contour of a part in the *x*-*y* area) and on top of another to verify the 3D contour (*z*-level). This also is called sandwich method.[2]

Based on these, the build material and its 3D shape are created simultaneously which implies that the properties of the material are defined within the manufacturing (printing) process. AM methods do not require part-related tools.

Additive manufacturing process chain: Starting point for all manufacturing processes is a complete 3D description of the product. It mainly comprises geometrical information but increasingly additional information concerning part properties and production details. It is mainly based on a 3D computer-aided design (CAD) system and hence is done in the cyberspace. The final result is a 3D data set.

Further, and still in the virtual space, the part CAD models are sliced in the build-up (z) direction into equidistant layers of which the final parts will be composed. The layer thickness is determined mainly by final parts desired accuracy, the overall dimension, and the build material.

Usually the layer thickness is order of tenth of a millimeter. The approximation of the real and the virtual geometry fits better by choosing a smaller layer thickness. On the other hand, the layer thickness is limited by the properties of the input material such as surface tension of the liquid build material or powder particle size, if sintering is used.

The contouring of each layer and the shape of the entire part in *x*-*y* direction, respectively, is defined by triangulation using a so-called STL (Standard Triangulation Language, sometimes also called stereolithography format) data set. As a result, on its surface, the part shows stair stepping in *z*-direction and facets in the *x*-*y* plane.

The transition from the virtual to the physical model takes place within the AM machine, the 3D printer. As indicated before, manufacturing is done layer by layer

---

[2] Today layered manufacturing is the only industrially used AM method, but it is not the only imaginable one. (Compare to Gebhardt, fifth edition: Sonoluminescence, Elastoviscosity, Ballistic Particle Imaging, Stratified Object Manufacturing).

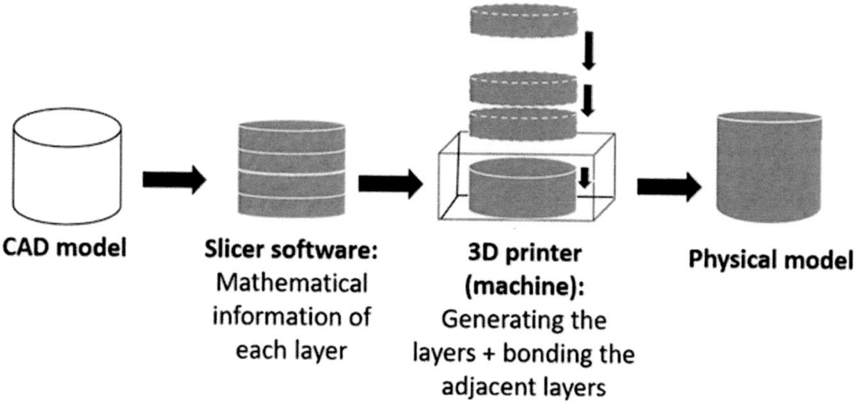

Fig. 1.1 Additive manufacturing process chain including the executed steps

(see Fig. 1.1). This discussion leads to the definition of "Additive manufacturing": *"Additive manufacturing" defines the automated layered manufacturing of arbitrary contoured 3D parts from its defining 3D data. AM therefore is the key to direct digital production in the sense of the "Industry 4.0" strategy.*

In the beginning, when AM was called "rapid prototyping," it solely was about geometry. Therefore, the approximation of the continuous contour by simple triangulation was sufficient.

Today and increasingly in the future, besides geometry other properties such as color, elasticity, and transparency are designated to the incremental elements and must be integrated into the data set accordingly. Therefore, new powerful data structures like AMF or 3MF are developed and integrated into the software suites.

The principle of AM (the layered technology) can be applied to various different material classes such as plastics, metals, or ceramics. Therefore, the appropriate working principle must be implemented by an automated machine (3D printer). Usually, this includes two simultaneously executed process steps:

- Creating a contoured physical layer according to the sliced virtual one
- Bonding it to the top of the proceeding one

As pointed out, the quintessence of the AM technology is to find a material combination and a fitting mechanism that simultaneously supports the local solidification and the desired contouring according to a 3D data set of the part.

1.2 Additive Manufacturing Processes

The technical execution of AM processes is carried out by means of direct layer manufacturing process. The developed technologies for AM in the market can be categorized into five main processes. Each of these processes uses different methodologies in order to form a solid layer and fabricate the final part by connecting adjacent layers. All five process families are introduced in this chapter.

1.2.1 *Stereolithography (SLA)*

Stereolithography (SLA) is the oldest and until today the most accurate AM process within these five families. SLA was invented by Chuck Hull in 1984 and patented in 1986. It was commercialized by his company 3D Systems, Rock Hill, SC, USA. SLA parts are generated by polymerization of a liquid monomer. With the help of a Ultraviolet (UV) laser beam, the liquid resin is transferred into a solid state through polymerization. Thus, corresponding to the part's contour defined by the 3D CAD file, scaled solid layers are formed in a layer-wise manner. According to the layer-wise structure, the part shows stair steps on its outer contour.

A laser-based stereolithography machine consists of a build chamber, filled with liquid resin (build material), and a laser scanner unit, arranged on top of it, which generates the contour in *x*-*y* direction (build area). The build space contains a movable build platform, which can be lowered in build direction (*z*-direction).

The part and the additional support structures (which are needed for the parts fixation and to prevent distortion) are built on a platform which is connected to a leveling system mounted in a resin bath. The laser beam generates the contour simultaneously providing the curing of each layer and its connection to the previous one. The movement of the laser beam is directed by a laser scanner according to the contour data of each layer. As soon as the laser beam penetrates the surface of the resin, an instantaneous solidification by polymerization is done. Depending on the reactivity and transparency of the resin, the layer thickness can be adjusted by the laser power and the traveling speed of the laser beam.

After solidification of each layer, the build platform, including the partially generated part, is lowered by the amount of one layer thickness. Subsequently, the next layer of resin is applied (recoating). Due to the viscosity of the resin, recoating is supported by a leveling system, e.g., viper of 3D Systems. It should be noted that the build platform is constructed in a mesh format, in order to avoid a local vacuum and ease the resin flow and layer deposition. The new layer is then solidified according to its contour using a UV laser. The procedure is continued from bottom to top until the part is completed. Subsequently, the part is cleaned and undergoes a post-curing step in a UV oven. Post-curing parameters of the manufactured parts such as curing time and energy vary based on the print material and print volume.

The fully cured parts can be sand blasted, polished, and/or varnished, if required. SLA fabrication process steps and a SLA fabricated sample are shown in Figs. 1.2 and 1.3, respectively.

Using this methodology, completely closed hollow geometries cannot be built directly; therefore, resin evacuation through a milled hole is needed as an extra step.

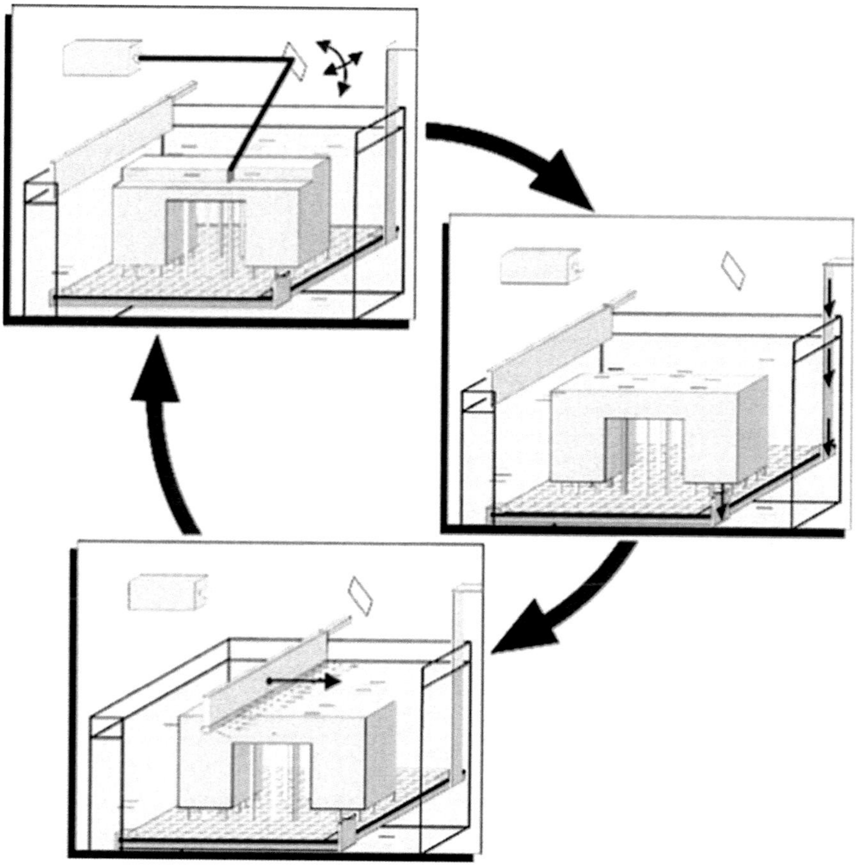

Fig. 1.2 Laser stereolithography process steps

A recent developed technology by Formlabs builds the part upside down, in which the build platform moves against gravity. Doing so, after emerging the build part from the resin, the resin pours out of the part.

Besides the laser-based SLA, other AM techniques such as Digital Light Processing (DLP) and Polymer Jetting (Poly-Jet Modeling (PJM)) use the polymerization phenomenon by the help of a UV lamp.

In DLP, a projector is used in order to project the images of each sliced layer for curing the resin by a UV lamp. In this case, whole layer is generated in one step with no need for scanning components.

In Polymer Jetting, build and support materials are dispensed simultaneously with a help of a 2D print head equipped with nozzles. Injected materials are cured by the help of UV lamps which are attached on the sides of the print head. Depending on the type of the nozzles, this technique enables manufacturing of multimaterials in different colors.

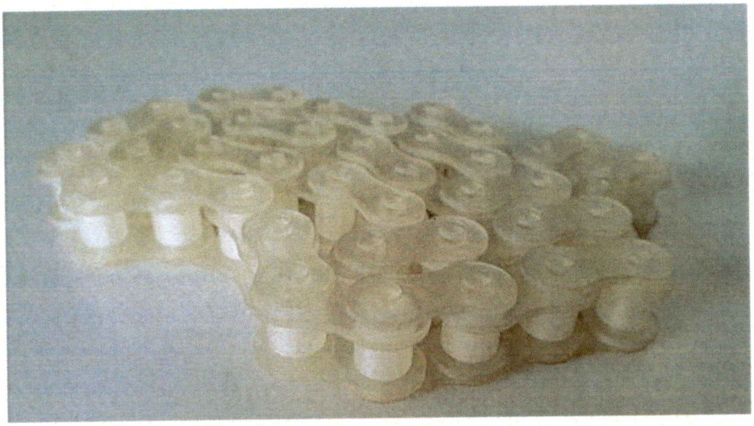

Fig. 1.3 Fabricated sample using stereolithography process (source: CP GmbH)

1.2.1.1 Pros and Cons

Polymerization results in very good surface quality and fine details. But it uses support structures that need to be removed, and the parts mostly need to undergo the post-curing. The plastic material and its comparably low glass transition temperature must be regarded as cons if tooling applications are needed.

1.2.1.2 Machines and Manufacturers

Laser stereolithography machines (Viper SLA and iPro SLA Series) and polymer printers (ProJet Series) are developed by 3D Systems. Polymer Jetting machines (Eden- and Connex Series) are offered by Objet. Machines using DLP projectors (Perfactory Series) come from EnvsionTec.

1.2.2 Selective Laser Sintering (SLS)/Selective Laser Melting (SLM)/Laser Powder Bed Fusion (LPBF)

The terms selective laser sintering (SLS), selective laser melting (SLM), and laser powder bed fusion (LPBF) use laser energy in order to partially or completely fuse the powder particles together. SLS was initially developed at the University of Texas in the mid-1980s.

The SLS/SLM/LPBF machines consist of a build space, filled with powder of a grain size of approximately 20–50 μm. A laser scanner on top generates the *x*-*y* contour in which each layer is outlined and particles are locally sintered or molten by a Nd:YAG or CO_2 laser beam and subsequently solidified due to the heat loss. Thus, a solid layer is generated. The bottom of the build space is designed as a movable piston, which can be adjusted to any z-elevation. After the solidification of a layer, this piston is lowered by the amount of one layer thickness, which also means that the complete powder bed and the growing part is lowered simultaneously. The free space above the powder bed is refilled with fresh powder which is taken from the adjacent powder reservoir and flattened by means of a powder recoater and/or a roller.

After the recoating the piston is lowered by the amount of one layer thickness, and the build process starts again generating the next layer. The complete process is continued layer by layer until the part is finalized (see Fig. 1.4). It should be noted that metallic parts have to be manufactured under inert gas atmosphere.

After the fabrication process is completed, the complete part including the surrounding powder is driven out of the build volume. The removal of the part from the powder (the so-called break out) is done manually by using a vacuum cleaner and brushing. In SLM/LPBF, metal parts are fixed with the help of support structures mainly in order to avoid the part's deformation. Removing the support structures is part of the post-processing which is mostly a manual and time-consuming step. An example of manufactured part with support structures, after support removal and polishing, is shown in Fig. 1.5.

Due to the nature of this process, SLS/SLM/LPBF allows processing of a wide range of materials such as metals (see Figs. 1.5 and 1.6), glasses (see Fig. 1.7), and ceramics.

Plastics are usually limited to only SLS due to their high rate of shrinkage. SLS of plastics do not require support structures as during sintering they are supported by the surrounding powder.

The recoating systems are specially adapted to handle the different materials, e.g., roller systems as well as hoppers for plastics processing and preferably wiper-type devices for metal processing systems. The leftover powders can be reused for SLS/SLM/LPBF application after going through the sieving process.

In contrast to SLM/LPBF fabricated metallic parts, SLS fabricated plastic parts are porous and need to be infiltrated. Final parts can undergo surface treatments such as manual polishing and varnishing or coloring. Fabricated samples using SLS of plastic are shown in Fig. 1.8, respectively.

Laser-based cladding, using a single nozzle system and a powder feeder, is a variation of sintering or melting in the powder bed. Herewith, the powder is applied directly into the laser beam and molten simultaneously. This process is used rather for repair than for manufacturing a whole part. The process is also suitable when higher layer thicknesses are applied. Applications are functional layers, e.g., on valve seats and local improvement or repair of brake disks and big tools.

1.2.2.1 Pros and Cons

Sintering processes can handle all materials that behave like thermoplastics which means they melt and solidify if heated and cooled down. Additionally, they must be available as powders. Sintering and melting enables making interior hollow

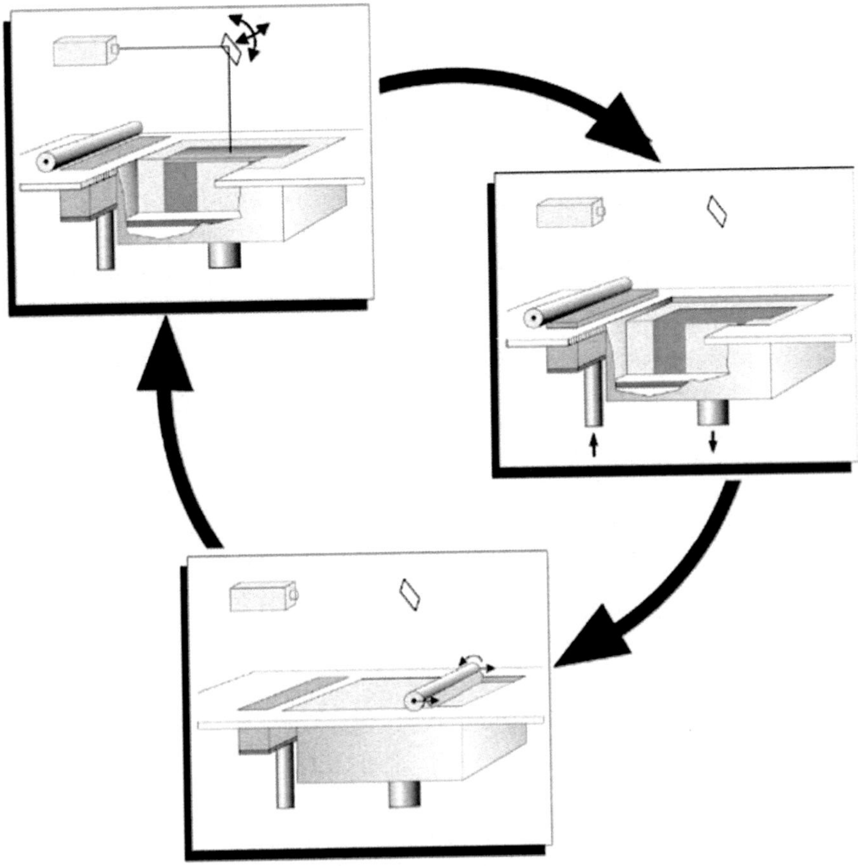

Fig. 1.4 Selective laser sintering/melting process steps

geometries. This opens up the possibility to make hollow parts with internal ducts to realize conformal cooling strategies.

The comparably rough surface structure is the biggest disadvantage of the process. Especially, if metal parts are concerned, support structure removal is an additional step. Also, smoothening (polishing) the surface is laborious and additionally may affect its accuracy.

1.2.2.2 Machines and Manufacturers

Laser sintering machines are offered by 3D Systems (Sinterstation sPro SLS Series, Sinterstation HiQ SLS, Sinterstation Pro SLM Series) and EOS (EOSINT P, M, S). Concept Laser (M1 cusing, M2 cusing, M3 linear) and MTT Group (MTT SLM 125, 250, 250 HL) name their products melting machines. Generating with a powder nozzle is used by Optomec (LENS 705, 800R) and POM (DMD 105D, DMD

Fig. 1.5 Fabricated parts using selective laser melting: metallic part with support structures (left); part after removing the support structure (middle) and after polishing (right) (source: FH Aachen)

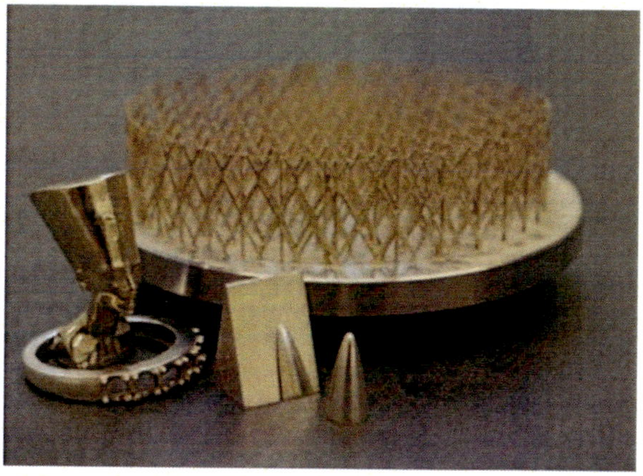

Fig. 1.6 Fabricated parts using selective laser melting of silver (source: FH Aachen)

44R, 66R). Low-cost variants are offered by LMI GmbH, Desktop Metal, und Trumpf.

1.2.3 *Fused Layer Modeling (FLM), Commercially: Fused Deposition Modeling (FDM)*

Fused deposition modeling (FDM) is registered as the protected brand name for a fused layer modeling (FLM) process, offered by Stratasys company, Eden Prairie, MN, USA.

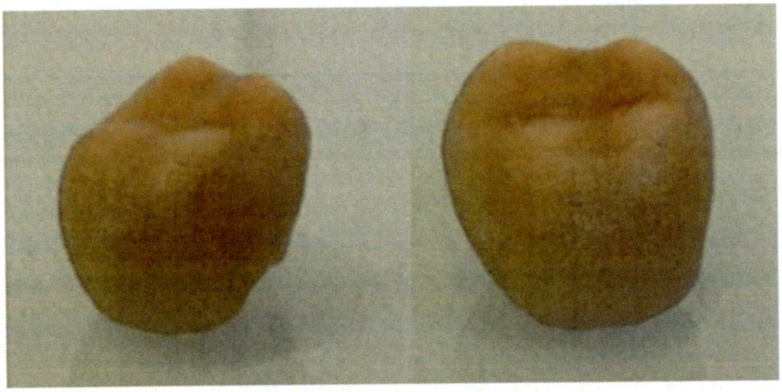

Fig. 1.7 Fabricated parts using selective laser melting of soda-lime glass [1]. (source: FH Aachen)

Fig. 1.8 Fabricated parts using selective laser sintering (SLS) of plastic. (source: EOS GmbH)

An FDM machine consists of a sealed and heated build space, equipped with an extrusion head and a build platform.

In FDM, the build material exists a prefabricated, thin plastic filament of generally 1.75 or 2.85 mm which is continuously fed into an extrusion head. The extrusion head (nozzle) heats up the feed material over its softening temperature by means of an electrical heater. The softened material is then deposited atop a platform and creates the profile in *x*- and *y*-directions.

The platform is subsequently lowered in *z*-direction by a predefined layer thickness or layer height of typically 0.1–0.3 mm. Afterwards, the nozzle starts extruding the next layer atop of the pre-deposited profile, and then the build platform is lowered again. This process continues until the part is finalized (see Fig. 1.9).

In order to fabricate specific freeform geometries, the build process needs support structures. These supports are generated by an additional nozzle, which simultaneously extrudes the support material that differs from the build material.

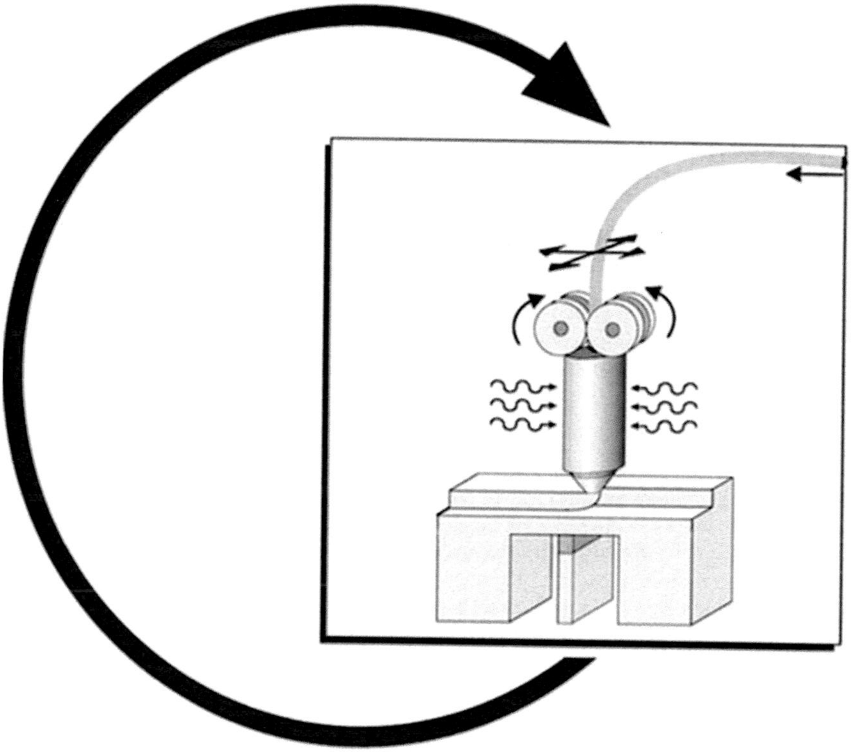

Fig. 1.9 Fused layer modeling process principle

The support materials could be washed away using a corresponding solvent or broken away manually. In order to dissolve the support material, parts are usually left in an ultrasonic bath filled with the specific solvent. Some support structures require to be left in a heated solvent. After supports removal step, parts could undergo the post-processing which includes surface finishing. This step could to be carried out by manual sanding and polishing, which is a time-consuming process. Alternatively, parts could be also polished by being exposed to their corresponding solvents which merges the print layer surface together. Parts polishing using their solvents could be done by a liquid or vapor bath process. Common plastic materials for FDM applications are acrylonitrile butadiene styrene (ABS), polylactide (PLA), and polyvinyl alcohol (PVA). Their corresponding solvents are acetone, tetrahydrofuran, and water, respectively.

In FDM, final products might show deflection and distortions. This is mostly caused by the shrinkage due to the temperature gradients between the freshly deposited softened material on top of already solidified layers. These deflections might also lead to parts detachment from the build platform. Towards these issues, the build platform is usually preheated. Ideally, the entire build chamber volume needs

Fig. 1.10 Fabricated sample using fused layer manufacturing process (source: FH Aachen)

to be sealed and heated up in order to minimize the temperature gradients during the print process. Available commercial machines in the market offer heated up chambers up to 80 °C. The build platforms could be also coated by polyvinyl acetate (PVA/PVAc) in order to provide a higher adhesion between the build material and the build platform. Doing all this, the process can be handled properly and result in usable parts.

Characteristic properties of FDM parts resemble those of injection molded parts. They exhibit anisotropic behavior which can be reduced by adapted design and adjustment of suitable build parameters. The manufactured parts are used as concept models, functional prototypes, or (directly manufactured) products. A sample of manufactured part using FLM/FDM process is shown in Fig. 1.10.

1.2.3.1 Pros and Cons

The process and the machines are easy to run even in an office environment. The process is mostly limited to plastics. The parts can be used within the product development process. The surface shows a texture due to the extruded strings that changes according to the orientation of the part within the build space. If used as master model for follow-up processes, the surface has to be post-processed intensively. The process requires supports.

1.2.3.2 Machines and Manufacturers

There are hundreds of inexpensive machines available even in hardware stores that need a big amount or manual adjustment. Stratasys (USA) offers a wide range of different professional machines following the FDM extrusion principle. It covers build spaces with different capacities that allow to process different materials (3D Production Systems, Fortus 360mc, 400mc, 900mc,). The Dimension series offers moderately priced so-called 3D printers (1200es, Elite, uPrint).

1.2.4 Powder-Binder Bonding (3DP)

The layer-wise bonding of powders by selective injection of liquid binders into the surface of the powder bed is called 3D printing (3DP). Sometimes it is named "powder-binder process," and referring to the entire process, it also is named "drop-on-powder" process. The process was developed and protected in the early 1990s by the Massachusetts Institute of Technology (MIT), Boston, USA, in the 1990s. Licenses have been granted to Z Corporation (now 3D Systems) and others who have marketed the process.

In the process, powder is spread atop of the build platform using a roller. The lower part of the machine contains the build chamber and stores the powder, having a movable piston for the adjustment of the layer. On top, a plotter-type device with a commercial print head is arranged, which is similar to a 2D office printer. The print head travels across the build area and injects the binder according to the actual contour onto the powder. The layer-forming particles are bonded by the binder and become one layer of the part. The loose powder supports the part. As opposed to sintering, this process requires neither heating nor shielding gas.

After solidification of the layer, the piston including the complete powder cake and the incomplete part is lowered by the amount of the layer thickness. The emerging space is refilled with fresh powder, taken from the adjacent powder storage by using the roller (recoating). After finalization of the last layer, the build process is completed (see Fig. 1.11). The loose powder is removed by vacuum cleaner, soft brushing, and blowing with air at low pressure.

In a next step this so-called "green part" part needs to be infiltrated by wax or epoxy resin to improve its strength. The result with respect to stability depends not only on the material but also on the quality of the infiltration. Frequently, bubbles are enclosed. Therefore, parts manufactured by 3D printing are not meaningful in stress tests.

Today, various materials such as plastics, metals, or ceramics are available for this process. In case of metal, binder is used and has to be removed by thermal treatment and by subsequent sintering to achieve the final mechanical properties.

As suitable binders are to be found for each kind of powder, the range of materials is nearly unlimited, including applications in the fields of food and pharmaceutical products—though only a small segment currently is commercialized.

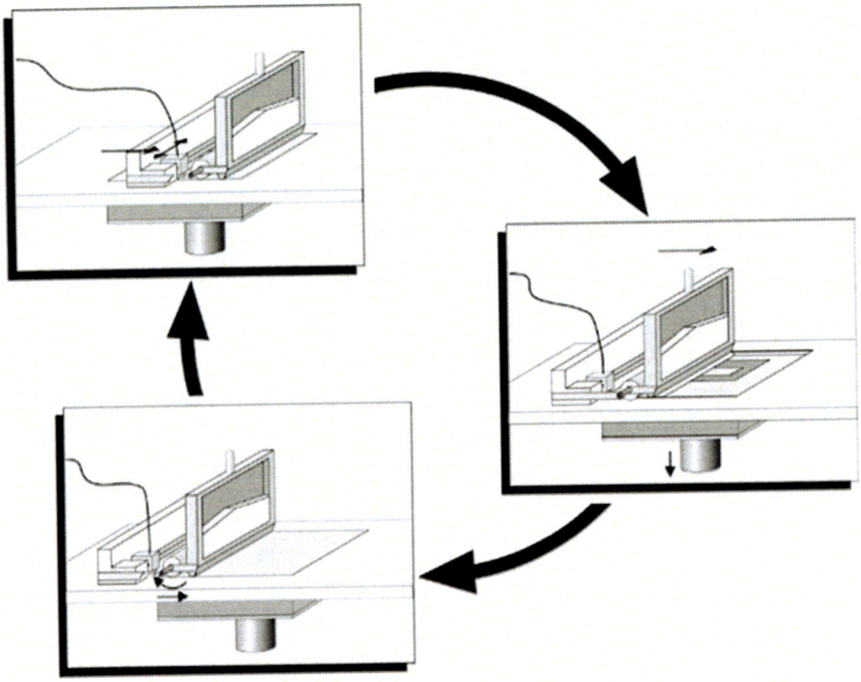

Fig. 1.11 Powder-binder bonding process steps

"Three-dimensional printing" or "3D printing" is becoming the synonymous term for all AM processes. This is because 3D printing is easily understandable by the general public familiar with conventional paper printers.

But, the use of two identical terms with different meaning sometimes causes confusion. Therefore, especially beginners should avoid mixing the two meanings of the term "3D printing." A sample of manufactured part using powder-binder bonding (3DP) process is shown in Fig. 1.12.

1.2.4.1 Pros and Cons

3D printing processes can handle all materials that are available as powders. No supports are required. Metal processes are multistep procedures that require a subsequent post-sintering process and an infiltration with bronze or another low melting point metal. The process works at room temperature, thus avoiding thermal distortion. The comparably rough surface quality is a disadvantage of the process. Especially metal processes require an intensive post-processing which additionally affects the accuracy.

Fig. 1.12 Fabricated sample using powder-binder bonding process (source: FH Aachen)

1.2.4.2 Machines and Manufacturers

The majority of the AM manufacturers offer 3D printing machines for making plastic parts, for example, Z Corp., USA (ZPrinter 310plus, 350, 450, 650), or voxeljet, D (VX800, VX500).

Prometal developed a family of metal machines, called *Direct Metal Printers* (R1 und R2). A so-called densification furnace to drive out the binder and to support infiltration is available as an extra as well as a cleaning station.

1.2.5 Layer Laminate Manufacturing (LLM)/Selective Deposition Lamination (SDL)

Cutting contours out of prefabricated foils or sheets of constant thickness according to the 3D CAD data set and bonding it to the surface of the previous layer is known as Layer Laminate Manufacturing (LLM) (see Fig. 1.13).

The foils or sheets may consist of paper, plastic, and metal materials. As cutting tools, laser, knifes, or mills can be used. Bonding of consecutive layers is done by gluing, ultrasound, soldering or diffusion welding. The majority of the processes

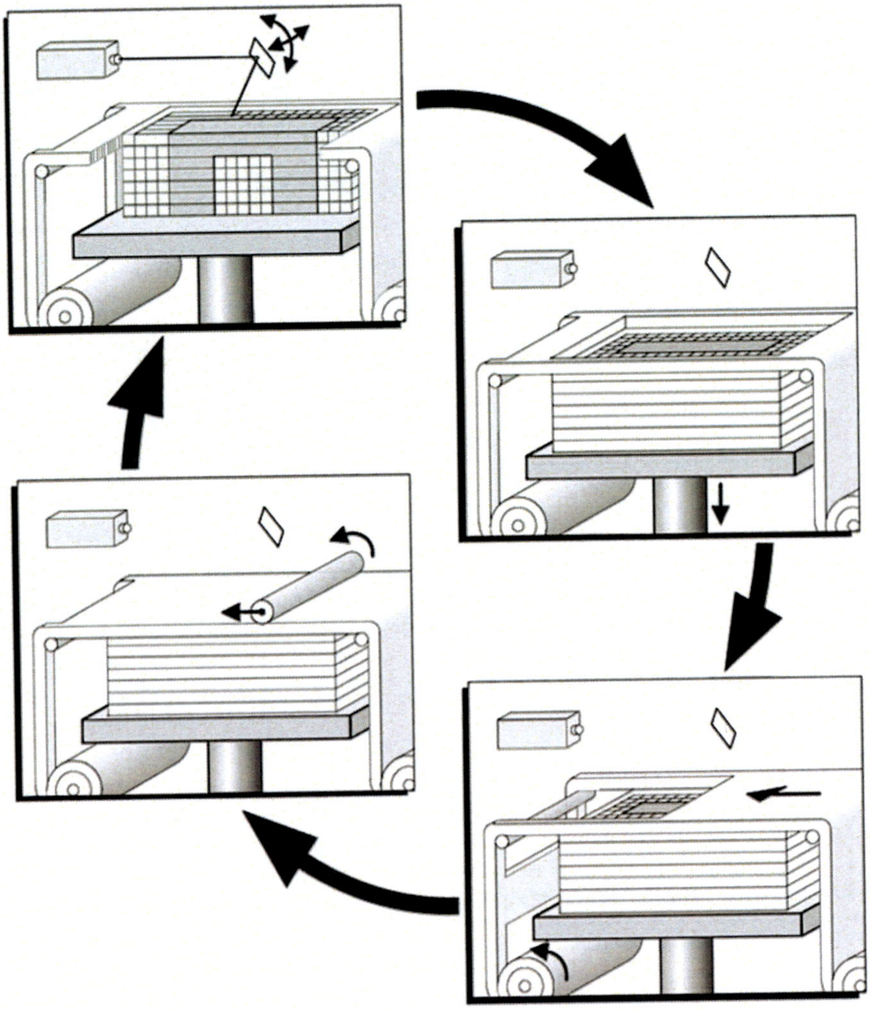

Fig. 1.13 Layer laminate manufacturing process steps

only need one manufacturing step; a few require post-processing such as compacting by sintering in an annealing furnace.

The LLM process was named laminated object manufacturing (LOM) when introduced in 1991 and frequently still is. In LLM process, the surrounding areas of the desired contours have to be cut layer-wise in order to enable the extraction of the print geometries. Thus, in LLM, relatively high amount of waste material is produced as depicted in Fig. 1.14. As such, use of LLM for commercial purposes became limited to use of cheap build materials such as paper. In this context, the LLM process has been modified further for commercial purposes by MCOR.

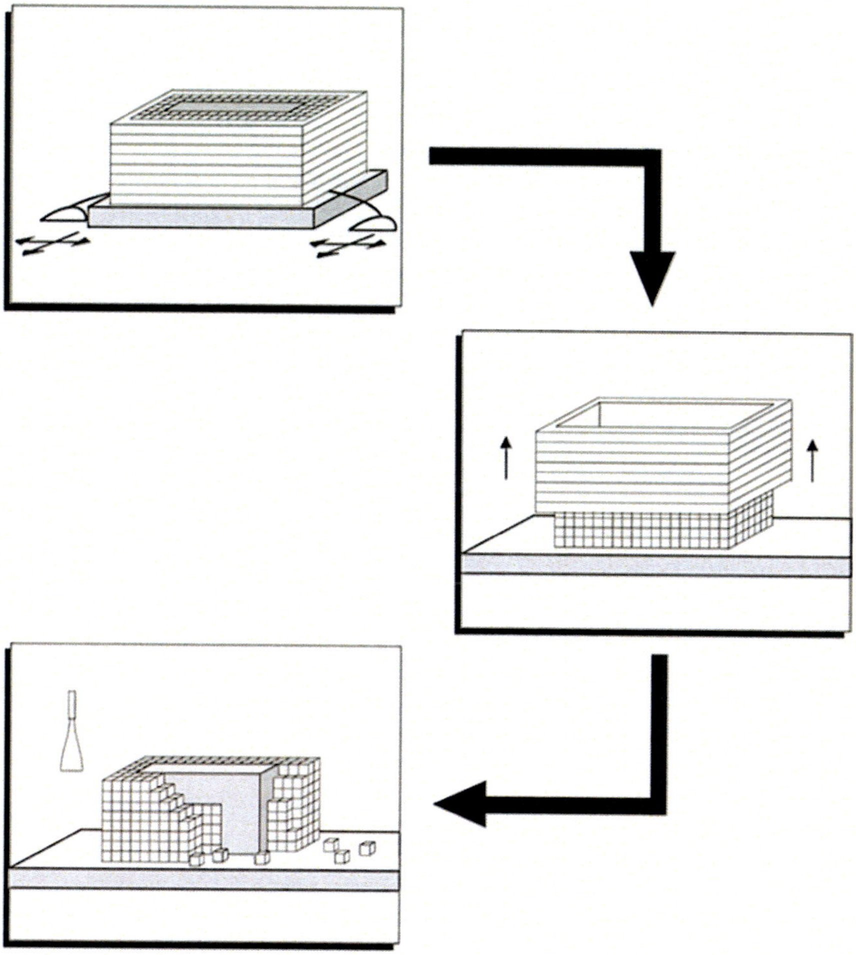

Fig. 1.14 Layer Laminate Manufacturing post-processing steps

Manufactured samples using Layer Laminate Manufacturing (LLM) process are shown in Fig. 1.15.

The current process (by MCOR), called Selective Deposition Lamination (SDL) (see Fig. 1.16), uses (standard A4) paper instead of special coated paper on rolls. It is contoured by cutting plotter using a blade instead of a laser. The sheets are bonded using standard glue (PVA) by a proprietary procedure of Mcor. Fully colored parts can be built by an advanced printer with and integrated 2D color printer. A sample of manufactured part using Selective Deposition Lamination (SDL) process is shown in Fig. 1.17.

Fig. 1.15 Fabricated samples using Layer Laminate Manufacturing (source: CP GmbH)

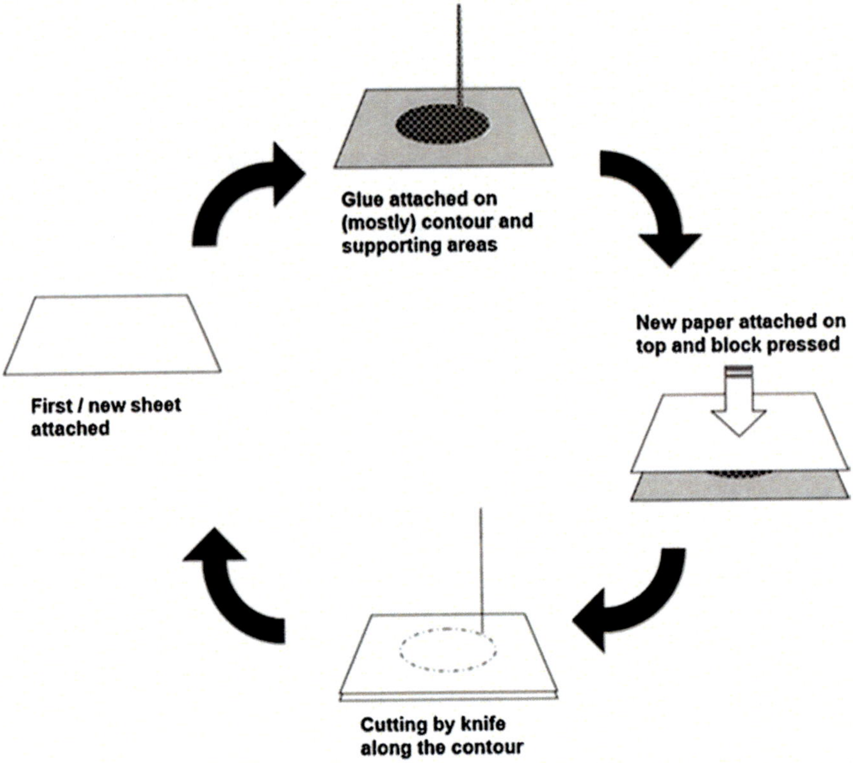

Fig. 1.16 Selective Deposition Lamination process steps

Fig. 1.17 Fabricated sample using Selective Deposition Lamination (source: MCor)

1.2.5.1 Pros and Cons

Layer laminate processes can process all types of materials that can be processed thermally or mechanically. The big amount of waste compensates this favor at least partly and forces the designer to optimize the part's shape and orientation. The advantage of LLM processes in general is fast manufacturing of massive parts.

1.2.5.2 Machines and Manufacturers

Cubic Technologies (LOM 1015plus, 2050H); MCOR Technologies, Kira (PLT A3, A4, Katana); and 3D Systems (LD 3D Printer) offer machines for making parts from paper or plastics.

Milling machines that process wood-like model materials are developed by Zimmermann (LMP), while software solutions (and hardware recommendations if wanted) are offered by Stratoconception/Charlyrobot (rp2i).

Solidica (MI; USA) introduced a hybrid milling and ultrasonic welding machine that makes dense aluminum parts based on their so-called Ultrasonic Consolidation process.

1.3 Processing Materials

Using AM a wide range of materials can be processed. Plastics may be unfilled or filled with glass or aluminum spheres or egg-shaped geometries.

Unfilled plastics are mostly commodities like semicrystalline polyamides from the PA11 or PA12 type or amorphous plastics such as polystyrene (PS). Further engineering plastics such as PP, ABS, or PEEK have been introduced and are close to industrial use. Flame retardant versions are available.

Table 1.1 Suitable processing materials for different additive manufacturing methods

AM Process	Material				
	Paper	Plastic	Form sand	Metal	Ceramic
Stereolithography (SLA)		X			X
Selective Laser Sintering (SLS)		X	(X)	(X)	(X)
Selective Laser Melting (SLM)				X	
Fused Layer Modeling (FLM)		X		(X)	
Poly-Jet Modeling (PJM)		X			
Powder-binder bond (3D printing)		X	X	X	X
Layer Laminated Manufacturing (LLM)	X	X		X	
Digital Light Processing (DLP)		X			X

X common material, *(X)* uncommon, but possible

New developments use nanoscale particles to improve properties such as durability and thermal stability.

Ceramics of nearly any type can be processed. Parts or molds can be made from Al_2O_3, SiO_2, ZrO_2, and SiC. Completely sintered parts are made from Si_3N_4. Controlled or graded microstructures, for example, zirconia toughened alumina (ZTA), can be made by depositing a ZrO_2 slurry onto an Al_2O_3 substrate.

Foundry sands can be sintered directly like plastics as they are prefabricated with a polymer coating. For off-Earth manufacturing proposes, sintering of lunar and Martian regolith (sand) has been investigated [2, 3].

For AM of glass components, quartz and soda-lime glass powders have been studied [4, 5].

Different one- or multicomponent metal powders are available. Available powders are made from steel, mild steel, tool steel, CoCr steel, titanium, and aluminum. The particle size varies from 20 to 50 μm. Small particles require a careful handling, as they may be respirable and explosive.

Reactive powders such as aluminum and titanium require a completely sealed build space and a closed material handling system flooded with shielding gas. Suitable processing materials for each AM method are listed in Table 1.1.

1.4 Characteristics of Additive Manufactured Parts

One of the AM advantages is that parts can be arbitrarily complex. As such, undercuts or interior hollow structures can be manufactured. When AM is applied for producing such parts, term of rapid tooling (RT) could be used.

RT was created in the early 1990s. At this time rapid prototyping (RP) was the umbrella term (or generic name) for all processes. RP solely was used to make samples or prototypes, both being positives. When the first negatives were made, mainly by inverting the data, they were used as cavities, dies, and molds or mold inserts.

In RT, mostly undercuts are not the problem for tools because they must be demoldable, but hollow interior structures are the prerequisite of new strategies to shorten the cycle time such as conformal cooling.

The best suited processes for RT are laser sintering and laser melting which avoid waste by using only a small amount of additional material than needed for the intrinsic part. Prominent geometries such as isolated domes, which cause a lot of waste chips if made by milling, can be made in a cost- and material-effective manner using AM/RT (see Fig. 1.18). RT can be applied for a wide range of metallic materials.

With regards to RT of transparent objects, LightFab company produces complex, transparent glass components by using selective laser-induced etching (SLE) process (see Fig. 1.19).

Fig. 1.18 Geometry complexity demonstrated by an interior hollow structure (source: FH Aachen)

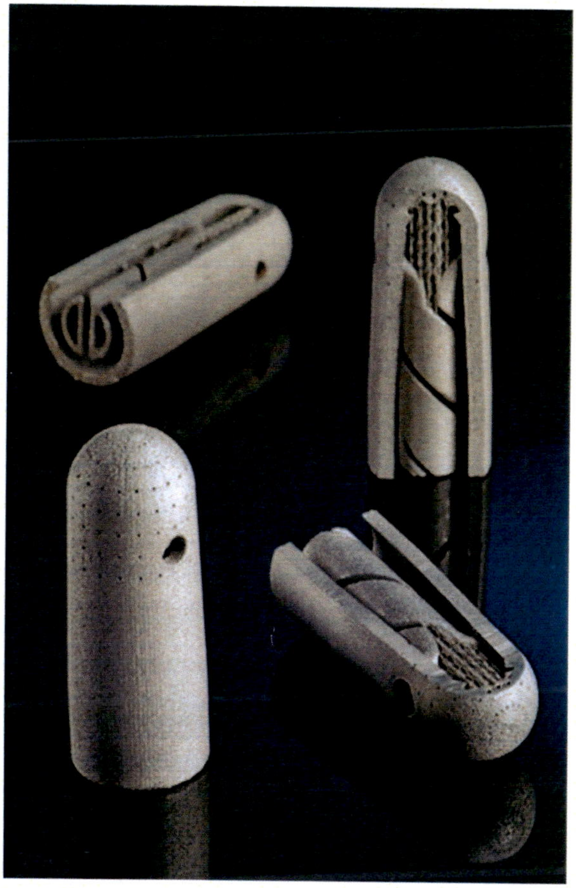

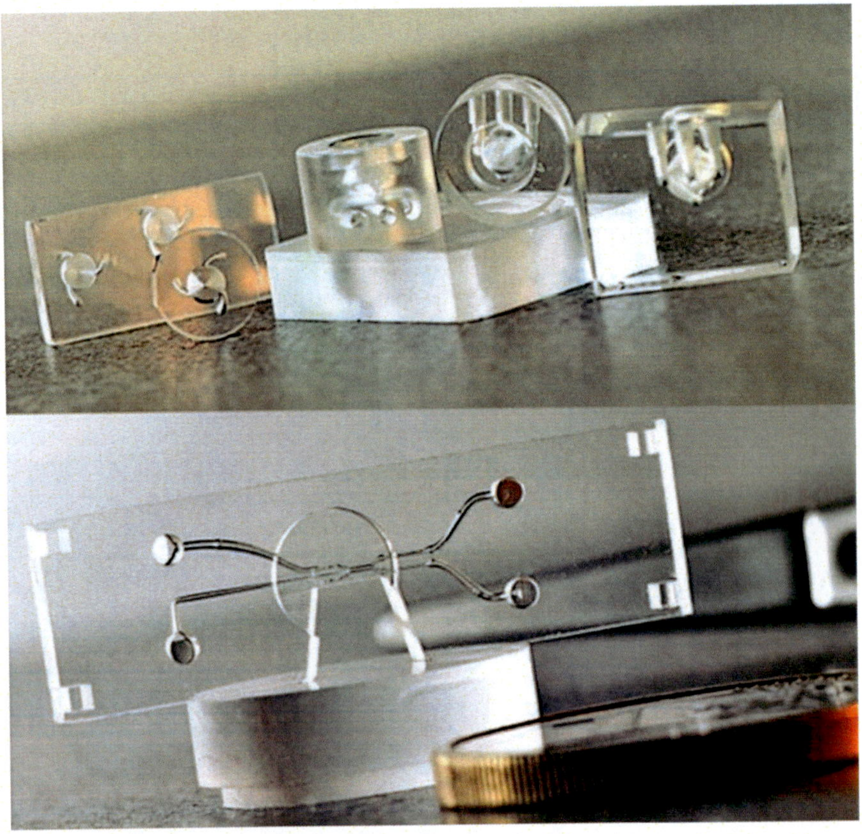

Fig. 1.19 3D nozzles in quartz glass (top); 3D microfluidics inside fused silica for cell sorting (bottom) fabricated using laser-induced etching process (source: LightFab)

References

1. Fateri, M., Gebhardt, A., Thuemmler, S., & Thurn, L. (2014). Experimental investigation on selective laser melting of glass. *Physics Procedia, 56*, 357–364.
2. Fateri, M., & Gebhardt, A. (2015). "Process parameters development of selective laser melting of lunar regolith for on-site manufacturing", applications. *International Journal of Applied Ceramic Technology, 12*, 46–52.
3. Fateri, M., Kaouk, A., Cowley, A., Siarov, S., Vera Palou, M., Gobartt González, F., Marchant, R., Cristoforetti, S., & Sperl, M. (2018). Feasibility study on Additive manufacturing of recyclable objects for space applications. *Journal of Additive Manufacturing, 24*, 400–404.
4. Khmyrov, R. S., Grigoriev, S. N., Okunkova, A. A., & Gusarov, A. V. (2014). On the possibility of selective laser melting of quartz glass. *Physics Procedia, 56*, 345–356.
5. Fateri, M., & Gebhardt, A. (2014). Selective laser melting of soda-lime glass powder. *International Journal of Applied Ceramic Technology, 12*, 53–61.

Chapter 2
Selective Laser Melting of Reflective Optics

Georg Leuteritz, Marcel Philipp Held, and Roland Lachmayer

Abstract Reflective optics are tremendously useful optical components due to their negligible chromatic aberrations when guiding light or forming light distributions. Using Additive manufacturing to realize freeform reflectors with additional functionality increases the potential of these components for highly efficient optical systems. In order to maintain the optical function of such reflectors, the relation between process parameters and part function needs to be accessible. This work describes a method to identify such relations, and the resulting efficiency is simulated and validated with additively manufactured samples.

Keywords Additive manufacturing · Selective laser melting · Laser polishing Reflective optics · Total integrated scattering · RMS roughness · Reflectivity Functional integration · Design parameters · Design limitations

2.1 Adjusting Optics Manufacturing

Optical technologies play a key role in the economy of the twenty-first century. Due to their enormous field of application, for example, in medicine, data transmission, metrology, or material processing, the production of optical components, such as lenses or waveguides, and systems is also being technologically pushed forward. Here, precision during production is particularly important. The standard shapes of lenses are first pressed, and then the surface is reworked in such a way that average surface roughness is achieved in the subnanometer range in order to achieve corresponding imaging properties. However, this post-processing includes, among other things, polishing steps which are often carried out manually and are therefore limited to the expertise of the respective processor. In addition, the standard shapes of the optical components offer no further scope for geometric adjustments or freeform geometries, which can be advantageous for many applications.

Therefore, miniaturized and function-integrated freeform optics are designed to reduce the assembly and alignment of many individual components to a minimum

G. Leuteritz (✉) · M. P. Held · R. Lachmayer
Leibniz University Hanover, Institute of Product Development, Garbsen, Germany
e-mail: leuteritz@ipeg.uni-hannover.de

A. Heinrich (ed.), *3D Printing of Optical Components*, Springer Series in Optical Sciences 233, https://doi.org/10.1007/978-3-030-58960-8_2

while increasing the functionality of optical systems and drastically reducing their size. However, no conventional manufacturing process can be used to realize these individual components in particular, as the production is usually tied to tools and the production of numerous freeform optics is therefore linked to an equally large number of forming tools and is not economically feasible.

In order to avoid the necessity of tools, the focus will be on Additive manufacturing, as it stands out due to its freedom in design of components and is also not bound to other tools. However, it is foreseeable that Additive manufacturing will not supply any components that do not require post-processing. The question therefore arises as to how these two aspects of production can be used in a targeted manner for reflective optics production.

At this point the legitimate question arises to what extent it may be necessary to present reflectors as freeform geometry if they are mostly used for light guidance. Individuality occurs in particular for lighting systems where individual light distributions have to be generated and corresponding optics have to be manufactured. With regard to scanning systems, which use a laser beam, for example, to scan a surface to be measured, this freedom of geometry seems to be misplaced. For such areas of application, potential for function integration on the reflector is sought apart from the optically functional surface. Thus, the reflector should not only assume the function of a light-guiding component, but can also be provided with additional cooling structures to counteract absorption-related heat generation. Internal structures can also be used to reduce weight or to avoid specific mechanical vibration modes. Especially with regard to internal structures, additive manufacturing becomes necessary, as undercuts or closed volumes are usually produced without further efforts.

Against this background, the core aspects of optical quality and functional integration for additive manufactured reflectors are considered in the following chapters. Which evaluation criteria have to be considered in order to evaluate the optical quality, and which influencing variables are available in order to be able to adjust them? Which functions can be implemented on a reflector, and which production-related restrictions are imposed? These and other questions will be discussed in more detail in the following sections.

2.2 Requirements for Reflective Optics

In this section a brief overview of different applications for reflective optics is given, which are used to derive requirements for such optical components. These requirements are used to specify the optical function of the component that is correlated to its design parameters. This relation will also be shown. Furthermore the possibility for functional integration using Additive manufacturing is presented.

2.2.1 Applications for Reflective Optics

In order to assess which value exists for reflective optics, studies by other research teams are considered below. In addition, examples are considered in which the reflectors were not manufactured additively in order to obtain a comparison with conventional manufacturing processes.

Sweeney et al. examine different metals with regard to the Additive manufacturing of reflectors. As an example geometry, a concave mirror is used, whose entire geometry was weight-optimized and is thus characterized by internal grid structures instead of solid material. It is found that the material AlSi10Mg, which is used for Additive manufacturing, has similar component functions, both mechanically and optically, compared to Al-6061-T6. Al-6061-T6 is used as standard for the conventional production of optical reflectors. The samples manufactured by Sweeney et al. have been reworked with precision milling and have an RMS roughness Rq of 7.5 nm [1].

In a work by Jiang et al., a reflector for a headlamp system is developed, the geometry of which is based on an optical simulation of the desired light distribution. The requirement for the reflector is to direct the light of the light source, here a HID D2S xenon lamp, efficiently and with an adapted light distribution onto the road. The surface of the reflector is facetted, i.e., the total surface of the reflector is divided into several individual surfaces, which are spatially separated by edges. In this case, the reflector is manufactured using exclusively conventional methods and has an RMS roughness Rq of 94.1 nm after production. The geometric deviations are approximately 500 nm. The light distribution produced by this method complies with the regulations of the ECE (European standard) and the SAE (American standard) [2].

The group around Hilpert et al. deals with the design and Additive manufacturing of reflectors for aerospace applications. The challenge here is to ensure not only the preservation of the optical function but also the minimization of the mass of the reflector while maintaining the same mechanical properties. The inner volume of the reflector made of AlSi12 is replaced by a spongelike structure with the help of Voronoi cell modeling. The same procedure is used by the group around Mici et al. [3]. The mass of the reflector is reduced by up to 64.3%, while the requirements for mechanical strength are still given. Using milling and coating processes, the optical quality of the reflector is finalized. The RMS roughness Rq in this case is 0.5–1.5 nm, and the form deviation is approximately 80 nm [4].

It is noticeable here that the optical design, i.e., the geometric design of the optically functional surfaces, is not limited by the manufacturing process and that Additive manufacturing is only used in the sense of a mechanical adaptation of the components. The guarantee of the optical function is further ensured by the use of conventional manufacturing processes.

A central knowledge gap can be derived from the applications described in this section. Only the design parameters of the reflector after passing through the entire manufacturing chain are reported, but not the values of the design parameters of the reflectors directly after Additive manufacturing. An evaluation of the capacities of Additive manufacturing with regard to the production of optical reflectors has therefore not been carried out. In addition, Additive manufacturing is mainly used for mechanical modifications, whereby the optical surfaces are designed with respect to conventional manufacturing processes.

2.2.2 *Geometry*

The geometry of reflective optics primarily depends on where they are installed in an optical system, since their function always remains the same—to guide light and form light distributions. If the reflector is located directly at the light source, it usually encloses the light source and guides the light collimated in the direction of the optical axis. These standard optics have parabolic, hyperbolic, or spherical geometries and thus have a three-dimensional optical functional surface. The light source is usually located in the focus of the optics in order to achieve collimation [5].

Reflectors, which are placed directly at the light source, not only have the effect of collimation but also increase the efficiency. An example is an LED with a Lambertian emission profile. If a collimating lens is used directly after the LED, only a part of the light according to the NA is collimated by the lens. The rest propagates past the lens. With additional reflective optics, light can be captured at large beam angles and guided to the collimating lens, thereby increasing the optical efficiency.

For reflective optics, which are arranged after the light source, flat surfaces usually result, since the light path is directed in the desired direction by reflection of the light rays. Concave or convex optics can also be used to introduce additional collimation or divergence effects.

2.2.3 *Relation Between Design Parameters and Functionality*

In order to be able to evaluate a component independently of its functionality and thus render it comparable to other parts, design parameters are introduced. These design parameters are [6]:

- Number
- Shape
- Tolerances
- Material
- Topology
- Surface roughness

On the basis of these parameters, first investigations can be carried out, which can be connected in the next steps with the functionality of the component. The reflectivity of a component will be influenced by the roughness and the material of the component. For optics, four different surface processes can be characterized without considering nonlinear optical effects. These are transmission, absorption, reflection, and scattering. For each of these quantities, the corresponding degree can be determined, which indicates the percentage of light influenced by the respective effect. For most reflective optics, transmission is not required, as these are not transparent materials, but mostly metals or coated polymers. Ceramic optics can also be used. For reflective optics, the target values of reflectivity (mirrors) and scattering (Lambertian samples) remain. Absorption is thus declared as a disturbance variable. The target quantities are to be adjusted according to their application, and the disturbance quantities are to be minimized. Above all aluminum and silver are used as reflectors for their high and relatively constant reflectivity of over 90% (when properly polished) in the visible range of electromagnetic radiation. The rest of the light is converted into thermal energy by absorption. For reflective optics it is now necessary to investigate the influence of the design parameters on its functionality. Therefore the approach of total integrated scattering (TIS) is introduced. In this case, the proportion of scattered light on a reflective surface is calculated. The equation for this is [7]

$$\mathrm{TIS} = R_0 \cdot \left(1 - e^{-\left(\frac{4\pi \cos(\theta) Rq}{\lambda}\right)^2}\right) \tag{2.1}$$

TIS total integrated scattering, R_0 reference reflectivity, Θ incident angle, Rq RMS surface roughness, λ wavelength of the incident light.In order to work with the TIS model, a reference measurement must be available which describes the reflectivity of a perfectly flat surface of the same material of which the component is to be made. This reflectivity represents the maximum value of the material. In Fig. 2.1 the spectral reflection for a perfectly smooth aluminum surface is shown.

In this measurement, both reflectivity and absorption of the material are determined and further used for the model. From the above equation, the roughness and its reference reflectivity can now be used to determine the proportion of scattering for rougher surfaces. Using a hypothetical surface with an assumed reflectivity of 85% at a wavelength of 550 nm, the following relationships can be calculated at an angle of incidence of 45° (Fig. 2.2).

The design parameters "material" (influences the reference reflectivity) and "surface roughness" are therefore decisive for the reflectivity. In addition, it quickly becomes clear that the surface roughness requirement for high reflectivity is very demanding. In order to convert 50% of the light into scattering, an RMS roughness of 50–100 nm is required, depending on the considered wavelength. This places high demands on the production of reflective optics. The extent to which Additive manufacturing is able to realize this will be examined in more detail in the following sections.Furthermore, the optics may have to have an imaging character and generate predefined light distributions. Reflectivity continues to play an important role

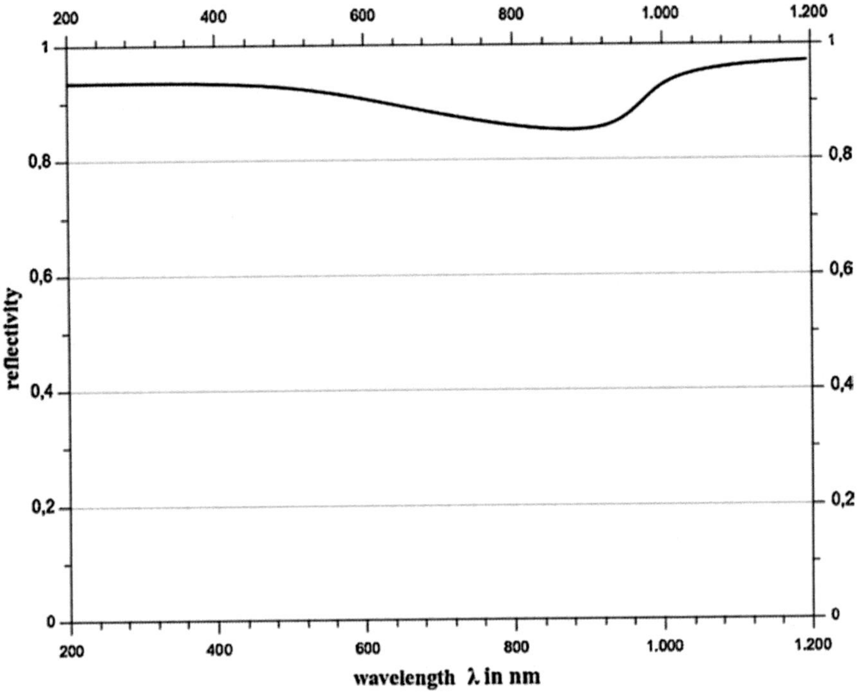

Fig. 2.1 Spectral reflectivity for an ideal aluminum surface; the reflectivity is nearly constant for a large area of the visible range according to [8]

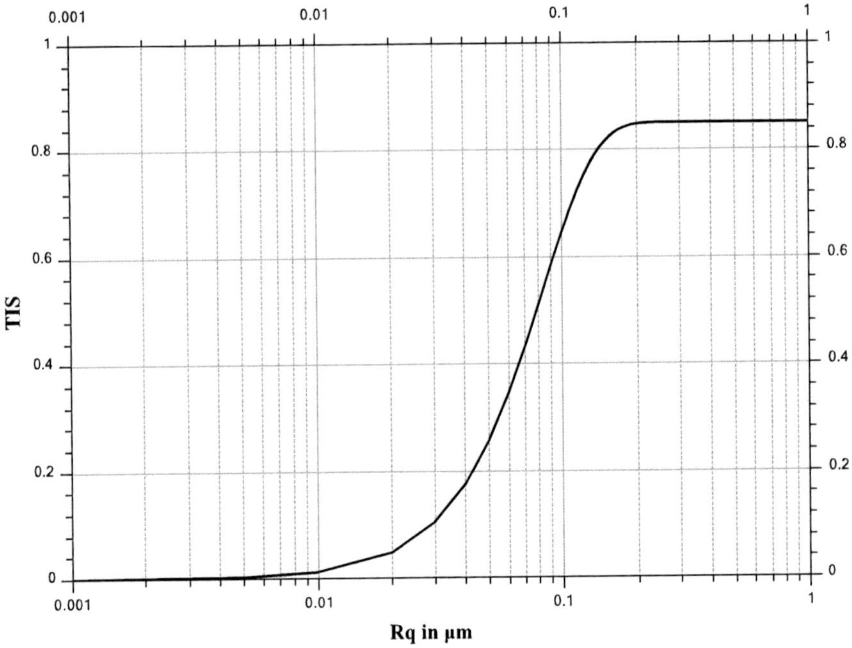

Fig. 2.2 Total integrated scattering for a hypothetical surface with a maximum reflectivity of 85%

here, but manufacturing tolerances also need to be taken into account, as these affect the light distribution. For conventional optics, tolerances of 0.1–0.2 mm are specified, which results in a second target value that must be met by Additive manufacturing.

2.2.4 Reflector Design for Additive Manufacturing

Additive manufacturing is mainly used for the layer-by-layer generation of components. Different processes are used depending on the material and aggregate state of the starting materials. When using Additive manufacturing, however, the question always arises as to the extent to which this technology is suitable for a particular component with specified application. In order to weigh up the relevance of this technology, the design objectives for which Additive manufacturing can be used are listed below. These design objectives (Table 2.1) are listed below [9].

If one—or better several—of these design objectives are achieved by Additive manufacturing, the component is particularly suitable for this process. For optical reflectors, function integration and material savings should therefore be named as main design objectives. If these are also imaging optics, the design of the reflector can also receive a higher weighting.

Figure 2.3 shows a flashlight reflector that has been equipped with several functions as a result of geometrical adaptations in order to show the potential of functional integration.

For this reflector, interfaces (e.g., with the housing) were first defined and the possible functions with which the reflector can be additionally equipped identified. The several functions can be structured in a design catalogue and transferred to other applications [10]. Whether the reflective surface of the component can be affected by functional integration depends on the individual application.

2.3 Additive Manufacturing: Selective Laser Melting

In order to render the content of the following chapters more comprehensive, in this section the process of selective laser melting (SLM) will be examined in more detail.

A powder with the previously defined layer thickness is applied via a re-coater. A laser beam with appropriate scanner optics selectively melts the applied powder and thus creates the first layer of solidified material. The building platform is lowered by one layer thickness, and the next layer of powder is applied. The process is repeated until the component is finished. All weldable metals are suitable as materials for this. Polymers can also be processed using this process, but in this case, the material is sintered, which makes the term selective laser sintering more fitting.

In order to be able to adjust the melting process, various process parameters are available. These include laser power, scanning speed, layer height, hatch distance,

Table 2.1 General design objectives for additive manufacturing

Design objective	Description
Material savings	Reduction of processed resources based on geometrical adaptations and material redistributions
Functional integration	Increasing the number of functions for a single component
Force flow adaptation	Optimized material distribution according to the load applied on a component
Thin walls	Reduction of processed material by using inner structures
Mass customization	Customer-based individualization of a component
Design	Realization of freeform components
Net shape geometries	Manufacturing of components based on simulation results
Integrated channels	Implementation of inner channels for different applications, e.g., cooling

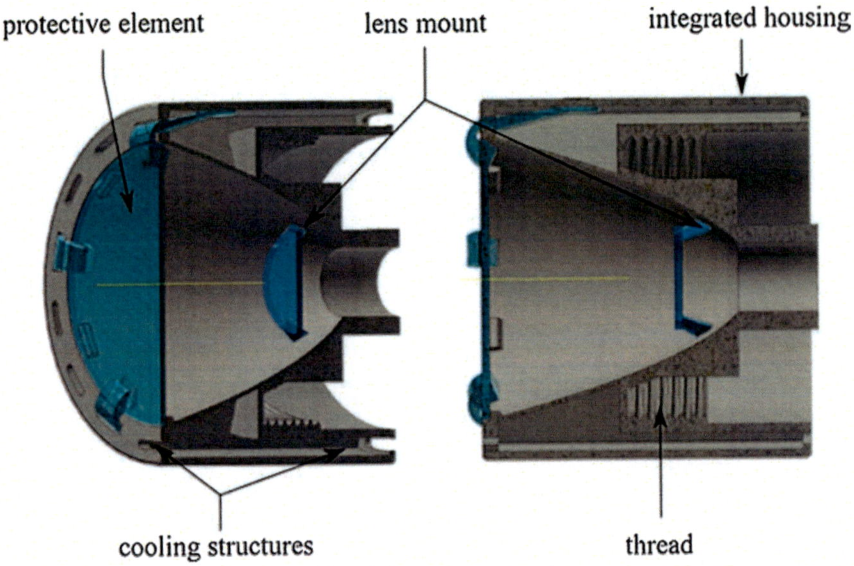

Fig. 2.3 Function-integrated flashlight reflector; the parabolic reflector surface is protected by an additional transparent window. Additional mounting options for collimating lenses are added as well as channels for passive air cooling. A thread serves to mount the reflector without additional housing elements

and many more. These process parameters influence the design parameters of the component and thus the functionality of the optical component. In this way, a chain of effects can be established which must be investigated.

In order to examine the process parameters specifically for their influence on the surface roughness, the process window must first be defined. By determining the process window, the selection of settings for the process parameters can be drastically restricted.

Care must be taken to ensure that such a process window changes not only with the materials used but also with the respective production machines. A change in the process gas, for example, also changes the process window. In order to determine the process window despite the large selection of process parameters, certain process variables can be combined to form a new variable. This variable is the energy density Ψ and is described below (Eq. 2.2) [11]:

$$\Psi = \frac{P}{v \cdot d \cdot h} \tag{2.2}$$

P laser power in *W*, *v* scan velocity in mm/s, *d* hatch distance in mm, *h* layer height in mm.

Using the energy density, it is now possible to set up a process window, which is generally divided into three areas [12]:

1. Too high energy densities mean that the material is liquefied too much, and a metal drop bulges due to its surface tension in such a way that it becomes higher than the actual layer thickness (balling effect). During the next coating process, the coater collides with this already solidified drop and thus destroys the printing bed, possibly by tearing out the component.
2. Too low energy densities lead to only partially melted particles which are not connected to each other, and therefore the component cannot be produced.
3. Sufficient energy density to melt the metal powder and process it into a mechanically stable component.

An example process window is shown in Fig. 2.4.

In order to identify the influence of the parameters on the surface roughness of the component, parameter studies are carried out, and the resulting components are examined for the necessary design parameters. The geometry of the components to be examined should be taken into account. These can be simple test specimens whose results are projected onto the actual components.

In this case, a sample geometry is used which takes up the essential aspects of the reflector geometry. This includes the assumption that reflectors do not take up more than one hemisphere and thus only angles of up to 90° to the building platform are required with appropriate alignment. In addition, it should be avoided that the optically functional surfaces have to be provided with support structures in order to minimize the post-processing effort and to be able to quantify the evaluation of the surface directly after additive manufacturing. In addition, the radius of curvature of the reflectors is approximated by several discrete steps. Figure 2.5 shows a sample geometry with seven differently sloped steps at 15° each, thus providing one plane each for measuring the samples.

The measured roughness can be composed of several effects of different orders of magnitude. Depending on the process, the surface has trenches that result from the sequence of melted lines. This depth and width of the trenches thus depends to a large extent on the layer height and the line spacing. The maximum depth that can be reached by the trenches is the layer thickness itself, when individual lines are

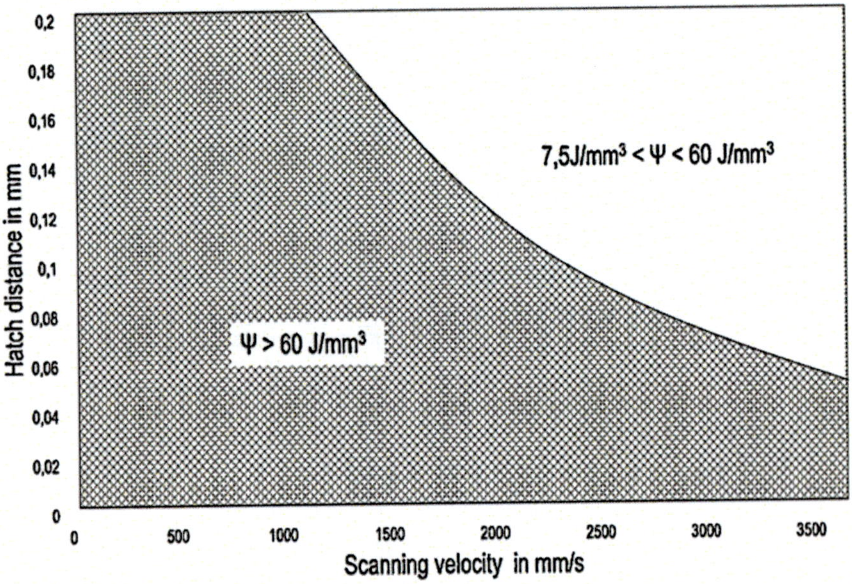

Fig. 2.4 Process window for the used SLM machine; areas of too high (> 60 J/mm^3) and suitable energy density can be identified with a constant laser power of 370 W and a constant layer height of 30 μm

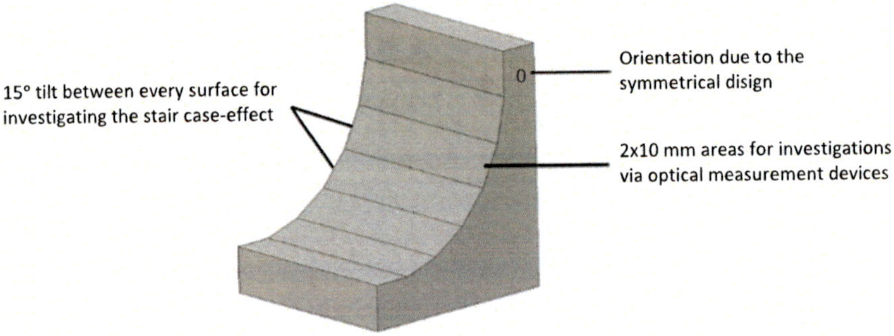

Fig. 2.5 Geometry of the samples for the general investigation of the surface properties of additive manufactured components

tangent to each other. Typical layer thicknesses for SLM are approximately 30 μm and thus represent the upper limit of this roughness effect.

The only partial melting of the metal powder particles exerts a further influence [13]. The particles form elevations on the surface and vary with their grain size and degree of melting. The particle size is in the range of 1–20 μm. Therefore, the maximum contribution of partially melted powder to surface roughness is lower than that of grid lines.

With sloped surfaces, the staircase effect is present, which is geometrically caused by the layered structure [14]. Since the component cannot be rotated in the installation space using the SLM process, sloped surfaces must be approximated by steps which have an immense influence on the surface roughness. The influence of the steps depends on the layer thickness and the angle of the surface and can lead to a deviation from the target geometry of more than 50 µm.

An example of results of the parameter studies can be seen in Fig. 2.6.

One quickly realizes that an enormous amount of measurement data is collected to characterize the process and its effect on the component surface. As a result, it is difficult to ensure that the process parameters can be set to produce a global minimum for surface roughness. At the same time, there is the challenge of basically meeting the requirements for the component, i.e., RMS roughness in this example.

For these reasons, a database is set up to serve as a process configurator. The results of the parameter studies are fed in, and missing values are approximated linearly. This rather course approximation is seen fit due to the high standard deviations. Functional correlations cannot be identified without introducing high errors in the fit parameters.

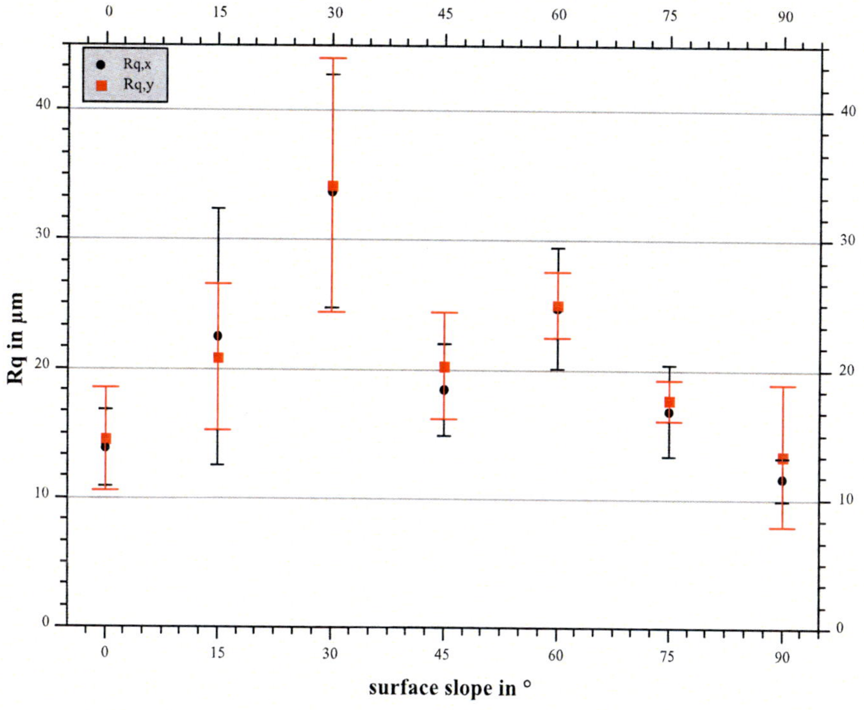

Fig. 2.6 Results of the parameter studies; the surface angle has an increased influence on the RMS roughness *Rq* due to the staircase effect. In addition the anisotropy of the additive manufacturing process leads to different roughness values for different directions

The database allows process parameters to be set in such a way that any roughness between the minimum and maximum measured roughness can be implemented. In addition, the process can be inversed, so that the input is a given roughness and the database calculates the best process parameters in order to configure the process according to the requirements.

The roughness value can also be coupled with a reflectivity via calculation by TIS. Since this reflectivity changes with each individual roughness value, the simulation of the components is made more difficult because only standardized surfaces are available for simulations. With the database, however, it is possible to additionally create a file as a coating, which contains the predicted reflectivity value, and thus the components can be simulated according to the manufacturing results. The flowchart for the process configurator can be seen in Fig. 2.7.

With the help of the database, it is thus possible to map the production capacities of SLM and to specifically adjust the process. The more parameter studies are carried out, the more accurate the database becomes, as the linear approximations become more accurate. It can also be extended by any target value. Thus it is possible to integrate form and position tolerances or even to introduce other production plants in order to be able to carry out the production of other optical, mechanical, and other components in a targeted manner. However, it quickly becomes apparent that additively manufactured components do not correspond to the required roughness values without further treatment, which are intended for reflective optics. Therefore, additional process steps must be carried out to ensure the optical quality.

Especially the post-processing is of special importance for the Additive manufacturing of optics. In contrast to mechanically stressed components, reflective optics mainly depend on the quality of the surface, which means that any

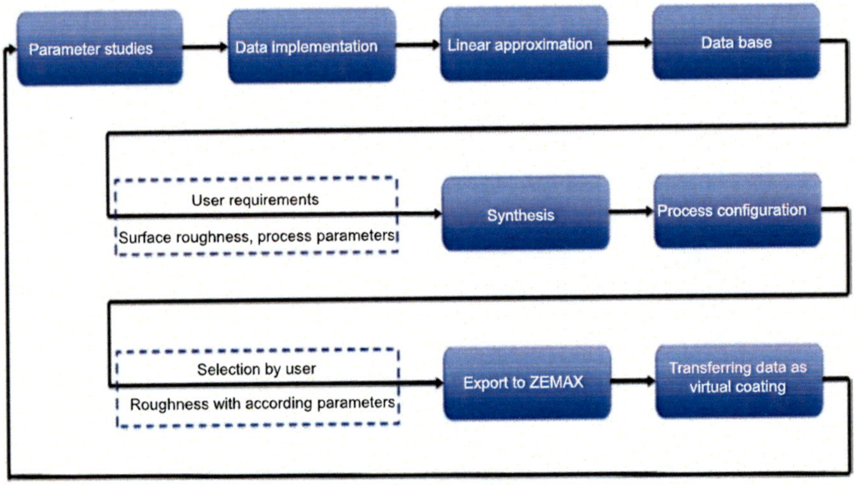

Fig. 2.7 Flowchart for the implementation of a process configurator based on a database with the capability of creating suitable files for further optical simulation

post-treatment of the surface is in the foreground of this process step. If the Additive manufacturing is considered alone, RMS roughness values of 20–25 μm result, which lead to a reflectivity far below 1%. In order to remedy this deficiency, additional post-processing methods are considered, the influence of which on the surface of the material will in turn be part of the process database.

The post-processing methods can essentially be divided into three different categories: subtractive/abrasive, additive/coating, or forming. Subtractive processes are traditional mechanical finishing processes such as milling or polishing. Additive processes provide for the application of additional materials, e.g., during coating. Forming processes, such as laser polishing, use energy input to change the material properties without removing material or applying new material.

Based on the requirements placed on the reflector, the post-processing methods can now be selected and applied. However, this chain of argumentation is accompanied by an enormous risk potential: Is the component geometry accessible for the post-processing procedure? The extent to which this question affects the process chain and the geometries of the reflectors is described in the following section.

2.4 Additive Manufacturing of a Reflector Array

In order to demonstrate the feasibility of the process configurator, the results gained from the parameter studies shall be transferred to a reflector array that is applied on an existing LED array.

2.4.1 Design of a Reflector Array

To validate the process chain, a paraboloid reflector array for front projection is designed for a high-performance LED matrix. The light sources of the LED matrix are 48 high-power Wicop-C LEDs from Seoul Semiconductor. These are arranged in a 6 × 8 matrix with a pixel pitch of 4.6 mm, have a light emission area $d \times d$ of 1.12 × 1.12 mm^2, and generate approx. 225 lm per LED. Assuming that the reflector contacts the printed circuit board flat, the height of the LED can directly determine the required focal length of the reflector to $f = 0.8$ mm (see Fig. 2.8). Each LED can be controlled individually so that all reflector geometries can be measured individually.

Apart from the LED array itself, which presents a given set of boundary conditions for the design of the reflector array, it has also to be taken into account that the post-processing method influences the design of the reflector array.

Figure 2.8 shows two different reflector array designs for the mentioned LED array, which strongly differ in their geometry. The left design is suitable for a milling treatment after the SLM process. Here, the process restrictions are set by the SLM process rather than by the finishing process, which means that there are

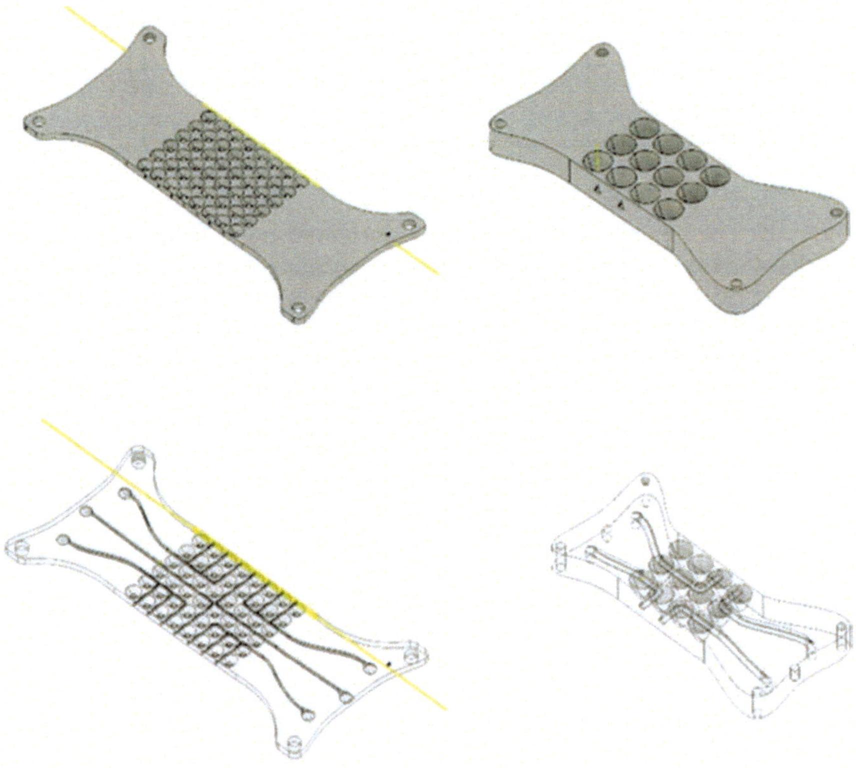

Fig. 2.8 Comparison between two reflector array designs with inner cooling structures based on the finishing method. Left: array suitable to be processed with further milling treatment. Right: array suitable for further laser polishing treatment (cw mode)

minimal wall thicknesses of 0.3 mm and the cooling channels are designed in a way to avoid support structures. In general the design restrictions for SLM apply to this design, since the restrictions induced by the milling process are lower in their magnitude regarding geometrical limitations. Using the given wall thicknesses, it is possible to create reflectors with a length of 0.58 mm for each LED that are cooled by the inner structures. The cooling channels lead not only to an additional function of the array but also to a reduced weight and the sheer necessity for Additive manufacturing, since they cannot be efficiently manufactured otherwise.

The right design shows an array that has been made suitable for laser polishing with continuous wave (cw) radiation. Since laser polishing is based on the remelting of the surface and the followed smoothing as a result of the surface tension, there are other design restrictions to be considered. The penetration depth of the remelting process is about 0.36 mm with the used set of parameters, but to ensure a sufficient heat flow, the wall thickness should be set to an empirical value of 1 mm in order to prevent heat accumulation. It should be kept in kind that this value does not represent the minimal value, but a value that ensures a smooth operation of the laser polishing process. The minimal wall thickness depends not only on the process parameters for laser polishing but also on the geometry and material of the component.

However, with a minimum wall thickness of 1 mm, it is not possible to implement a reflector for every LED. Therefore only every second LED in every second row is equipped with a reflector, which reduces the overall amount of reflectors to one fourth of the original amount. The benefit of having less reflectors is the possibility to increase the length of each reflector, here up to 5.28 mm, with regard to the wall thicknesses. Additional cooling structures again increase the usefulness of Additive manufacturing.

In order to exploit the design freedom of SLM, the possibility of functional integration is considered for the reflector array, using the design catalogue from [10]. Most of the listed functions cannot be implemented due to the low height of the array (corresponds to the length of the reflector elements) of 1.25 mm and the high density of reflector elements; otherwise the manufacturing restrictions for wall thicknesses, as well as overhangs and associated support structures, cannot be maintained. Ignoring these restrictions can result in component defects, such as cracks, holes, or other defects that endanger the function of the array.

Two functions of the design catalogue are used for the reflector array with 48 elements. Firstly, mounting aids such as drill holes can be provided to ensure the connection between the LED and reflector array. On the other hand, cooling channels are integrated to reduce the power dissipation of the LEDs.

14 side openings and thus 14 cooling paths for the array are implemented, which lead to 6 separate outputs inside. These outputs are each located at holes in the LED array, which are used to feed coolant into the cooling channels. The cooling channels have a drop-shaped cross-section to be manufactured without support structures, provided that the array is built with the reflector surfaces in building direction.

The cooling channels are deliberately not laid in a meandering pattern around the paraboloids in order to be able to free them from residual powder after production. The integration of further functions for the reflector array is not planned due to the small thickness of the array, since, for example, channels for fiber guides cannot be integrated into the array due to their size. Electrical functions are not important due to the already existing LED array, since all necessary electrical functions are already covered by the LED array.

The functional integration on the variant of the array with 12 elements is adapted so that the number of cooling channels is reduced to 4 and the holes for mounting the previous array are retained.

The channels are dimensioned larger and are provided with a triangular cross-section instead of a drop-like cross-section, since this fills the spaces between the paraboloids more effectively in order to ensure cooling near the surface. Once again, a meander-shaped course is avoided in order to be able to clean the channels better after production.

In addition, cut-outs are implemented which are used for the LEDs which cannot be provided with a paraboloid to embed them. These 36 notches are located on the bottom of the array in a matrix pattern.

2.4.2 Validation of a Process Configurator

The reflector geometries can be mounted directly on the board as shown in Fig. 2.9 to ensure an exact alignment. In the following, only 2 × 2 LEDs in the center of the matrix (Fig. 2.9, framed in blue) are operated in order to keep simulation and measurement times short.

The maximum reflected fraction of the luminous flux Φ_R can be described similar to the transmissivity fraction [15] with Eq. 2.3:

$$\frac{\Phi_R}{\Phi} = 1 - \Phi_F = \frac{1}{2}\left(1 + \cos 2\varepsilon_1\right) \tag{2.3}$$

The angle ε_1 defines the fraction of the light emission not deflected by the reflector (see Fig. 2.10).

Using the angle ε_1 the reflector diameter (Eq. 2.4) and the reflector length (Eq. 2.5) can be determined according to

$$d = \frac{4f}{\sin \varepsilon_1}\left(1 + \cos \varepsilon_1\right) \tag{2.4}$$

$$l = \frac{d}{2\tan \varepsilon_1} \tag{2.5}$$

Figure 2.11 shows the effective fraction of the reflected luminous flux Φ_R for reflector length l and reflector diameter d. Both parameters have been normalized to the focal length. In the present case, the maximum diameter d^* is geometrically limited by the pixel pitch of 4.6 mm and the focal length $f = 0.8$ mm to $d^* = 5.75$ mm, which results in a maximum length for a parabolic reflector system of $l^* = 1.04$ mm. The maximum length for a parabolic reflector system is therefore the same as the maximum length for a parabolic reflector system. This corresponds to a useful share of the luminous flux of $\Phi_{R1} = 11.7\%$. However, certain minimum wall thicknesses are required for post-processing the reflectors, so that the maximum reflector diameter must be reduced to $d^* = 5$ mm. As a result, the maximum reflector length is now only $l^* = 0.58$ mm, so that the useful fraction of the luminous flux is reduced to a maximum of $\Phi_{R2} = 5\%$.

In order to adjust the geometry of the array so that it is suitable for laser polishing in cw mode, both the distances between the paraboloids and the wall and base thickness within the array must be increased to at least 1 mm. As a result, the array no

Fig. 2.9 Image of the measurement setup consisting of a 6 × 8 high-power LED matrix and printed reflector array

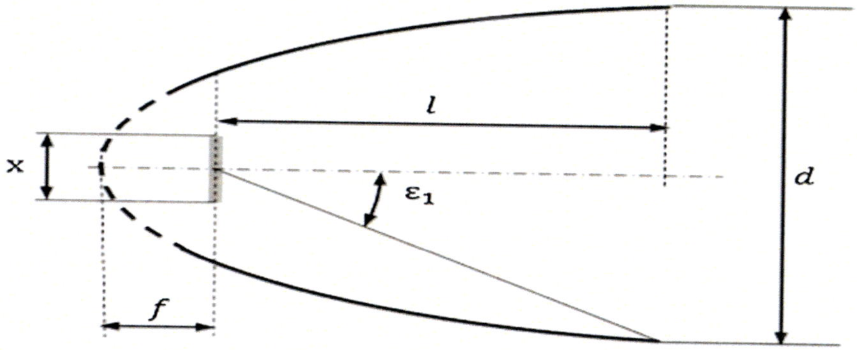

Fig. 2.10 Geometric characterization of the paraboloid reflector. x = edge length of the LED, ε_l = nonreflected angle of a beam, f = focal length, d = reflector diameter, l = reflector length

longer has the original 48 reflector elements, but the number is reduced to 12 reflector elements.

By reducing the number of reflector elements, their size can now be increased in order to redirect a larger portion of the luminous flux through the array. With the condition that the minimum distance between the reflector elements should be 1 mm, a maximum diameter d^* of 8.22 mm is thus achieved. The resulting length l^* of a reflector element is thus 4.28 mm, and the undetected angle range ε_l is reduced from 77.32° to 43.84°. Thus the portion of the collected luminous flux is increased from 5 to 52%.

By increasing the reflected part of the luminous flux, it is shown that this reflector geometry is better suited to increase the efficiency of the optical system. However, it must be taken into account that only 25% of all LEDs can be operated

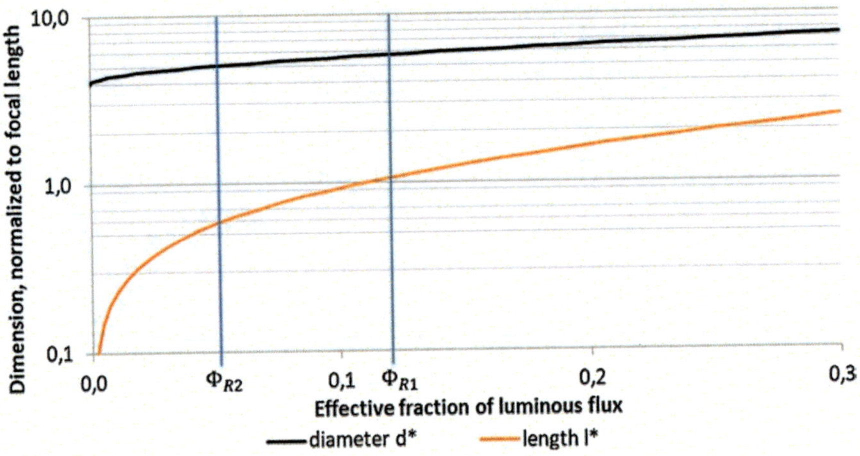

Fig. 2.11 Reflector diameter and length as a function of the reflected fraction of the luminous flux

with this reflector array. Compared to the array with 48 elements, the proportion of the detected luminous flux is thus increased by a factor of 10.8, and the number of reflector elements is reduced by a factor of 0.25. In terms of power dissipation, the 12-element version is therefore to be preferred, but the function of the LED array is significantly impaired, since only a quarter of all light sources can be used.

For the prediction of the reflection properties of new parameter studies, the already recorded roughness values of previous studies are transferred to the optical simulation program Zemax OpticStudio as a coating.

The previously described reflector geometry in Fig. 2.10 is imported into Zemax in STL format. The LEDs of the matrix are represented by an ideal square emission area with corresponding dimensions and ten million beams each. To simulate the reflection properties, the reflector array is provided with a surface coating which simulates the measured reflectivity. The resulting light image is projected on a detector in a distance of 1 m.

For the following example, an unprocessed reflector array with a roughness value of $Rq = 32$ μm and a roughness value of approximately $Rq = 1$ μm is simulated. The latter corresponds to a treatment using a ball milling tool.

The maximum possible reflected fraction of the luminous flux corresponds to 5%, which corresponds with the geometry shown in Fig. 2.10. Therefore, the following Eqs. (2.6) and (2.7) can be used to determine the reflectivity of the reflectors:

$$dR = \left(\frac{\Phi_S - \Phi_{\text{Absorbent}}}{\Phi_{\text{Reflective}}} \right) \cdot 100\% \tag{2.6}$$

$$R = \frac{dR}{dR_{\max}} \cdot 100\% \tag{2.7}$$

Table 2.2 shows the simulation results for an ideally absorbing, an ideally reflecting, and three manufactured reflectors. The resulting values of the reflectivity gained from the parameter sets and the simulations coincide. The simulated reflectivity and the values gained from the database mostly match. Only for the untreated reflector, the values deviate from each other since the reflectivity is very low and the simulation cannot reproduce such small values. All in all the data from the database can be transferred to a simulation environment without significant errors.

In Table 2.3 the measured values of the luminance camera are compared with the simulation values. It can be clearly seen that for the untreated reflector arrangement, there is a deviation between simulated and measured luminous flux of 0.04%. The agreement between the simulation and measurement results is high. It must be taken into account that only a small proportion of the light can be deflected through the array by this reflector array variant (48 elements), which means that measured values close to the minimum luminous flux can be expected and thus the functioning of the process configurator cannot be clearly checked. Furthermore, the consideration of the simulated reflectance negates the fact that the process configurator calculates a value of 10^{-18}% and this value cannot be validated due to the measurement inaccuracy of the test setup.

The reflector array refined with a spherical milling cutter is very well represented by the simulation, as the measurement deviates from the simulation results by only 1%. The dependence of the surface inclination is not applicable in this case, because the milling process homogenizes the surface roughness over the reflector surface and therefore there is no dependency on the surface slope. However, analogous to the unmachined array, it should be noted that only very small portions of the luminous flux can be deflected by the array and thus the influence of the array is small.

The laser polished array with 12 elements is measured with a deviation of 55% for simulation. Analogous to the spherically milled array, there is no surface slope dependence of the roughness and thus of the reflectivity in this case. However, an oxidation of the surface is caused by the polishing process. Even if this layer is treated by laser cleaning, residues of this oxide layer on the surface can be identified. There is an increased absorption on the reflector surface due to this residual layer, which leads to this deviation of the simulated and measured luminous flux by 55%. Due to the changed geometry of the paraboloid compared to the array with 48 elements, this array is also more affected by the simplification that the reflectivity is only valid for a single angle of incidence according to the TIS model.

The simulation of the parameter sets can be performed very well with a deviation of 1% and less in comparison to the measurement results. It is important that the number of simulation beams is not too small, as otherwise the random generation of the beams in combination with a small number of rays provides too large variance in the results. This can cause untreated reflectors with single-digit reflectance to be represented incorrectly. Alternatively, a measured ray file can be used. In the presented case, a reliable result could be achieved with a number of ten million beams per LED.

Table 2.2 Simulation results for different reflector samples

Post-processing (# reflectors)	Reflected fraction *dR*	Reflectivity *R*	Parameter set reflectivity R_M
Ideal absorbent	0%	0%	–
Ideal reflective	5.05%	100%	–
Untreated (48)	0.06%	0.01%	10^{-18}%
Ball-milled (48)	0.75%	17.7%	15.9%
Laser polished (12)	20.7%	8.5%	8.8%

Table 2.3 Measurement results

Post-processing (# reflectors)	Simulated luminous flux	Measured luminous flux	Deviation
Untreated (48)	852.3 lm	852 lm	0.04%
Ball-milled (48)	859.4 lm	851.2 lm	1%
Laser polished (12)	691.7 lm	380 lm	55%

2.5 Challenges for SLM of Reflective Optics

As described in the last few chapters, the usage of database process configurator proves helpful in having precise predictions about the outcome of optical functionality of reflective optics. The configurator gets more accurate with each measurement taken with manufactured samples, and the simulation of the optics creates promising and reliable results. However, the requirements for reflective optics have not yet been met without having an increased effort in post-processing. It is not yet possible to produce reflective optics without having several separate process steps that lead to demounting and remounting the components and the corresponding additional effort in alignment. But there is still potential in improving the SLM process in order to decrease surface roughness and therefore increase the optical quality of the components.

Laser polishing is a very promising finishing method since it is not of a subtractive kind, but it transforms the surface by remelting. There is no loss in material and no contamination by removed material, nor is there any need for additional materials. But due to the process restrictions of cw-mode laser polishing, it is not possible to manufacture miniaturized optics, since the material thickness at the optical surface has to be too thick. Using pulsed laser radiation would solve this problem, because the penetration depths for this kind of laser polishing are lower than the penetration depths for the cw mode, and in addition, the surface roughness decreases even more, since the polishing effect only occurs within the focal volume, which is only a few microns. The pulsed laser radiation is only suitable for surfaces with an *Rq* of approximately 1 μm, which is why it often follows the cw-mode treatment. If this sequence is kept, the process restrictions of the cw mode still have to be considered.

Having this in mind, it is only plausible that the SLM process has to be optimized in a way that the surface roughness is reduced to an amount that allows a direct treatment with pulsed laser radiation. The three main causes for the high surface roughness are partially melted particles, the staircase effect, and the structures originating from the spacing between the molten lines. The effect of partially melted particles can be reduced by using powder compositions with decreased grain size distribution, only allowing maximal grain sizes of about 10 μm. The staircase effect can be limited, when the component is movable within the building space. The radiation can always propagate perpendicular to the components surface and therefore reduce the approximation of tilted surfaces.

Considering these approaches, the building platform and the scanning system of an SLM system can be improved by providing more kinematic degrees of freedom, for example, by using hexapods as building platforms. However, the SLM process still uses a powder bed, which is easily destroyed by moving parts in the building room. A suitable concept for such an integrated SLM system is still missing, but when found it promises to introduce us to a new level of optics manufacturing.

References

1. Sweeney, M., et al. (2015). Application and testing of Additive manufacturing for mirrors and precision structures. In M. Krödel, J. L. Robichaud, & W. A. Goodman (Eds.), *Material technologies and applications to optics, structures, components, and sub-systems II* (p. 957406). Bellingham: SPIE.
2. Jiang, J. B., et al. (2006). Design and fabrication of freeform reflector for automotive headlamp. In *2nd international conference on power electronics systems and applications* (pp. 220–224). Piscataway: IEEE.
3. Mici, J., et al. (2015). Optomechanical performance of 3D-printed mirrors with embedded cooling channels and substructures. In A. E. Hatheway, editor. Optomechanical engineering SPIE. S. 957306 .
4. Hilpert, E., et al. (2018). Precision manufacturing of a lightweight mirror body made by selective laser melting. *Precision Engineering, 53*, 310–317.
5. Wolf, A.. (2016). Laserscheinwerfer für Kraftfahrzeuge. Garbsen: TEWISS—Technik und Wissen GmbH Verlag. isbn: 978-3-95900-121-2.
6. Roth, K. (2000). *"Konstruieren mit Konstruktionskatalogen", Band 1: Konstruktionslehre* (3.Aufl. ed.). Berlin: Springer-Verlag.
7. Elson, J. M., Rahn, J. P., & Bennett, J. M. (1983). Relationship of the total integrated scattering from multilayer-coated optics to angle of incidence, polarization, correlation length, and roughness cross-correlation properties. *Applied Optics, 22*, 3207–3219.
8. Bass, M. (1995). *Handbook of optics*. New York: McGraw-Hill.
9. Lippert, R. B., Leuteritz, G., & Lachmayer, R. (2017). An approach to implement design for Additive manufacturing in engineering studies. In *Proceedings of the 21st international conference on engineering design (ICED17), Vol. 5: Design for X, Design to X, Vancouver, Canada, 21–25 August, 2017*. issn: 2220-4342.
10. Leuteritz, G., Rohling, M., & Lachmayer, R. (2018). Functional integration for additively manufactured reflectors. In *DGaO proceedings*.
11. Zghair, Y. A., & Leuteritz, G. (2017). Additive Repair von Multimaterialsystemen im Selektiven Laserstrahlschmelzen. In R. Lachmayer & R. B. Lippert (Eds.), *Additive Manufacturing Quantifiziert—Visionäre Anwendungen und Stand der Technik* (pp. 195–215). Springer Vieweg Verlag.

12. Kruth, J.-P., et al. (2005). Binding mechanisms in selective laser sintering and selective laser melting. *Rapid Prototyping Journal, 11*, 26–36.
13. Liu, B., et al. (2011). Investigation the effect of particle size distribution on processing parameters optimisation in selective laser melting process. *Additive Manufacturing Research Group Loughborough*, 227–238.
14. Wang, W., et al. (2016). Improved surface quality in 3D printing by optimizing the printing direction. *Computer Graphics Forum, 35*, 59–70.
15. Held, M. P., et al. (2020). *Hochauflösende LED-Scheinwerfer für Kraftfahrzeuge*. Garbsen: TEWISS—Technik und Wissen GmbH Verlag. isbn:978-3-95900-439-8.

Chapter 3
3D Printing of Optics Based on Conventional Printing Technologies

Manuel Rank, Andre Sigel, Yannick Bauckhage, Sangeetha Suresh-Nair, Mike Dohmen, Christian Eder, Christian Berge, and Andreas Heinrich

Abstract This chapter describes the potential of 3D-printed optics based on conventional printers. The materials used in the Additive manufacturing of plastic optics are first presented using the example of photopolymerization. The characterization of additively manufactured components is then discussed. For samples, "post-processing" proves necessary, especially with regard to their surface structure. Another aspect is the resulting inhomogeneity in the refractive index, which is presented in detail.The key focus of the chapter is a discussion of completely different examples of additively manufactured optics to illustrate the potential and limitations of additive manufacturing in this area. Starting with light-guiding elements, the advantages and disadvantages of a range of 3D-printed imaging systems are presented and discussed. The potential of 3D-printed liquid lenses, freeform lenses, volumetric displays, and mirror elements is also described. The discussion of these macroscopic optical elements is followed by a presentation of the 3D printing of microlenses. Both spherical and aspherical microlenses can be realized additively based on the inkjet process. The question of how this might allow a greater functionalization of additively manufactured optics is explored by discussing the printing of organic LEDs, as well as the additive manufacturing of random lasers and of photoluminescent optics.

The final focus of this chapter is the additive manufacturing technology itself. 3D printing specifically optimized for optics production is discussed. Three different concepts are presented: a robot-based additive manufacturing platform that achieves resolutions in the subpixel range, an additive manufacturing method for microlenses combined with electric fields, and the multimaterial printing approach based on stereolithography.

M. Rank · A. Sigel · Y. Bauckhage · S. Suresh-Nair · M. Dohmen · C. Eder · C. Berge
A. Heinrich (✉)
Center for Optical Technologies, Aalen University, Aalen, Germany
e-mail: Andreas.Heinrich@hs-aalen.de

A. Heinrich (ed.), *3D Printing of Optical Components*, Springer Series in Optical Sciences 233, https://doi.org/10.1007/978-3-030-58960-8_3

Keywords Additive manufacturing · Optics · 3D printing · Individualized optics Illumination · Microlenses · Liquid lens · Fisheye lens · 3D-printed mirror Random laser · Multimaterial printing · Double patterning · Quantum dots Organic light-emitting diodes · OLED · Volumetric display · Inkjet printing Dispenser printing · Robot-based printing

3.1 Introduction

Standard 3D printing systems are highly attractive for the additive manufacturing of optics. One approach to classifying the existing methods is, for example, to group them by the material that they use. Thus, a distinction can immediately be made between metallic materials and plastics. For optical applications, this gives a distinction between reflective and transmissive optics. Aluminum can be used for reflective optics, for example. Selective laser melting (SLM) is a method that allows powdered metals to be processed. Common additive manufacturing methods for plastics include fused deposition modeling (FDM), multijet modeling (MJM), and stereolithography (SLA).

The FDM process involves melting one or more plastic threads with one or more extrusion heads that can be moved through space. This allows 3D components to be built up layer by layer. The layer thickness is typically around 100 μm (depending on the print volume and the actuator system). For applications in the field of optics, however, this process produces components that are too inhomogeneous, leading to high levels of light scattering within the component. Furthermore, the surface quality achieved by this process is relatively low.

The MJM process is comparable to inkjet printing. A liquid UV-reactive synthetic resin is applied directly by a print head in layers. After a layer is deposited in droplets, the material is leveled with a roller and cured with UV light. The minimum layer thickness possible with this process is a few dozen microns. To print overhanging structures, a second material (support material) is often printed in parallel as a support alongside the build material. The support material is then removed from the component, for example, with a water bath or solvent.

The SLA method primarily uses epoxy resins that are applied in layers and then hardened with a laser or a mask image (e.g., using a DMD projector). In some cases, this can achieve a minimum layer thickness of less than 10 μm. To realize overhanging structures, the build material is also used as the support material. The support structure is generated as a delicate structure that connects to the build area at "predetermined breaking points." This leaves behind a rough component surface when the support material is removed, requiring post-processing.

The general workflow of an additively manufactured optical element begins with its design in an optical design software program to specify the desired optical functionality. The element is then exported as an "STL file" that can be read directly by the printer. This is where so-called slicing is performed, which involves breaking down the model into individual layers that can be printed one by one. Post-processing is then typically required for the surface of the printed component.

Conventional 3D printing technologies are highly attractive for implementing optical components, since they allow new degrees of design freedom, for example, in terms of their shape or the combination of materials. However, these manufacturing methods also have various disadvantages that directly affect the optical properties of the components. This means that both an understanding of the manufacturing process and an intensive analysis of the printed components are essential prerequisites to successfully developing a custom additively manufactured optical element. Another benefit of adapting the manufacturing process is that it allows greater functionalization of the 3D-printed samples. This enables further differentiation relative to conventional manufacturing methods.

The next few pages present both the potential and the limitations of additively manufactured optics based on standard 3D printers. Various approaches to more advanced printing technologies that compensate for the inherent disadvantages of printing optical components on conventional 3D printers are also introduced.

3.2 Materials Used for the Additive Manufacturing of Optics Using Polymerization

3.2.1 Photopolymerization Categorized According to the Reacting Species

The fundamental principle used by most 3D printers for additive manufacturing with photoreactive resins is the polymerization of monomers, oligomers, or prepolymers, a process that is initiated in the presence of photons of a specific energy range. Being able to control this chemical reaction, which starts with the fluid resin composition and ends with a solid three-dimensional network of cross-linked polymer chains, is of prime importance when implementing this technique to realize rapid prototyping. In general, a photopolymerization route follows the three major steps of initiation, propagation, and termination. The photopolymerization dynamics can be optimized to achieve the desired control by varying the polymerization route, the reactive species, and the resin composition. The two major kinds of photopolymerization of interest in the field of additive manufacturing are explained in this section.

3.2.1.1 Radical Photopolymerization

When photopolymerization proceeds through the resin medium with a radical reactive species, it leads to free radical/radical photopolymerization [1]. The initiation route of a free radical photoinitiator is described in Sect. 3.2.2. Since free radicals are extremely reactive in nature, they benefit from high reaction rates. Using free radicals as the polymerization propagators results in a faster reaction at the initiation and termination stages. This is crucial when considering the swiftness with which the polymerization reaction can be "switched on and off" to form the constituent photopolymer drops/films/layers that make up the component. The possibility of continued radical formation is why acrylate-based resin compositions are used in radical polymerization-based 3D printing. But the extremely reactive nature of the radical species may also undermine the polymerization in some cases. One very common situation with acrylate-based or acrylated oligomers that undergo free radical polymerization is their passivation to photopolymerization in the presence of oxygen. The passivation occurs because the reaction between the radical species from the resin composition and oxygen forms unwanted reaction intermediates, such as side products or very stable peroxide radicals, thereby ending the reaction. Despite being an unwanted sidetrack when photopolymerization is desired, this passivation feature is utilized in some stereolithography 3D printers (see Sect. 3.7.4) to facilitate continued layer-by-layer accumulation to create the components on the print platform. A layer of a polymer network like polydimethylsiloxane (PDMS) that contains diffused oxygen in its porous surface is kept in close contact with the liquid resin layer in the resin container (vat). When the resin is irradiated through this PDMS layer, the liquid resin in immediate contact with the oxygen-diffused PDMS layer is inhibited from photopolymerizing due to the unavailability of a free radical population. However, this oxygen passivation can only inactivate the liquid resin up to a certain height. As a result, the liquid resin directly above this passivated liquid layer of resin (called the oxygenated dead zone) polymerizes onto the print platform.

3.2.1.2 Cationic and Anionic Photopolymerization

When the reactive species that initiates and maintains the propagation of photopolymerization is a cation or an anion, the process is called cationic or anionic photopolymerization, respectively. Cationic photopolymerization is a common method to photopolymerize epoxide and vinyl monomer compositions. Photoinitiators capable of producing cationic reactives include species such as onium salts and metal carbonyl compounds [1, 2]. The main restriction of cationic polymerization in the context of rapid prototyping processes is the extended time required to create and inhibit the reactive species needed for photopolymerization. In other words, the delay before initiation derails the layer-by-layer procedure. The delay in the inhibition of the reaction in the absence of photon energy leads to dark polymerization and extends anywhere from seconds to several hours [1]. Diffused oxygen has a negligible effect on the reaction rates of cationic photopolymerization, but the presence of moisture still hinders the reaction rate. Anionic photopolymerization is currently more or less in its infancy, owing to the absence of corresponding resin compositions and fabrication techniques. The upside to anionic photopolymerization routes is that the reaction is not inhibited by oxygen or moisture in the reaction environment.

3.2.2 *Resin Composition*

As the raw material of additive manufacturing, it is essential for the photoreactive resin entity to be versatile with respect to the various stages of the photopolymerization, including initiation, controlled propagation, and reaction termination. Spatial or temporal variations in the resin behavior in any of these steps may result in fabrication artifacts like blooming, changes in the printed part's dimensions due to nonuniformity in layer stacking within the part, or even complete print failure. It is therefore important to understand the basic constitution of the raw material, to test a reasonable hypothesis in the event of variations, or to reverse engineer specific artifacts in the raw material or its behavior. The basic ingredients required to synthesize a modest photoreactive resin mixture are outlined in Fig. 3.1.

3.2.2.1 Photoinitiators

The photoinitiator is the most prominent light absorber in a photoreactive resin mixture. Depending on the molecular composition of the photoinitiator and the chemical environment surrounding it, photoinitiators absorb energy and initiate a chain of chemical reactions (polymerization reactions) that bond the prepolymer strands into long polymer chains with cross-links, leading to their solidification, which results in a three-dimensional network of cross-linked polymer chains. The Jablonski diagram that depicts the transition of an electron in the photoinitiator molecule after its photon absorption is shown in Fig. 3.2. Upon photon absorption, photoinitiator molecule electrons move to an elevated singlet state, from where the

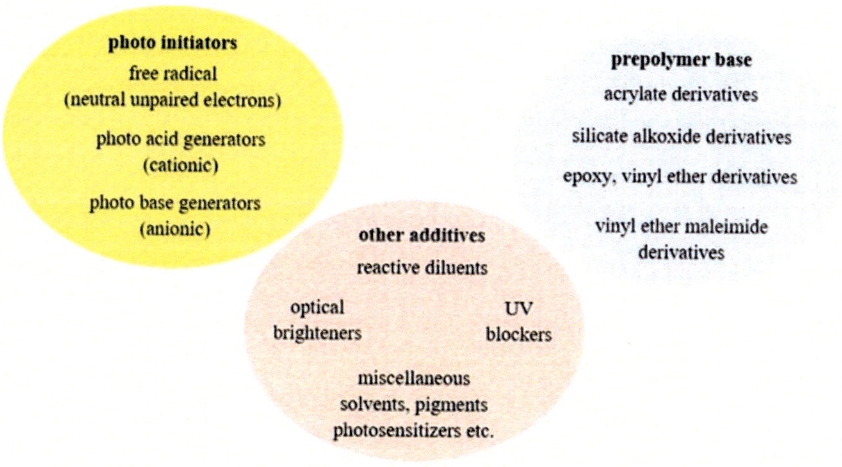

Fig. 3.1 A schematic diagram indicating the most commonly found constituents in a prepolymer liquid resin mixture reactive to photon initiation

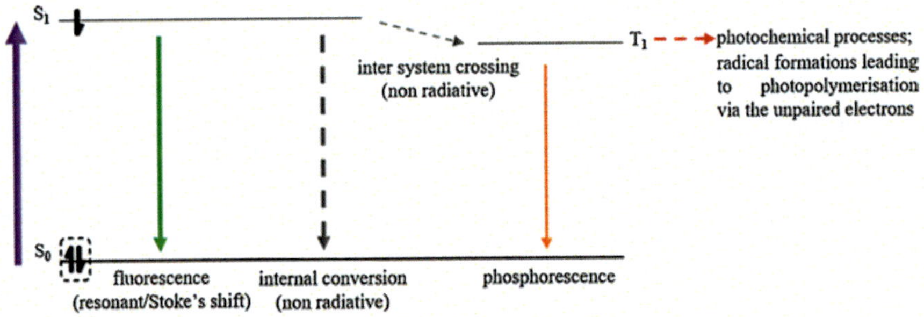

Fig. 3.2 A schematic drawing of the Jablonski diagram indicating the possible relaxation routes that an excited electron may follow in response to photonic excitation in typical photoinitiators—the relevance of an unpaired excited electron retained in the triplet state to initiate the radical photopolymerization reaction

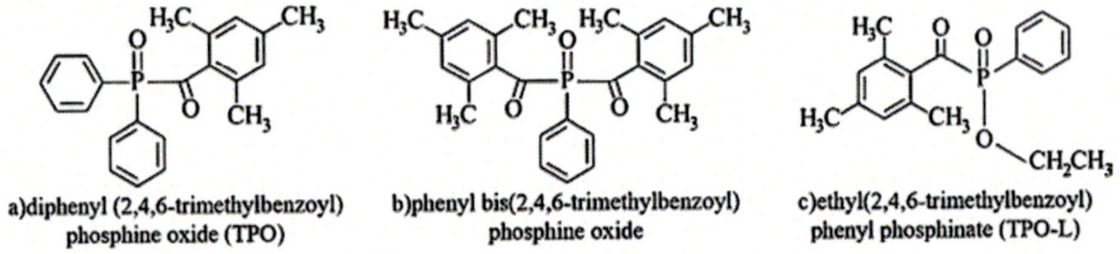

Fig. 3.3 Schematic showing certain commonly used derivatives of acyl phosphine oxide that absorb photons with specific wavelengths based on their molecular structure, thereby initiating photopolymerization in acrylate-based resin compositions

electron may either relax into its ground state via radiative or non-radiative transitions or take on a slightly lower-energy triplet state by reversing its spin (intersystem crossing). In most photoinitiator molecules, this free electron in the triplet state is the starting point for the polymerization reaction. Depending on the surrounding chemical environment, either an energy transfer or an electron transfer occurs in this longer lifetime triplet state, creating the reactive species that attacks the prepolymer molecules and thereby initiating the 3D network formation.

It should also be noted that only photons absorbed by the molecules (and not all irradiated photons) can cause a photochemical event (the first law of photochemistry) and that all photons absorbed by the molecules will cause a physical/chemical event (Stark-Einstein Law). In other words, only a certain proportion of the photons of the total irradiation on the photoreactive resin can be absorbed by it, and only a certain proportion of this quantity can initiate the desired photochemical event of photopolymerization initiation (the quantum yield of initiation).

A typical molecule capable of photon absorption and its derivatives used as photoinitiators in photoreactive resin compositions are shown in Fig. 3.3. These derivatives fall under the category of type I photoinitiators, where the initiation reaction is a result of radical reactive species due to electron transfer from the photoinitiator molecule. The absorption wavelength and the physical properties of the photoinitiator molecule vary as a function of the chemical moieties present as side groups. For

Table 3.1 Collective representation of common industrial photoinitiators and their absorption wavelengths

Photoinitiators (PI) and photosensitizers (PS)		Peak of absorption wavelength usable for rapid prototyping (nm)
Type I PI for radical polymerization	TPO L	274 [3]
	TPO	382 [4]
	BAPO (Irgacure 819)	371 [3]
Type II PI	Camphorquinone	467 [5]
PS for cationic photopolymerization	2-Isopropylthioxanthone	383 [6]
	Irgacure 784	460 [6]

instance, the derivative TPO-L (Fig. 3.3b) is liquid in form, which makes it easier to integrate into the fluid resin composition, while TPO (Fig. 3.3a), which has a solid form, has an absorption wavelength that is redshifted relative to TPO L. A summary of the absorption wavelengths of different photoinitiators is shown in Table 3.1.

Type II photoinitiators with an energy transfer step before the radical formation due to electron transfer are also used in the free radical polymerization route followed by the acrylate resins used for additive manufacturing with photoreactive resins. The main advantage with such systems is the longer wavelengths of absorption, which are attractive for applications like dental fillings on a live patient, where blue light radiation to solidify the fillings is more desirable than subjecting the live tissues to ultraviolet radiation.

With elevated research in the direction of increasing the efficiency of photoinitiation and simplifying the process, certain interesting photoinitiating systems are also in use. One concept being developed is that of an iniferter, where the same constituent in the resin composition works to initiate the polymerization process, induce chain transfer during the process, and terminate the photopolymerization. Progress is also underway for panchromatic photoinitiators for a wide range of prepolymer bases (cationic polymerization of epoxides, radical polymerization of acrylates, and thiol-ene polymerization) over a wide range of absorption wavelengths (405 nm, 470 nm, 520 nm, 594 nm, 636 nm) to yield various degree of curing (~44–67% for an epoxide and ~30–39% for an acrylate) [7].

3.2.2.2 Prepolymers

Prepolymers are the resin constituents that primarily undergo polymerization to form the long polymer chains that cross-link and become physically entangled with each other to form the 3D polymer network. In this respect, they could be thought of as the backbone of the structural formation of the layers fabricated within the 3D-printed component. The choice of oligomer or prepolymer used in a particular resin composition depends primarily upon the expected properties of the end product, as well as compatibility with the rest of the mixture.

The most popular choice of prepolymer in the context of additive manufacturing based on layer-wise model fabrication comes from the branch of acrylate oligomers or acrylated variations (e.g., epoxy acrylates). This is because the acrylate family of oligomers exhibits much faster reaction times to photon excitation. This is a very critical factor for the initiation and termination of photopolymerization when a specific dosage of photon energy is supplied to fabricate each layer of the component. With the cationic photopolymerization mentioned in Sect. 3.2.1, which applies to epoxy resins, dark polymerization (the prolongation of photopolymerization for a small period of time after the removal of the photon energy) is prominent, which obstructs rapid prototyping. Multifunctional urethane acrylate oligomers are commonly seen in commercial photoreactive resins used for stereolithography. Within this class, the functionality of the oligomer (the number of other oligomer molecules that one oligomer molecule can bond with) can change the properties of the formed network and sometimes even the degree of curing of the photopolymerized layer/film. For instance, the curing speed of tetrafunctional oligomers was measured in one study to be faster than that of hexafunctional acrylate oligomers [8]. In other words, the oligomers occupied different states within the curing cycle (gelation to vitrification) for the same energy dosage. The physical properties like elongation, tensile strength, yellowing, and degradation are also influenced by the choice of oligomer. In most cases, multicomponent systems where similar proportions of different prepolymer acrylates are mixed together are used instead of a homogenized monomer mix that undergoes bulk polymerization, as is the case with the methyl methacrylate that forms PMMA. To summarize, the choice of the prepolymer base follows a logistic thought process where the pros and cons for a customized product need to be weighed to make the best choice.

3.2.2.3 Reactive Diluents

While the prepolymer base is the main component in the resin composition by weight, it is often a very viscous mixture. This is an important parameter when considering a potential raw material for rapid prototyping, where the components need to have enough mobility for the components to be within mutual vicinity to allow the bonding reactions to continue. Reactive diluents are often acrylate monomers that are mixed into the resin composition to bring down the viscosity of the mixture to maintain fluidity between the components and to maintain the optimum fabrication conditions. It should also be noted that the proportion of reactive diluents in the mixture directly influences the molecule mobility within the resin mixture and therefore can be used to control the molecular weight distribution of the polymer network that forms and hence the network density of the polymer film [9].

3.2.2.4 Other Optional Additives

Optional components can be added to the resin composition, often in very small quantities, to fine-tune the basic characteristics of the resulting photopolymer films. One such important additive in a photoreactive resin composition used for additive manufacturing is the UV absorber. These additives absorb the excess UV irradiated onto the resin composition. The absorbed energy is converted into vibrational and rotational energy for the molecular constituents, thereby inhibiting further radical formations that could change the photopolymerization rates.

The presence and quantity of UV absorbers in the resin composition has an influence on the curing depth of the layer that forms from the photoreactive resin during 3D printing. In some cases, UV absorbers are used in combination with other stabilizers like hindered amine light stabilizers (HALS) to protect the fabricated components from light and oxygen triggers [10]. Fluorescent whitening agents are also added in some cases to combat the common problem of yellowing of the photopolymerized film due to extended exposure to ultraviolet-containing radiation in daily life by absorbing the ultraviolet radiation and re-emitting the energy at a longer, blue wavelength, giving the sample a bluish hue in appearance. Because of this brightening effect, these fluorescent agents are also called optical brighteners.

Fillers are included in certain compositions to increase the tensile strength, toughness, heat resistance, color, clarity, etc. Titanium dioxide is used as a filler in certain compositions to reduce the yellowing. Other plasticizers are used as an alternative or in combination with reactive diluents to increase the free volume between the polymer chains, thereby increasing the chain mobility, which in turn has a direct influence on the flexibility of the polymerized structures that form. Adding certain solvents to the resin compositions in very small quantities is preferable to controlling the reaction environment within the composition. For instance, BHT (butylated hydroxytoluene) acts as a polymerization inhibitor when added to certain monomers, to facilitate their safe storage [11].

Innovative optical components can also be fabricated using additive manufacturing techniques by adding optional additives like nanoparticles into the resin composition to fine-tune optical properties like the refractive index (e.g., dioxides of titanium, zirconium) or photoluminescence (e.g., quantum dot nanocrystals). One such nanocomposite application is explained in Sect. 3.6.3.

3.3 Analysis of Additively Manufactured Optics

The previous section considered a wide variety of materials and material systems for additive manufacturing. The following discussion is restricted to additive manufacturing based on radical photopolymerization.

Analysis is another factor that plays a key role in successfully realizing additively manufactured optics. Firstly, it is important to consider ways to characterize the printing process itself. But the final printed sample also needs to be analyzed, especially in terms of certain optically relevant properties. Finally, the analysis methods need to establish whether post-processing is necessary.

3.3.1 Analysis of the Printing Process

Prior to any analysis of the optical properties of an additively manufactured optical element, there are various key parameters that are relevant to the printing process itself. First and foremost, the curing wavelength must match the sensitivity curve of the resin, or more precisely the photoinitiator [12]. If the resin is only sensitive to wavelengths below 385 nm, an LED with a wavelength of 405 nm may not be able to achieve the desired curing, or possibly even any curing at all. To identify suitable exposure and layer thickness parameters for the process, the resins must be characterized in terms of both their penetration depth D_p and their critical energy density E_c. This energy density describes an energy dose per unit area. The dependence of the curing depth C_d on the penetration depth, the supplied energy density E_0, and the critical energy density E_c is described by Jacobs as follows [13]:

$$C_d = D_p * \ln\left(\frac{E_0}{E_c}\right) \tag{3.3.1}$$

The model predicts a logarithmic dependence of the thickness of a cured layer on the energy dose supplied per unit area. This is an empirical formula that does not consider the influence of other actors such as the photoinitiator on the reaction [14]. Nevertheless, in practice, it often provides a sufficient description of the parameters required to describe the process, as is shown below.

To obtain the measurement curves shown in Fig. 3.4, an area of 1 cm^2 was irradiated. The energy density can be adjusted by changing the irradiance and/or exposure time. The figure plots the curing characterizations of different resins. For each resin, the thickness of the cured layer was determined at four different energy densities. From the critical energy density E_c, the values form a straight line of slope D_p in log scale, which is consistent with the model. The slope D_p is defined as the penetration depth. A steeper slope means that slightly increasing the energy density will allow a significantly thicker layer to be cured, which is beneficial in the context of rapid build processes. Resins with lower D_p values and hence gentler slopes allow the desired thickness to be adjusted more finely by varying the time control as usual while maintaining a constant irradiance within the printer.

Different printing processes also have different requirements in terms of the rheological characteristics of the resin. Inkjet-based processes in particular require low viscosities of less than 100 mPas, whereas some resins for stereolithography (SLA) can exceed 1000 mPas at room temperature. Inkjet printers often include a heating feature that allows the viscosity to be adjusted by increasing the temperature of the resin. Figure 3.5 shows the viscosity of various resins as a function of temperature in the range from 20 to 90 °C. Over this range of temperatures, the viscosity

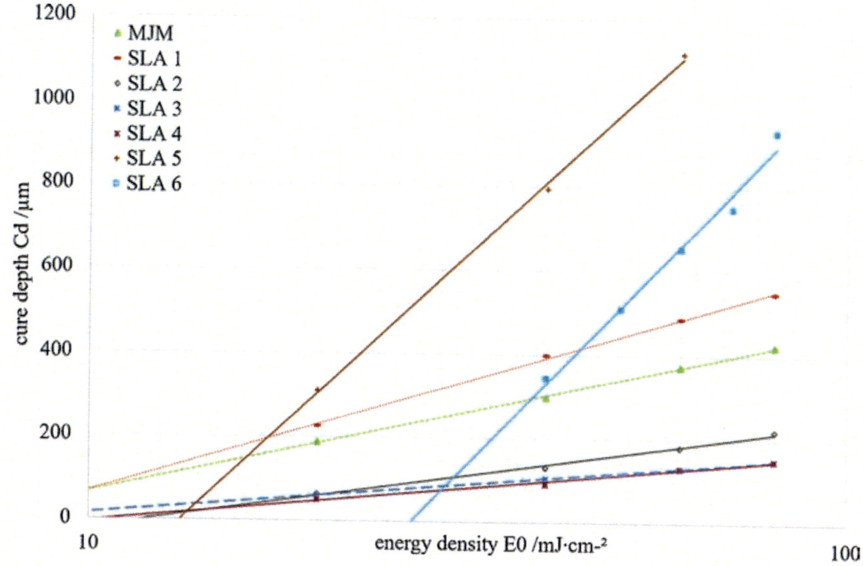

Fig. 3.4 Relationship between the cured layer C_d and the energy input E_0. According to Jacobs' model, the penetration depth D_p is given by the slope of the fit line

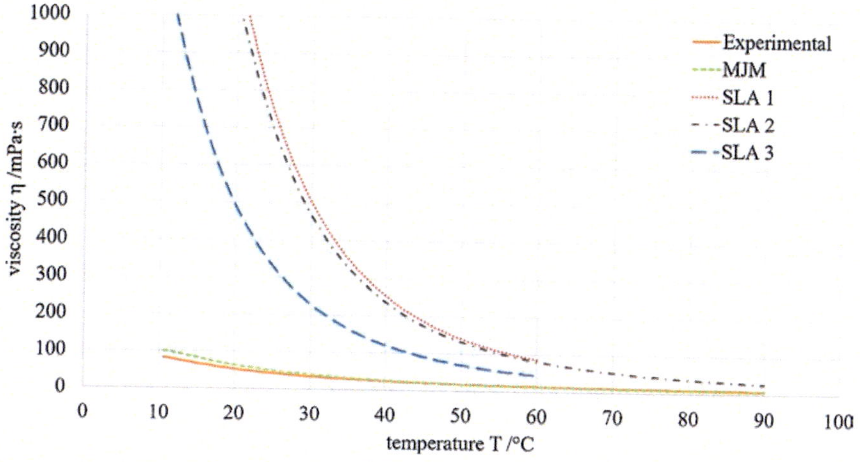

Fig. 3.5 Temperature dependence of the viscosity of different resins in the range from 20 to 90 °C

decreases significantly; accordingly, resins designed for the SLA process at room temperature can also be used with inkjet printing when heated. The resins shown in the illustration are all Newtonian fluids, so the viscosity does not depend on the shear rate. It should also be noted that photopolymerization is an exothermic reaction. The curing process of the material releases heat. For stereolithography, this changes the temperature of the resin bath. Especially when additively manufacturing structures in the µm range, this temperature increase can influence the kinetics of the curing process.

3.3.2 *Analysis of the Shape and Surface of Additively Manufactured Optics*

The functionality of an optical component is largely defined by its material, but also depends on its volume, shape, and surface. One of the great advantages of additive manufacturing is that it is capable of realizing complex geometries. Nevertheless, the layered structure of the process needs to be considered. Figure 3.6 gives an example. Figure 3.6a shows the optical shape measurement (point cloud) of an inkjet-printed spherical lens. The distinctive layered structure is clearly visible. Although each step is strongly rounded, the slicing process causes a plateau to form near the center. In other words, due to the lack of curvature, this area does not produce any optical effect. Artifacts resulting from the resolution of the print head or the printing process itself (e.g., the local application of individual droplets in the case of inkjet printing) are also visible. The artifacts and the formation of steps as described above mean that surface post-processing is required.

Figure 3.6b shows a flat plate (1 mm thick) that was printed vertically on the build platform. This upright orientation resulted in clearly visible structure in the component, consisting of many individual layers (each 30 µm thick). This creates roughness in the final surface.

As described above, decomposing the component into layers leads to the formation of steps when printing inclined or curved surfaces. Some printers allow the machine data to be viewed and analyzed after the component is decomposed into layers. A simple example of such an analysis is visualized in Fig. 3.7a. If the individual layers are reassembled into a 3D object (red step curve) and compared with the target data (blue curve with even curvature), there are clear deviations. The discretization approximates the curvature as best as possible with horizontal steps subject to the layer thickness and the lateral resolution of the printer, but a residual error remains. This residual error depends on the z-resolution of the printer, as well as the shape and orientation of the component during printing. At the extreme ends of the spectrum, the curvature is rendered as a flat surface at large radii, and the smallest

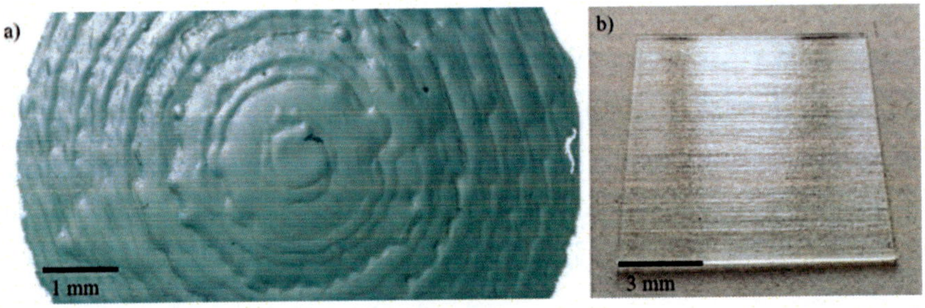

Fig. 3.6 The layer-by-layer build process of conventional Additive manufacturing can lead to (**a**) shape deviations when approximating curved surfaces and (**b**) rough surfaces in areas that consist of a large number of layers

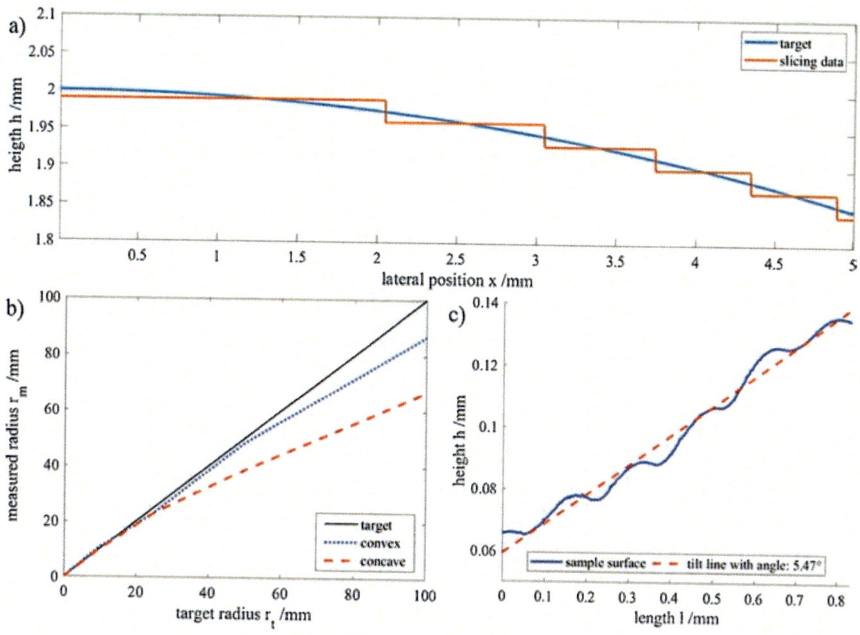

Fig. 3.7 (**a**) Visualization of the curvature approximated by a layered structure; (**b**) comparison of target and measured values of printed concave and convex spherical surfaces; (**c**) measured step structure of an inclined printed surface with a line of best fit

possible sphere is represented as a cube. If these steps were implemented perfectly, the stair-stepping effect would not result in any deflection for rays perpendicular to the layers, since there is no curvature, simply a sequence of flat levels with different heights. In practice, however, the edges are rounded, which has an unfavorable effect on light propagation.

To establish a better classification of the deviations, it is advisable to perform measurements on printed reference geometries. Figure 3.7b shows a comparison of the target and measured values at different radii. The radii were calculated by fitting circles to the topology profiles of real printed samples measured with a laser scanning microscope. To simplify the visualization, the value of the radius is plotted for concave and convex curvatures separately. The deviation between the measured radius and the target radius can clearly be seen to increase as the value of the radius increases. Concave curvatures deviate more strongly from the target than convex surfaces. One possible explanation for this could relate to the wetting effect, a phenomenon relevant in inkjet printing that determines the shape (or more specifically the contact angle) that forms when a droplet is applied to a substrate, governed by the surface tension and the surface energy. To prevent the applied material from flowing away, the printing process may place the paths further apart when printing concave geometries, which manifests locally as a smaller radius. The measuring equipment can only detect a very limited section of the curvature. The relative error in the discretized curvature becomes less important as the size of the printed spherical section increases. The paraxial area around the optical axis is therefore an extreme case.

The effect described above is not limited to curved surfaces, but also occurs on inclined surfaces, as shown in Fig. 3.7c. The red dashes show the line of best fit for the blue measurements of an inkjet-printed component with a 5° slope relative to the build platform. The ripples once again display the rounded-off stair-stepping effect described above. Over the section of just under 1 millimeter shown in the figure, the measured angle deviates from the target angle by about half a degree. In such a context, it is helpful to consider the frequency spectrum as a power spectral density or the simplified approach of a Fourier transform to conduct a detailed analysis. This allows the spatial frequencies of the process to be identified.

The stepping effect mentioned earlier is shown in Fig. 3.8 from above. Figure 3.8a shows how the slicer adds a small protrusion to the top of the highest flat surface to minimize the error in the target radius (marked with an arrow). Figure 3.8b shows a case where only a single step is generated due to a very large radius of curvature. Figure 3.8c shows a complex freeform with clearly visible discretization artifacts. As can be seen in the images of Fig. 3.8, there are rounded sections (black rings) at the edges of each step, meaning that a plateau with reduced optical effect only forms at extreme radii. However, since small bumps can act like independent microlenses, these elements and the resulting edges can contribute to light scattering.

In addition to the shape, the surface and surface roughness must also be considered to characterize the additive manufacturing of optical components. The layered build process creates surfaces whose roughness varies as a function of the angle, curvature, and orientation. The lowest roughness values and hence the best surfaces are obtained with surfaces that are flat parallel to the printing platform, since the surface is formed by a single layer. With projector-based SLA printing systems, even flat layers may exhibit ripples, since the curing unfolds locally with a projected mask. Figure 3.9a shows the measured topology of a supposedly flat surface that was cured with a DMG projector. The pixels of the projector, which have an edge length of 50 μm, can be clearly distinguished. Figure 3.9c–e shows transmitted light images of different surfaces of an inkjet-printed cube. As mentioned earlier, the top

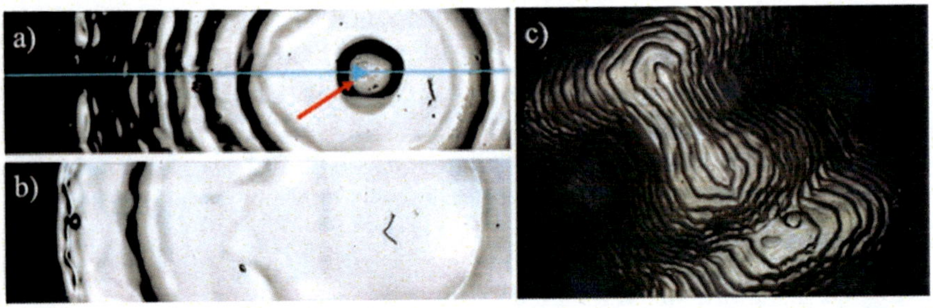

Fig. 3.8 Top view of surfaces with different curvatures: (**a**) spherical cap with a small radius with a protrusion (arrow) on the plateau surface; (**b**) spherical cap with a large radius whose final layer forms a plateau; (**c**) freeform with clearly visible step artifacts in all directions

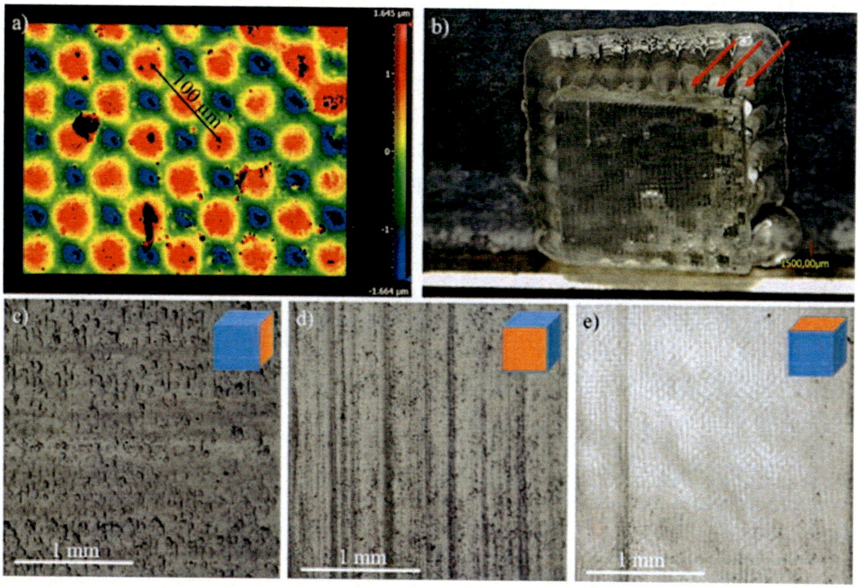

Fig. 3.9 Surface analysis for various additively manufactured samples; (**a**) projector-based curing produces a pixel-like structure within the same printed layer; (**b**) support structure on the underside of a plate; (**c–e**) transmitted light images of different surfaces of a printed cube. Different orientations result in different surfaces

side (e) is smooth, whereas the lateral sides have visible structure in some places. The roughness value R_a can reach up to 5–10 µm with unfavorably oriented surfaces, meaning that post-processing is unavoidable.

Different printers and printing technologies differ not only in terms of the achieved surface quality but also in terms of the support structures that they can implement. Support structures are used to create elements with strongly overhanging sections (e.g., >60°). They are in direct contact with the build material, so they significantly affect the surface roughness. Printers that realize supports from the same material as the build material usually leave behind small contact points on the component. To prevent these contact points from causing scattering, the surface requires post-processing. Alternatively, the orientation of the component within the printing process can be arranged in such a way that the support structures connect to areas that are not optically active. Figure 3.9b illustrates this problem. It shows a top view of a 3D-printed plate that was built on support structures (see arrows) because of its inclined orientation. The very flat but inclined surface is supported by a large number of tips, which leave behind a large number of protrusions on the underside of the component. These protrusions require post-processing.

As an alternative, some printing systems use a water-soluble support material. Nevertheless, the quality of surfaces with supports remains significantly rougher than unsupported surfaces. This example once again illustrates the importance of design and orientation for the conventional 3D printing of optical components.

There are various approaches for the post-processing of conventionally printed optical elements. The classical method is mechanically abrasive polishing. However, this restricts the degrees of freedom when designing complex geometries due to accessibility. The fluid jet polishing method can be used as an alternative to abrasive polishing that maintains the design freedom [15]. The material removal function of this method has its own set of characteristics, meaning that not all spatial frequencies can always be smoothed out equally. In general, these post-processing methods can achieve surface roughness values of less than 10 nm R_a. However, they are relatively time-intensive to implement.

A simpler and faster method is to apply a spray-on clear lacquer that smooths out the steps and surface structures. To ensure that the refractive indices of the lacquer and the build material are consistent, the build material itself can be used, for example, by dip coating. In both cases, the resulting shape deviations depend on the wetting behavior of the material on the component. The advantage of these methods is that they allow the coating and post-processing of inaccessible areas. The dip coating process is discussed in more detail in the next section.

3.3.3 *Dip Coating to Improve the Surface of Additively Manufactured Optical Elements*

The surface quality of additively manufactured components does not meet the specifications of optical elements. Consequently, post-processing is required. However, many objects have a high complexity that makes it difficult to access their surfaces. This can be especially challenging for conventional polishing processes.

The dip coating process is an inexpensive and quick alternative. The process involves coating rough surfaces of a sample with resin from the printing process in a single dipping step. The surface tension of the resin produces a self-smoothing effect in the liquid resin layer. The dipped part is then cured with ultraviolet radiation (e.g., 405 nm) within an inert nitrogen atmosphere. Since the refractive indices of the component and the coating are the same, the point of optical transition shifts from the original rough surface of the component to the smooth surface of the coating.

The dip coating process is illustrated below with the example of small, flat plates with an edge length of 10 mm and a thickness of 1.5 mm (see Fig. 3.10a). The samples were realized with the multijet modeling process.

Robot kinematics were used to ensure that the process is repeatable and flexible. The robot moves the sample between an ultrasonic bath (1), a container filled with coating material (2), an analysis module (3), and a nitrogen chamber (4) for curing (see Fig. 3.10b). The robot kinematics system allows both the angle and the speed of immersion to be varied for process optimization.

Three analysis methods were used to qualify the process development steps: a transmitted light camera setup to detect particles and air bubbles, a visual transmission test that observes a stripe pattern projected through the sample, and a measurement of the surface roughness using a white light interferometer (WLI).

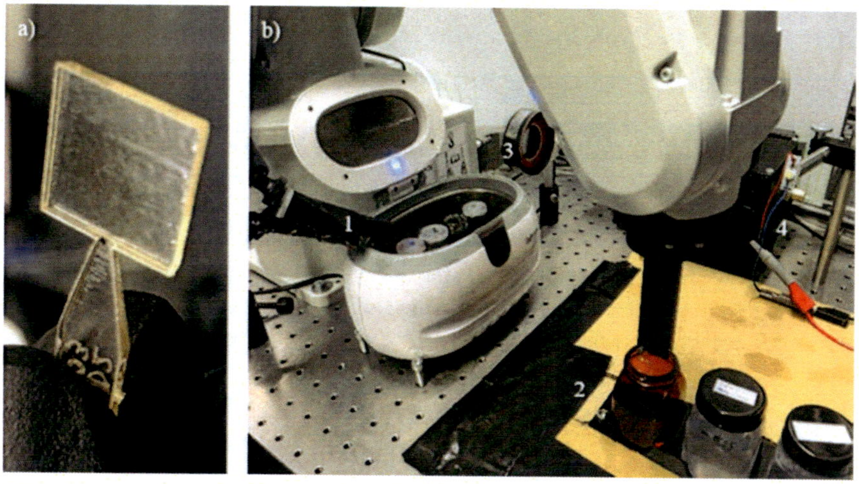

Fig. 3.10 MJM test sample with 1.5 mm thickness and 10 × 10 mm^2 edge length

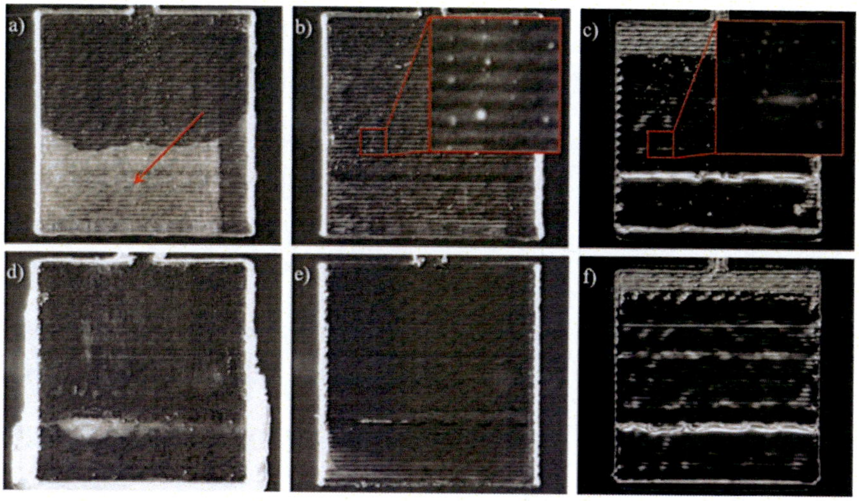

Fig. 3.11 Sample with deposits (**a**), particles (**b**), coating with trapped particles (**c**), support material (**d**), cleaned (**e**), coated (**f**)

The samples must be cleaned before the coating process itself. This is especially important if the printing process uses a water-soluble support material. Cloudy deposits may remain on the sample on any horizontal surfaces that come into contact with the support material during printing (see Fig. 3.11a). These deposits and particles are then trapped by the subsequent coating, or form nucleation points for air bubbles (see Fig. 3.11b and c). Thorough cleaning is therefore essential to achieve homogeneous and consistent wetting.

A wide variety of solvents or acids can be used for cleaning – depending on the build material and the support material. In the present case (Keyence Agilista multijet printer), the support material was removed in an ultrasonic immersion bath of 37% hydrochloric acid. Unlike cleaning with water, the support material is completely dissolved, whereas the build material is largely unaffected. The hydrochloric acid turns into a slightly reddish color, so it can be used as an indicator of dissolved support residues from the sample.

However, the cloudy deposits and particles on the sample do not dissolve in the hydrochloric acid. The sample was therefore cleaned with four successive isopropyl alcohol immersion baths. Since isopropyl alcohol has a degenerative effect on the material, the cleaning is accelerated by the ultrasonic bath and limited to 60 seconds per immersion bath. Figure 3.11d–f shows a sample before cleaning, after cleaning, and after dip coating. Residual ripples can still be seen on the coated sample, but there are no longer any particles or air bubbles.

There are two decisive parameters for the dipping process: the smoothing of the sample and its shape retention. The most influential process parameter for these factors is the pull-out speed. A low pull-out speed ($v_a = 0.1\ mm/s$) applies less resin and therefore achieves more favorable shape retention. This can be seen in the profile of the sample, shown in Fig. 3.12a. However, this also produces a weaker smoothing effect. By contrast, a thicker layer of resin ($v_a = 0.5\ mm/s$) provides better smoothing. As a trade-off, the resin causes a deformation as it runs off the edges of the sample (Fig. 3.12b). Drainage structures are a possible remedy for this problem. By adding such a structure to the lower edge of the samples, excess resin can be collected on the optically irrelevant lower edge to improve the shape retention (Fig. 3.12c).

Another approach is repeated dipping. Multiple very thin layers are applied in sequence until sufficiently good smoothing is achieved. Since the coating material will often struggle to wet favorably on layers that have already been cured, the intermediate layers are cured within an oxygen atmosphere. This limits polymerization on the surface (Fig. 3.12d) and improves the process as a whole.

Figure 3.13 shows the optical analysis of an uncoated plate (a) and a coated plate (b). The transmission of a stripe pattern with 100 line pairs per centimeter is shown

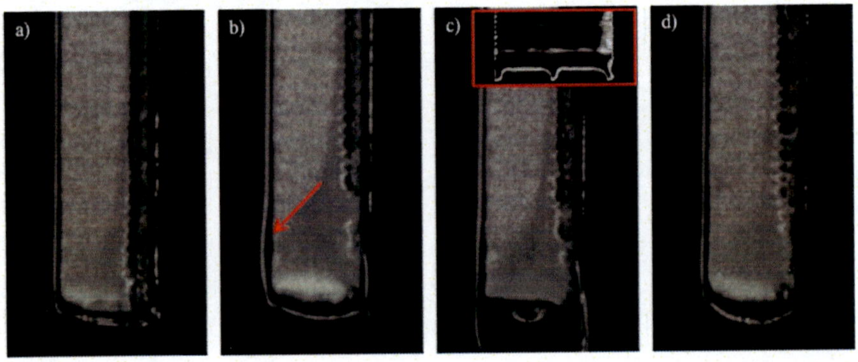

Fig. 3.12 Samples in profile: $v_a = 0.1$ mm/s (**a**), $v_a = 0.5$ mm/s (**b**), $v_a = 0.5$ mm/s + drainage structure (**c**), 3× layers with $v_a = 0.1$ mm/s (**d**)

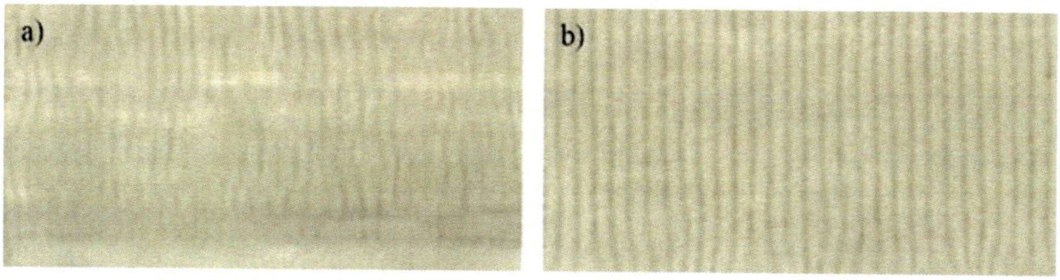

Fig. 3.13 Representation of 100 LP/cm through an uncoated sample (**a**) and a coated sample (**b**)

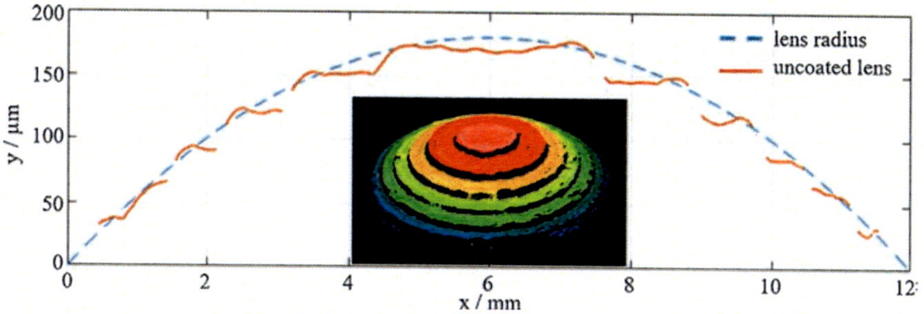

Fig. 3.14 Topology of the uncoated lens (WLI image)

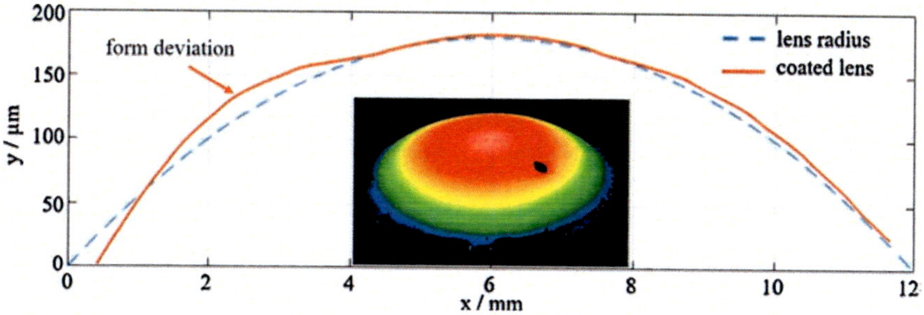

Fig. 3.15 Topology of the lens after dip coating

through the plates. Before the coating process, the line pairs cannot be resolved due to the rough surface after printing. However, the dip coating makes them clearly visible. The deformation in the lines at the bottom of the image is caused by the accumulation of resin at the bottom of the sample.

The dip coating of a plano-convex lens with a radius of curvature of 100 mm and a diameter of 10 mm is discussed below as another example. The surface of the uncoated lens is shown in Fig. 3.14. The curved surface is discretized into five layers of 30 µm thickness each by the printing process.

Figure 3.15 shows the same lens after dip coating ($v_a = 0.5\ mm/s$). The individual layers are no longer distinguishable after the coating is applied. However, there is some shape deviation caused by resin runoff.

There remain challenges when attempting to coat samples with very narrow and deep grooves. In such samples, insufficient wetting can occur, leading to air pockets. A possible solution is to reduce the viscosity of the coating material, for example, by heating it or electrically activating the surface of the substrate.

To reduce shape deviations, the runoff behavior of the resin can be anticipated when designing the designing the component geometry. After dividing the geometry into optically relevant and optically nonrelevant areas, the latter can be used to collect excess material from any optically relevant areas. Another way to achieve homogeneous coating is by controlling the path movement with robot kinematics, which allows six degrees of freedom.

3.3.4 Analysis of the Optical Properties of Additively Manufactured Elements

It only makes sense to discuss the shape and surface of a 3D-printed optical element if the material is already known to be suitable for optical applications. This section therefore discusses the optical properties of 3D printing materials, most importantly the transmittance and the refractive index.

The spectral transmittance of a material describes how the material transmits light at different wavelengths. As the classical material for optical components, glass usually has a high transmittance across the entire visible spectrum. For UV-curable 3D-printing resins, the transmitted spectrum is limited by the material's curing wavelength. In most cases, transmission starts at approx. 10 nm above the curing wavelength. The transmittance curves of various widely used resins are shown in Fig. 3.16. Five resins with a nominal curing wavelength of 405 nm are shown, as well as one resin with a curing wavelength of 365 nm. As a general rule, it can be observed that the transmittance first increases abruptly and then continues to increase more gently up to the infrared domain after a certain transition region. The measuring system introduces artifacts into the diagram at 650 nm that should be ignored by the analysis. As the example shows, it is not true in general that materials with a lower curing wavelength have higher transmittance values in the visible spectrum. The individual properties of each material are ultimately decisive.

The absolute power transmitted at specific wavelengths also depends on the thickness of the material. Whereas good transmittance values are achieved by thin layers of material over a wide spectral region, thicker samples or samples with longer light paths such as light guides exhibit greater differences between wavelengths. With values of 0.01 dB/min, the attenuation is significantly higher than in glass.

The transmittance values can often be improved by post-curing—either with another round of intensive UV radiation or with tempering.

It should also be noted that production factors can create light scattering within the component. Figure 3.17 shows an example of this. In Fig. 3.17a, a light-guiding element is shown immediately after the printing process (multijet modeling

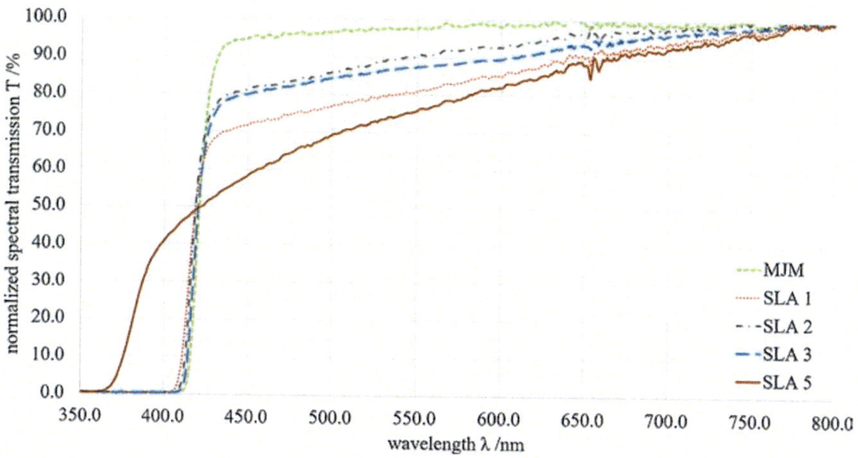

Fig. 3.16 Relative transmission for various 3D printing materials

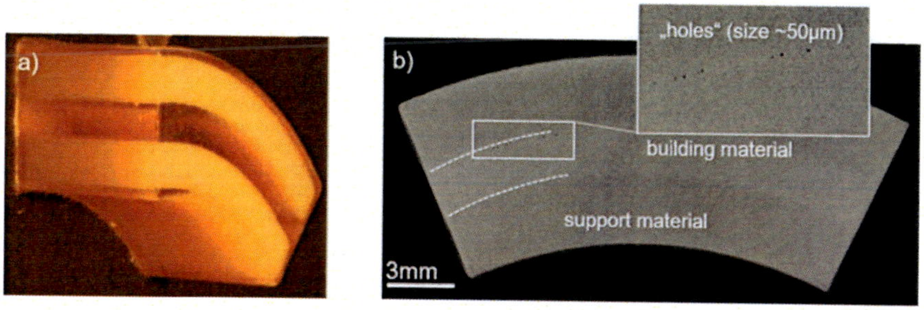

Fig. 3.17 (**a**) Additively manufactured light-guiding element immediately after the printing process; (**b**) computed tomography image of the light-guiding element

printing), still embedded in its support material (light yellow). A computed tomography image of the component is shown in Fig. 3.17b. This image shows that the printing process left behind holes with a diameter of 50 µm, mostly between the build material and the support material. These holes act as clear scattering centers.

The refractive index and the Abbe number are two other key characteristics of optical components. The Abbe number describes the dispersion within the element. Since glass is the default material for optical components, there is a large selection of glasses with different refractive indices and Abbe numbers. Glass manufacturers typically present their product portfolios in the form of a "glass chart," as shown in Fig. 3.18 [16]. The labeled areas separated by blue lines allow rough positioning relative to standard reference glasses. The orange areas connected by an arrow show the change in the refractive index between the liquid and cured states for a selection of transparent 3D printing materials. As the diagram shows, the refractive index increases during the curing process.

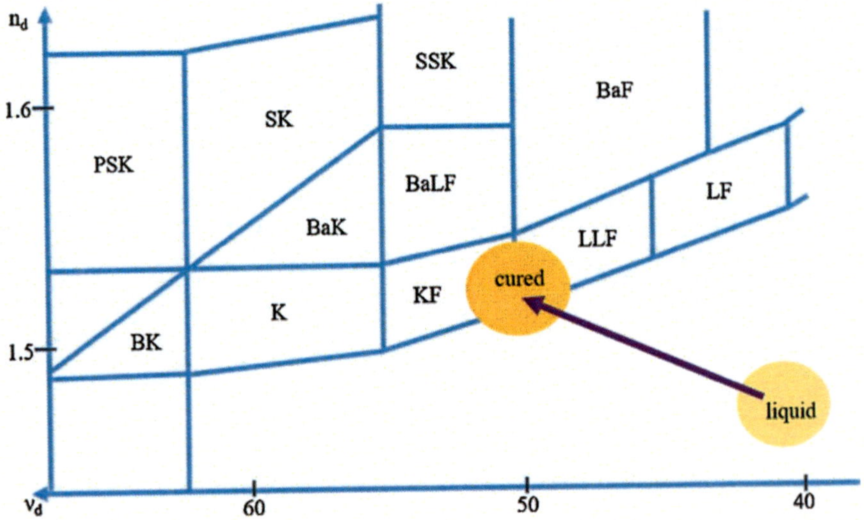

Fig. 3.18 Classification of various 3D printing resins in liquid and cured states relative to standard reference glasses

The standard 3D printing resins shown here, which are based on the polymerization of monomer/oligomer mixtures without additives, have refractive indices ranging from $n = 1.5$ to 1.53 and an Abbe number of approximately 50. This range can be expanded to $n = 1.35$ to 1.7 by adding nanoparticles or adapting the material composition. Typically, however, at particularly high refractive indices, the viscosity also increases very strongly with the nanoparticle concentration. This often makes inkjet printing systems difficult to use. Dispenser systems offer a way out (see Sect. 3.7).

The increase in the refractive index that occurs during curing as mentioned above plays a major role in local and layer-by-layer curing, as differences in the extent of curing can produce inhomogeneities in the material. To measure the local distribution of the refractive index in a 3D-printed sample, the "scanning focused refractive-index microscopy" or SFRIM method can be used [17]. With this method of analysis, the sample is placed on the surface of an equilateral prism (see Fig. 3.19a). A laser is focused on the interface between the prism and the sample (resin) with a high numerical aperture at an angle of 60°, i.e., perpendicular to the entry surface of the prism, and the totally reflected beam after the exit surface of the prism is evaluated with a camera chip (arrow). The focused beam has a wide-angle spectrum, so, depending on the refractive index of the sample at the focus/measuring point, rays of certain angles are either totally reflected or coupled out to the sample. Figure 3.19b shows a section of the camera image with a light-dark edge, as well as an evaluation of a profile section. The profile section allows the refractive index to be determined by fitting the curve describing the Fresnel reflection between the prism and the sample. After attaching the prism and the sample to an XYZ translation table, the sample can be scanned in two dimensions.

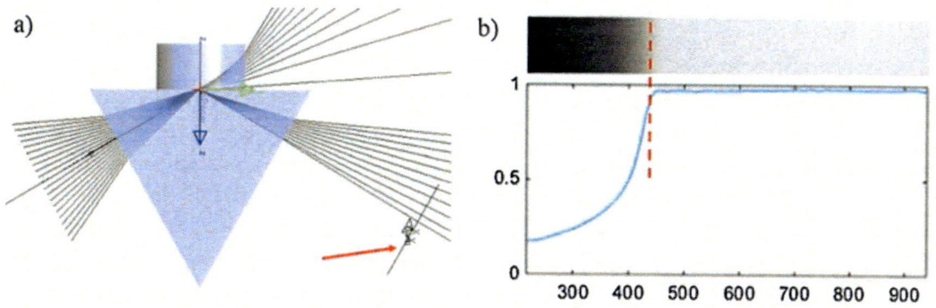

Fig. 3.19 Schematic representation of the SFRIM setup

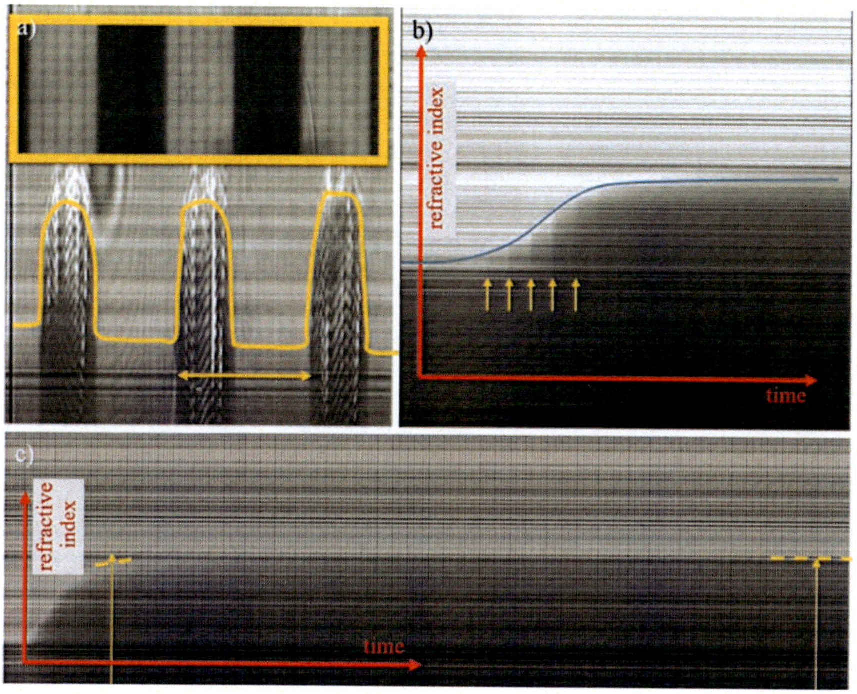

Fig. 3.20 (**a**) Locally resolved measurements of the refractive index for a structure cured with a UV projector; (**b**) increase in the refractive index with pulses of UV exposure; (**c**) long-term measurements of the change in the refractive index with curing over a period of 30 min

An example SFRIM measurement is shown in Fig. 3.20a. The sample was exposed to a 5x5 pixel grid from a DLP projector at 405 nm for curing. The pattern alternates between five inactive pixel columns and five active pixel columns along one direction, as shown in the overlay at the top of Fig. 3.20a. For the evaluation, the light-dark edge needs to be identified in each image column, which can be identified and extracted by referencing and least-squares fitting (shown schematically in Fig. 3.20a with an orange line). The higher the position of the curve, the higher the refractive index. With the scanning resolution of 5 μm shown here, the periodicity of the curing is clearly reproduced by the refractive index curve. The jump in the refractive index, i.e., the difference between the maximum and minimum values, is $\Delta n \approx 0.01$ in this example. This is smaller than the difference between the cured and uncured material states mentioned above ($\Delta n \approx 0.03$). The curing can therefore be concluded to be partial in this example.

Incorporating a globally curing UV LED into the SFRIM setup also allows the dynamics of the curing process to be investigated by visualizing the change in the refractive index over time during the curing process. A corresponding result obtained with a modified SFRIM setup is shown in Fig. 3.20b. The figure shows a sigmoid-like increase in the refractive index over time. This global increase fits well with the description of photopolymerization provided by kinematic models of chemical reactions [18]. More specifically, the exposure was arranged by intervals, with UV exposure for a duration of 0.5 s followed by a pause of 30 s. The yellow arrows in the figure indicate the times of the first UV "flashes" and the sudden jumps associated with them. Each UV flash relaunches the decaying curing, but the curing speed once again decreases as the maximum refractive index is approached. If a single UV flash is considered individually, the change in the refractive index associated with curing does not only occur during the period with active UV radiation; the curing process also continues to a lesser extent after the flash itself.

When the curing is considered over a period of 30 min rather than just a few seconds, as shown in Fig. 3.20c, the refractive index can once again be seen to continue to increase slowly and asymptotically after an initial phase of steep curing. To illustrate the change in the refractive index, two arrows of the same length were added to the figure.

To supplement the preceding quantitative characterization of materials for the additive manufacturing of optical components in terms of the transmittance and the refractive index, the topic of diffraction on 3D-printed samples is also discussed qualitatively below. If a collimated laser is directed through a printed and polished cube, different diffraction patterns may appear, depending on the printing process that was used.

Figure 3.21a shows a typical diffraction pattern obtained from an inkjet-printed sample. In this example, a red laser beam was coupled into the cube parallel to the layered structure, and the resulting diffraction image was recorded. Accordingly, diffraction maxima are visible perpendicular to the build direction. Starting from the zeroth-order maximum, the intensity of the secondary maxima decreases with the single-slit envelope, as is also observed in a diffraction simulation for this example (Fig. 3.21b). Fitting the diffraction model to the measurement data allows the underlying grating parameters to be deduced. The grating constant calculated in this way is 30 μm, which matches the z-resolution of the printer.

Figure 3.21c shows the diffraction pattern of a layer cured with a DLP projector. A white light laser aligned perpendicular to the layer was used as the source of illumination. The spectral band of the laser generates a color gradient due to the spectral dependence of the diffraction. The overlaid image in the figure shows the exposure mask, which features alternating rows of active and inactive pixels. The on-off pattern with a large grating constant (2 pixels) in the vertical direction leads to vertical diffraction orders with small distances. In the horizontal direction, a

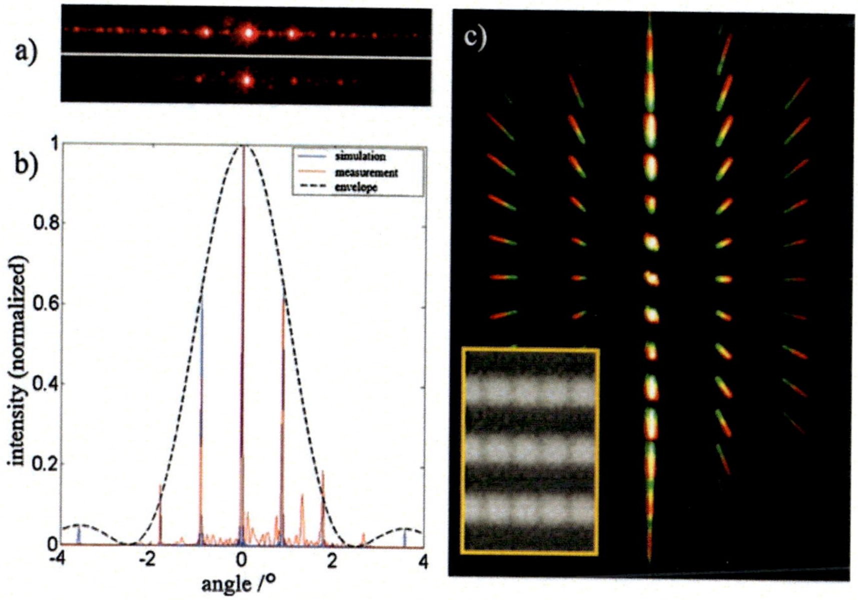

Fig. 3.21 Diffraction effects produced by 3D-printed optical components; (**a**) images of diffraction patterns from the layered structure of inkjet samples; (**b**) simulation of the diffraction in (**a**) with an envelope of the slit width; (**c**) image of the diffraction pattern of a sample with structured curing (see overlay)

double-frequency diffraction pattern is observed. This is caused by the dead area with reduced curing in the illumination pattern between the pixels of each row, which leads to a periodic reduction in the refractive index.

Some initial analyses of the impact of post-curing on diffraction effects have demonstrated that the imprinted refractive index structure can be "overwritten" or rehomogenized with homogeneous lighting if the curing process has not yet reached an advanced stage. However, once the curing process passes a certain threshold of cross-linking within the polymer, the mobility of the molecules becomes too limited to reverse the imprinted pattern.

3.4 Additively Manufactured Macroscopic Optics

3.4.1 Light-Guiding Elements

Additive manufacturing opens up completely new possibilities in the field of lighting technology. Whereas conventional 3D printing methods are only moderately suitable for imaging optics without post-processing, the realization of lighting optics with these processes is most certainly viable.

One way to realize lighting systems is with light-guiding elements. This takes advantage of the difference in the refractive index between the transparent printing material and the surrounding air. Since air has a lower refractive index than the printing materials [19], total reflection occurs at the surface of the element above a certain angle threshold. This couples the light beam into the light guide [20].

3.4.1.1 Complex Light-Guiding Structures

3D printing allows new light-guiding elements with complex shapes to be realized. These elements can also be used as illumination units in applications involving sensors. Many of the limitations of conventional manufacturing technologies no longer apply. Angled and difficult-to-access structures are not a problem with 3D printing. Furthermore, new degrees of design freedom can be implemented. For example, mechanical fittings or blinding elements can be incorporated directly into the lighting optics.

Figure 3.22 shows two examples of complex light-guiding structures. In both cases, the light-guiding elements are connected to surfaces. Illuminating these surfaces allows light to be coupled into the guide, for example, to spatially resolve the illumination of a component using different wavelengths.

3.4.1.2 Additively Manufactured Scattered Light Sensor for Quality Control

In this example, printed optics are used not only for the lighting itself but also to detect any light that scatters on the part. The goal is to implement a sensor to perform quality testing of component topology. The general idea is to illuminate the component homogeneously with the printed optics and locally detect any light scattered on the surface of the component using printed light-guiding elements. If the component has a defect on its surface, the signal response of the scattered light in the light guides changes, which is then registered.

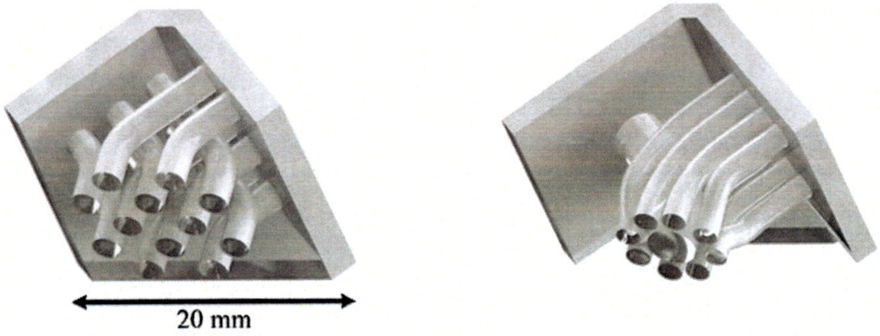

Fig. 3.22 Additively manufactured light pipe structures for lighting applications

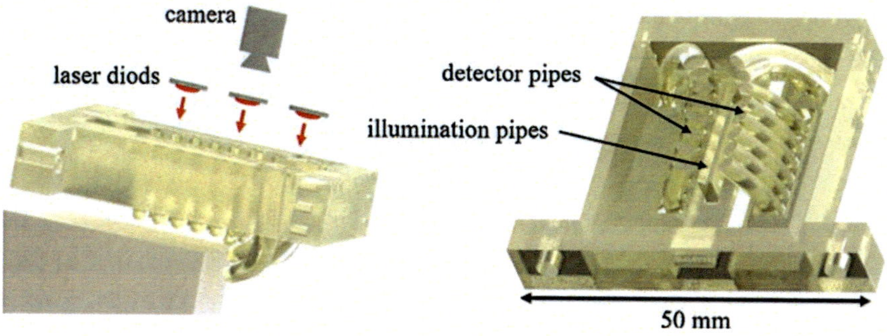

Fig. 3.23 Concept of an additively manufactured scattered light sensor. The test object is illuminated with three laser diodes, and the light is subsequently detected with a CCD camera. The light is guided to the object and camera through light pipes

Thanks to additive manufacturing, this sensor can be adapted to measure components with various surface shapes. The light-guiding structure ensures that the interface between the optics and the light source and between the optics and the detection unit (camera) is uniform across all variants of the component shape.

The concept of the sensor is shown in Fig. 3.23. The lighting is provided by several laser diodes, which ensure that the object is uniformly illuminated using printed optical elements. The scattered light is captured by detector light pipes that guide the light back to the top of the sensor. The signal from each light pipe is then recorded, for example, with a camera.

Figure 3.24a shows the simulation of the sensor lighting unit. The light is supplied through flat, refractive, and totally reflective surfaces. An example of a real object illuminated with this unit is shown in Fig. 3.24b. The light scattered on the object's surface and collected with the detector light pipes is then recorded with the CCD camera at the top of the sensor (Fig. 3.24c). If there is a local change in the topology of the test object, the intensities in the corresponding detector light pipe also change. If the system is installed symmetrically as shown here, any relative difference in the signal between two symmetrically arranged light guides is easy to determine.

The sensor shown in Fig. 3.23 only performs a very rough scan of the component. An improved version is shown in Fig. 3.25. This setup uses light-guiding elements with significantly smaller diameters. This allows a higher spatial resolution to be achieved. The individual light pipes are also embedded in the support material of the printer to reduce cross talk between adjacent light pipes caused by absorption in the support material. This also increases the mechanical stability of each light-guiding element. The sensor's mechanical fittings are also integrated directly with the optics.

In addition to sensor technology applications, light elements can also be used for decorative purposes. An example of a complex light-guiding element is shown in Fig. 3.26, together with each of the steps involved in its realization. The light is guided from a single LED through several elements to the exit surface at the top.

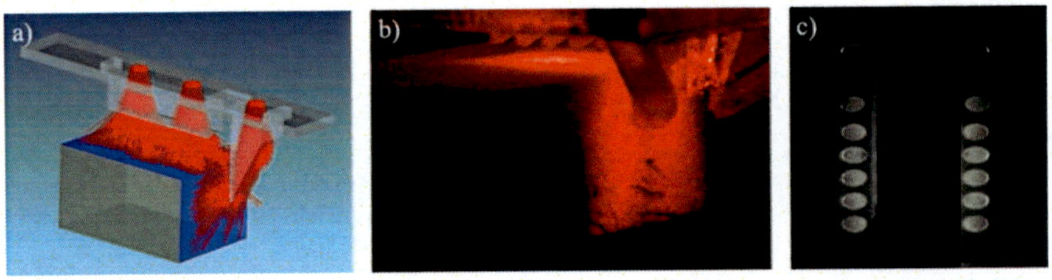

Fig. 3.24 Simulation of the lighting unit (**a**). Illuminated test object (**b**). Signal image from the CCD camera (**c**)

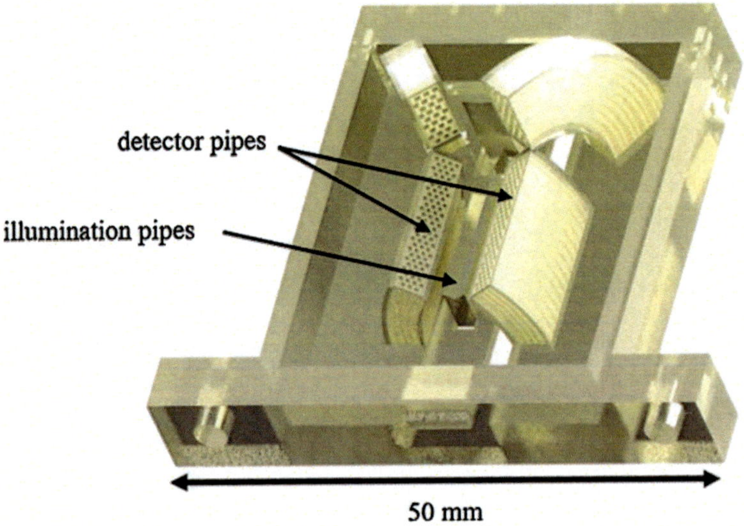

Fig. 3.25 Sensor variant with a finer detector light pipe resolution. The light pipes are embedded in the support material to avoid cross talk

Figure 3.26a shows the target structure illuminated by the light-guiding element on a screen, with the optical design shown in Fig. 3.26b. Each element of the light guide is assigned to an area on a circular LED array, as shown in Fig. 3.26c, allowing the different sections to be optimized relative to one another. Figure 3.26d shows a simulation of the irradiance in false colors on a screen immediately after the light-guiding structure. The lion's head shines more brightly than the other elements, whereas the foot is too dim. Figure 3.26e and f shows the illuminated light-guiding element from different perspectives after printing and further optimization. The unprocessed rough surface structure allows light to escape from the sides, but this is not a limitation in this particular application. The additional optimization allowed the printed complex light-guiding element to achieve a homogeneous illumination of the target structure.

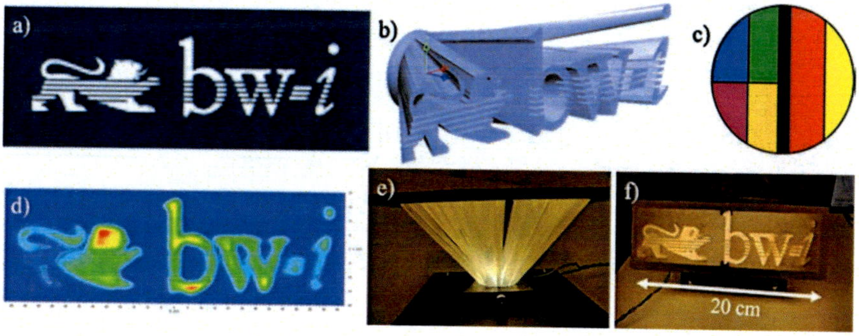

Fig. 3.26 Workflow for a complex light-guiding structure; (**a**) target; (**b**) optical design; (**c**) division of the LED into different light-guiding elements; (**d**) simulated irradiance distribution on a screen; (**e**) side view of the printed light guide; (**f**) front view of a screen illuminated by the printed light guide

This example shows that there are almost no limitations in the complexity of the component shape and, most importantly, that individual solutions with a batch size of one are easy to implement.

For shapes with simple rectangular cross-sections, the use of FDM printers is also conceivable to realize light-guiding elements. In the FDM process, filaments, i.e., plastic strands, are melted and applied by a nozzle in layers. The typical nozzle diameter and hence the smallest possible structure width is 0.4 mm, with a typical layer thickness of 0.1–0.3 mm. Transparent filaments can be used to reliably print planar light guides, i.e., light guides whose cross-section is always perpendicular to the build platform. Since the layer resolution is often coarse at more than 100 μm, the lateral surfaces of any light guides whose path runs diagonally relative to the build platform form into steps. The suitability of FDM for printing such components tends to be limited. Figure 3.27a shows an FDM-printed light guide with a simple radius, illuminated from the left side with a green laser. The light guide ends in an integrating sphere to collect all of the transmitted light. The light guide is visibly giving off light at the entrance and exit surfaces, as well as in the middle of the circular arc. The microscope image in Fig. 3.27b shows a polished cross-section of an FDM light guide. The zoomed-in overlay shows lines drawn on top of the measurement camera to bound the thickness of a melted filament strand. The measured result of 420 μm matches the melting nozzle, which has a diameter of 0.4 mm. Even after melting the material and attempting to find suitable parameters for the amount of extrusion, the individual sheets (circle) and layers are still recognizable. The cavities and inhomogeneities present in the sample are one of the causes of the increased scattering.

Typically, light guides convey light within the body of the material into the surrounding medium, in this case the air, by total reflection at the boundary layer of the light guide. This requires special settings for the printing parameters, since with FDM printing typically only the outer path is printed solidly, and the remaining volume is filled with a type of hatching, e.g., only to about 10% (infill). The slicer

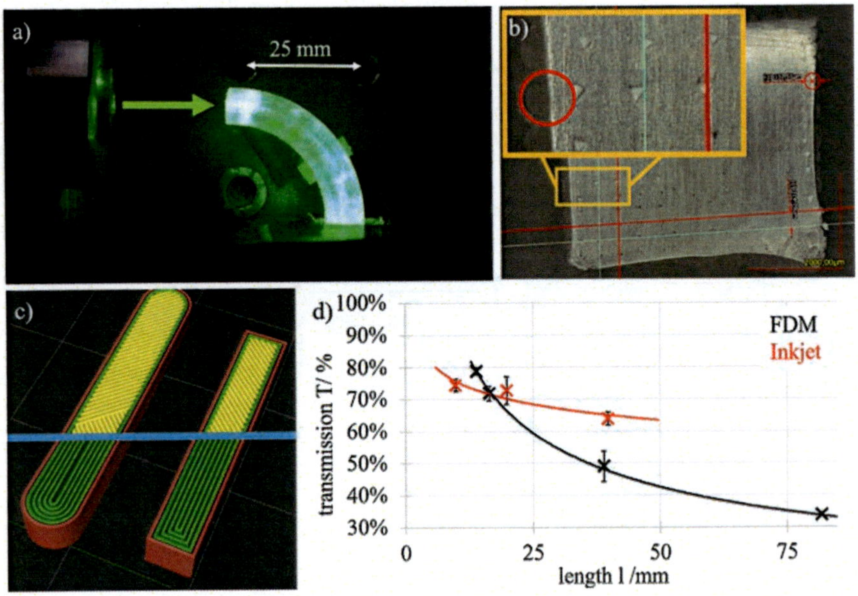

Fig. 3.27 Analysis of FDM light guides; (**a**) light guide with a laser; (**b**) microsection with visible filament strands; (**c**) preview of slicing data with hatching (top) and paths (bottom); (**d**) comparison of the transmission of FDM and inkjet as a function of length

settings must also take light propagation into account, as shown in Fig. 3.27c. The top section of the slicing preview shows a solid hatching pattern that is oblique relative to the desired direction of light propagation along the tube. This type of diagonal hatching would cause increased scattering due to the constant presence of boundary layers. Parallel tracks in the body of the material, as shown in the bottom part of Fig. 3.27c, instead act much like a light-guiding matrix that carries the light. However, at the end of the light guides, the paths once again cross the desired direction of light propagation. The ends were therefore cut off so that only parallel paths are used.

Another observation is that the unprocessed filament often appears "crystal clear" at first, but then becomes significantly cloudier after extrusion. This is presumably caused by air trapped in microinclusions during the melting process, leading to cloudiness and scattering.

Figure 3.27d shows the transmission of straight light guides of different lengths. A curve with an exponential decay is also fitted to the measured values. The graph compares FDM-printed light guides and inkjet-printed light guides. Whereas the FDM light guides achieve a higher transmission at short lengths of below 20 mm, their losses are otherwise significantly higher than those of the corresponding inkjet-printed solid light guides. The attenuation is therefore significantly higher in filament light pipes than in photopolymer components.

3.4.2 *Lens Systems*

It does not appear to make much sense to print single lenses, since the disadvantages of additive manufacturing with regard to the optical properties outweigh the disadvantages of conventionally manufactured lenses. Nevertheless, it is interesting to ask what possibilities additive manufacturing might offer for macroscopic imaging optics.

3.4.2.1 Additively Manufactured Doublet

Figure 3.28 shows the example of an additively manufactured doublet. Figure 3.28a shows a cross-section of the optical component. As can be seen in the figure, the doublet has a monolithic structure. The four optically active surfaces are labeled as lens surfaces 1–4. An aperture was also integrated into the doublet. This is implemented as a cavity in the design that is filled with nontransparent support material during the printing process (inkjet printing). Since the optical surfaces require post-processing, a way to access the inner lens surfaces 3 and 4 from the outside was also added to the design. This allowed the surface quality to be significantly improved using the dip coating process (see Sect. 3.3.3). Figure 3.28b shows the results. A resolution of approx. *10 lp/mm* was achieved.

3.4.2.2 Refractive Panoramic Lens

Two different approaches to refractive panoramic lenses are discussed below as further examples.

As a first example, a fisheye lens was designed as a retrofocus lens that inverts the telephoto design. Thus, the focal length of the system is smaller than its back focal length [21]. The fisheye lens consists of a group of converging lenses on the

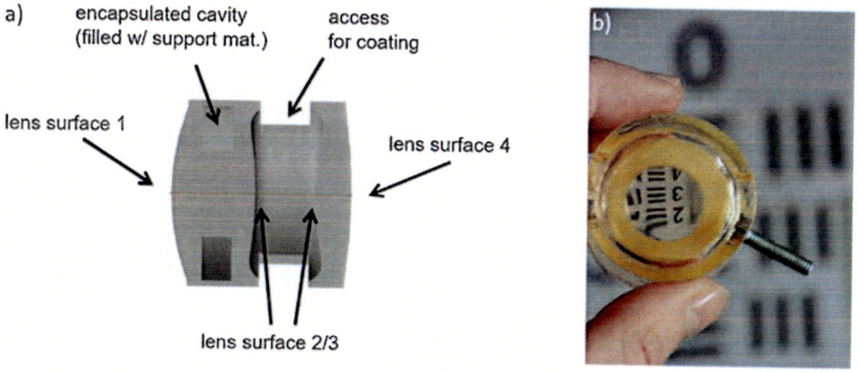

Fig. 3.28 Additively manufactured doublet. (**a**) Basic structure; (**b**) 3D-printed optical component

Table 3.2 System parameters for the designed fisheye lens

Field angle	126°
System length	73.9 mm
Focal length	5.5 mm
F-number	2.54

front side followed by a group of diverging lenses on the rear side to focus the beam into the detector.

To realize a lens system with a large field angle, two strongly curved negative lenses are used at the front side. The properties of the fisheye lens are shown in Table 3.2.

In addition to the optical components, the system contains mechanical components to mount the lenses (see Fig. 3.29). These components are attached to each lens. They make the lenses stackable without any other mechanical components. This makes it possible to work with or polish each lens separately. An aperture (black part in Fig. 3.29) was also introduced to reduce unwanted scattering effects.

The components were realized using the multijet modeling method (layer thickness, 15 microns). To reduce the total manufacturing time and to achieve the best surface quality, it is helpful to optimize the orientation of the lenses, as the surface characteristics may vary according to whether or not the model and support material are in contact. The yellow-marked areas in Fig. 3.30 indicate where the printed part has to be supported for two different orientations: horizontal and vertical. The vertical orientation is preferable, as then the surface on the top side does not need support material, which makes it shinier and less rough.

Figure 3.31 shows one of the lenses directly after the printing process (top side and bottom side). The surface of the bottom side is rougher than the surface of the top side.

To obtain sufficient optical quality in the lens surfaces, post-processing was necessary for each surface. The lenses were polished using the negative polishing technique on a standard polishing machine. Accordingly, for each lens, a corresponding polishing head was constructed with the opposite negative radius of the lens curvature. After polishing the surfaces with sandpaper (grit size 2500), the polishing process was continued by applying polyurethane polishing pads with different suspension agents for each polishing step. Initially, aluminum oxide was used with different grain sizes (20–6 microns), followed by cerium oxide with a grain size of 1 micron. The stacked lens system with the polished lenses is shown in Fig. 3.32. Reworking the printed optics with this polishing method leads to a surface roughness RMS of about 12 nm.

The experiments conducted to verify the optical performance of the fisheye lens were based on spatial frequency measurements. For this, a sine pattern with different frequencies and different contrasts, the Sine M-6 pattern (Fig. 3.33a), was used.

Figure 3.33b shows the appearance of line pairs with a spatial frequency of six line pairs per mm (red-marked part of the pattern) when viewed through one of the lenses. With the other lenses, a different number of line pairs per mm, from 6 to 32,

Fig. 3.29 Optical and mechanical components of the fisheye lens

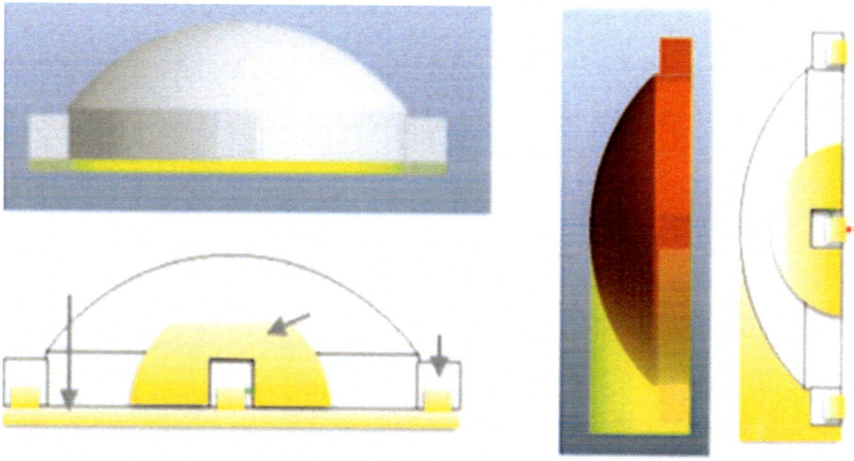

Fig. 3.30 Orientation of the lenses on the printing bed

Fig. 3.31 Surface quality of one lens directly after the printing process

Fig. 3.32 Stacked fisheye lens with polished lens surfaces

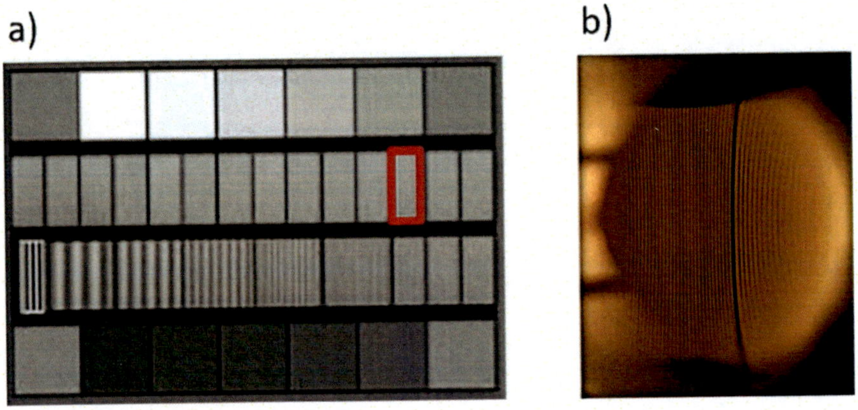

Fig. 3.33 Sine M-6 pattern with different frequencies and an image of six line pairs per mm viewed through lens no. 3

can be observed. The imaging quality is certainly lower than conventionally realized lenses. This is mainly due to the layered structure of the components and the post-polishing process.

The example of a catadioptric panoramic objective is also discussed below. The corresponding optical design is shown in Fig. 3.34a (based on [22]). The system consists of an additively manufactured panoramic lens (diameter 70 mm, f/# = 5) combined with a commercial achromatic relay lens. The objective collects the light from surface 1 in Fig. 3.34a refractively, reflects it via the reflective surfaces 2 and 3, and then conveys it through surface 4 (= circular nonreflective area in the center of surface 2) into the achromatic lens (surfaces 5–7).

The system was optimized so that the rays captured by the objective are mapped to as large a surface as possible on the sensor (2464 × 2056 pixels, 8.8 mm × 6.6 mm,

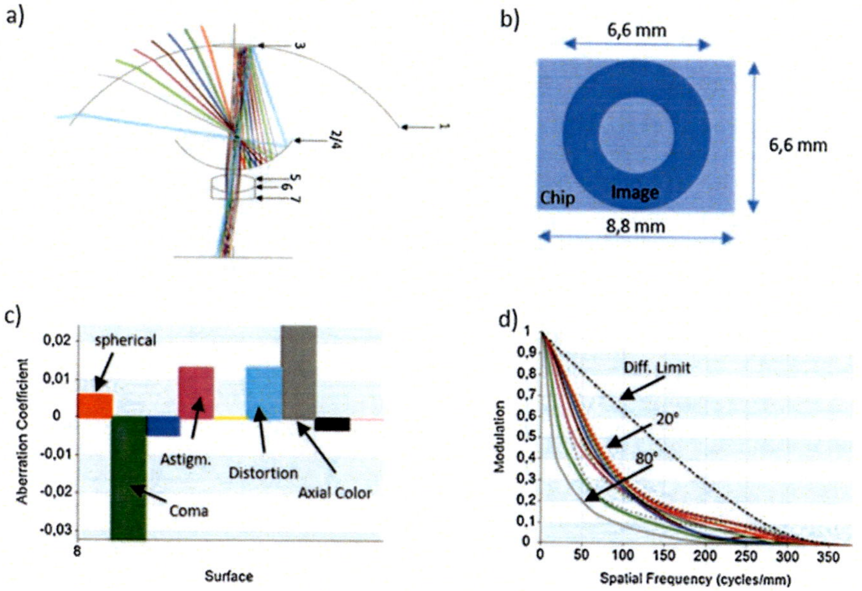

Fig. 3.34 (**a**) Design of the objective; (**b**) view of the image on the sensor surface; (**c**) aberrations in the design; (**d**) resulting MTF

pixel size 3.45 μm × 3.45 μm). The catadioptric system produces an image in the form of a ring on the sensor at a distance of 7 mm from the relay lens (see Fig. 3.34b). To optimize the utilization of the sensor surface, the design was optimized to ensure that the outermost ray, incident at 100°, hits the edge of the sensor surface. The rays incident at 20° form the inner boundary.

Surfaces 1, 2, 3, and 4 were released for optimization in the optical design, with surfaces 2 and 4 having the same radius. One particularity of surface 3 is that it was designed aspherically rather than spherically, as this allowed the spot sizes to be reduced and the aberrations to be improved. The system aperture is realized as a recess in the reflective surface 2 (circular recess in the center).

The resulting third-order aberrations and the MTF from the simulation are shown in Fig. 3.34c and d. The tangential coma and the axial chromatic aberration are key factors of the aberration of the system. The MTF shows the expected decrease in the system's resolution as the angle of incidence increases.

The simulated panoramic objective was then manufactured with the multijet modeling process and polished by mechanical abrasion with individual tools until the surfaces achieved optical quality. Surfaces 3 and 4 were then treated with silver by physical vapor deposition to make them reflective. The mount that attaches the panoramic lens and the achromatic lens to the camera was manufactured additively by fused deposition modeling. The housing was threaded to make the distances between the panoramic lens, the achromatic lens, and the camera chip continuously adjustable.

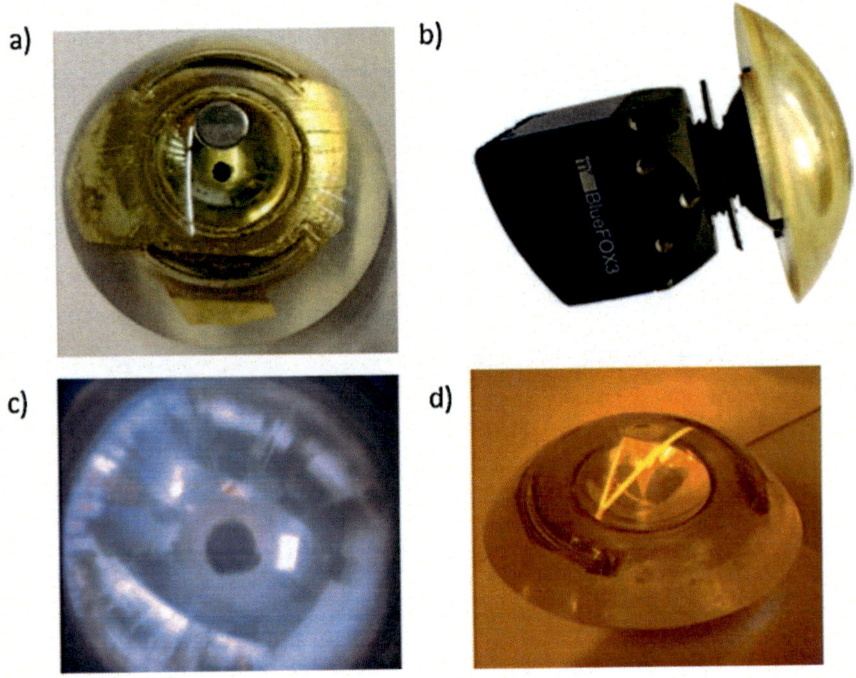

Fig. 3.35 (**a**) Objective after post-processing; (**b**) assembled system; (**c**) resulting image; (**d**) fluorescence when a green laser (532 nm) is shone through the system

Figure 3.35a shows the panoramic lens after polishing and reflective coating, and Fig. 3.35b shows the panoramic lens and the achromatic lens mounted on the camera. A picture taken with this system is shown in Fig. 3.35c. It depicts the laboratory, taken after attaching the camera fitted with the lens to the ceiling. An elongated table is visible in the middle of the picture, and the laboratory walls can be seen around the outside. As expected, the image quality of the additively manufactured panoramic lens is not comparable to the usual image quality achieved by commercial optics. The primary reasons are light scattering and light absorption within the body of the material. Fluorescence effects can also be observed, for example, when a green laser beam is transmitted through the sample (Fig. 3.35d). This also reveals the ray path within the catadioptric system.

A better image quality can be expected to be achievable with optimized materials.

3.4.3 Liquid Lenses

Another promising application for the conventional additive manufacturing of optical components is given by adaptive membrane lenses, also known as liquid lenses. Liquid lenses with adaptive focal length control are already used by some sensors

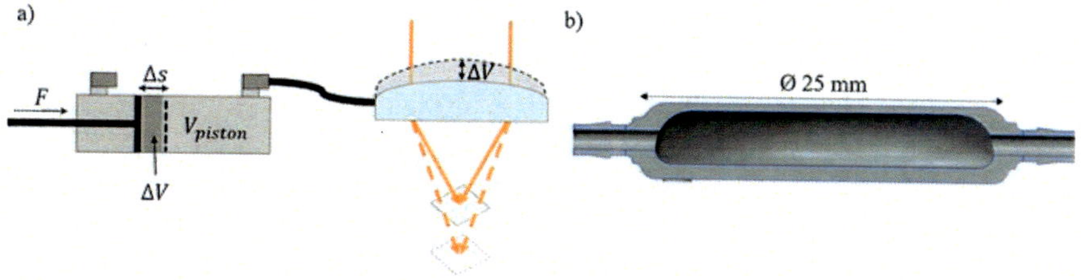

Fig. 3.36 (**a**) Sketch of the principle of a 3D-printed liquid lens; (**b**) cross-section of the CAD model of a 3D-printed liquid lens

with autofocusing features. The shape of the boundary layer and hence the optical effect is adjusted using either an electric field or piezoelements [107].

Lenses can also be realized using conventional techniques for the 3D printing of optics. Figure 3.36a shows the basic principle of a liquid lens system consisting of a lens chamber, a liquid reservoir, and a compression unit. The system uses oil as a liquid. Ideally, the refractive index of the oil should be equal to the refractive index of the membrane. The volume of the lens chamber is adjusted by moving the piston. The membrane changes shape accordingly, which modulates the optical effect.

A diagram of the cross-section of an additively manufactured liquid lens is shown in Fig. 3.36b. Water-soluble support material can be used to create a cavity to ensure that the liquid lens is monolithic. The design already incorporates mechanical fittings. The thickness of the membrane can also be resolved locally to allow predefined aspherical membrane curvatures to be realized in practice.

Other mechanical structures can also be realized together with the membranes. Figure 3.37a shows a few examples. The structures can only be in contact with one side of the membrane (star structure and circular structure), or they must span from the top membrane to the bottom membrane (cylinder in the middle of the liquid lens, on the left in Fig. 3.37a). The structures influence the deformation of the surface.

After the components are printed (see Fig. 3.37b), the surface requires polishing, and the support material must be removed from the cavities. In the examples, the polishing was performed conventionally by mechanical abrasion. Measurements of the surface roughness found values of around $R_a = 35$ nm.

Figure 3.37c shows the three designs from (a) immediately after the support material was removed and the tubes were connected. The lenses have not yet been filled with oil. Note that, thanks to the oil, post-processing is not necessary for the inner surfaces of the membrane. The oil has the same refractive index as the 3D printing material, so there an immersion effect that compensates for any roughness due to 3D printing.

Figure 3.38 compares the measured shape of the star-shaped lens to the target values in the CAD design. The membrane surface bulges slightly inwards (circular blue area). The bulging is reduced around the star.

Figure 3.39 shows the functionality of the third additively manufactured liquid lens (which has a connection between the top and bottom membranes). The image

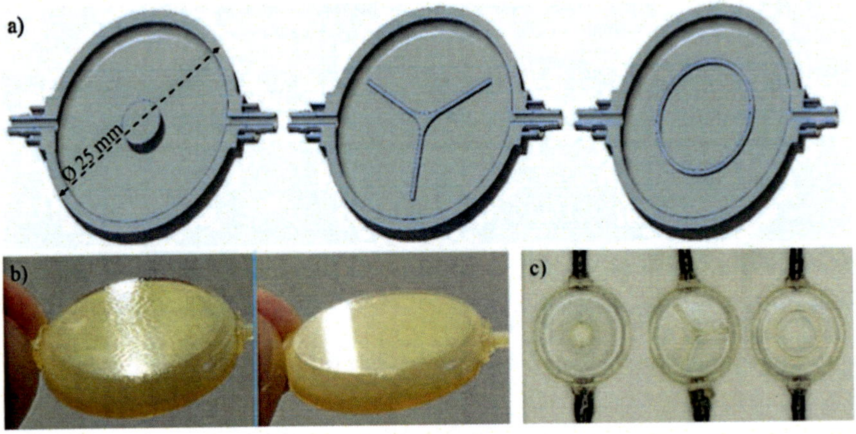

Fig. 3.37 (**a**) CAD models (without back) with various structures on the membrane; (**b**) liquid lens, still filled with support material, before (left) and after (right) the polishing of the membrane surface; (**c**) liquid lenses after cleaning and connecting the tubes

Fig. 3.38 CAD comparison of the liquid lens

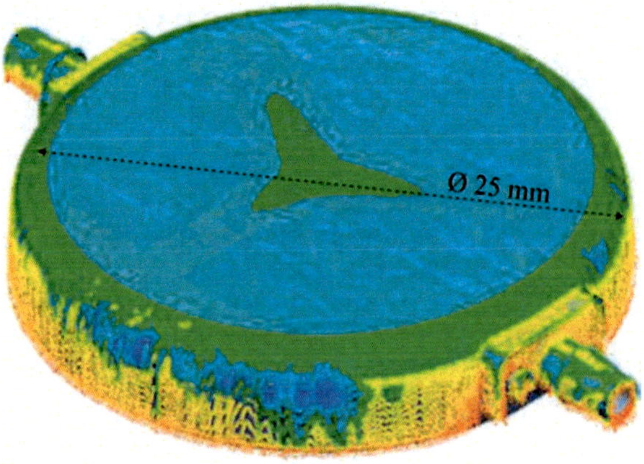

of a checkerboard pattern is shown, projected through the liquid lens in its relaxed state (Fig. 3.39a) and when the pressure is increased (Fig. 3.39b). The distortion in the central area in the relaxed state is caused by a slight overpressure in the system. The aberrations decrease towards the edges, where an undistorted checkerboard pattern is reproduced. The optical effect clearly changes when the lens pressure is increased to 1 bar of overpressure (Fig. 3.39b). As expected, there is no change in the optical effect near the mechanical connections between the two membranes. In the other areas, the checkerboard pattern is massively distorted.

This example demonstrates the possibilities of the custom refractivity configurations that can be achieved with additively manufactured liquid lenses. Shining a widened laser beam through this liquid lens in its initial state produces a

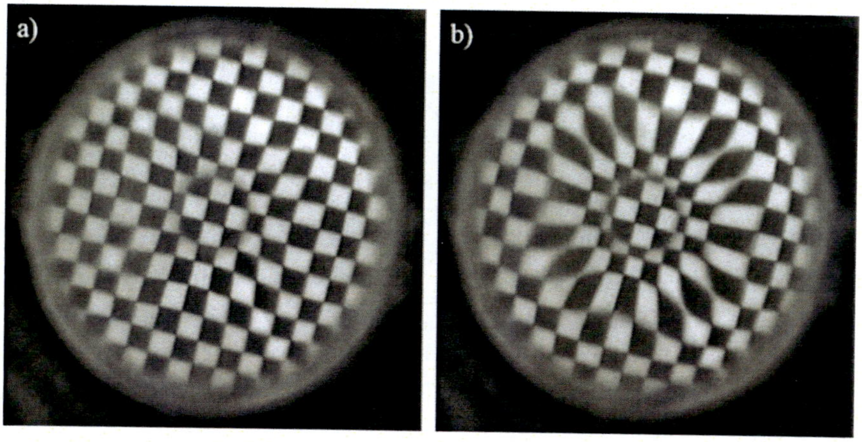

Fig. 3.39 Checkerboard pattern viewed through a liquid lens in its default state (**a**) and with 1 bar of overpressure (**b**)

homogeneously illuminated spot. When the pressure is increased and the curvature of the membrane increases accordingly, a bright ring forms around the central point of the laser. The focus of the ring varies with the pressure and can be adjusted to different distances.

As a general rule, when designing lens systems such as these, it is desirable to digitally map out the entire process. This means that a closed digital process chain must be established, consisting of the following steps: optical design-mechanical design-additive manufacturing-systematic experimental investigation of functionality. Appropriate interfaces are needed between fields.

The procedure is as follows. The desired optical effects are defined for each deformation state, and the required shape is calculated in the optical design software program. The target shapes of the membrane are themselves the input variables of an FEM mechanical simulation that optimizes the topology and membrane structure so that the target shapes are achieved at predefined pressure values. The CAD model that will be additively manufactured is then deduced. The analysis data collected when qualification is performed for the component is then fed directly back into the optical design to allow the system to be improved iteratively.

Figure 3.40 shows an example of this process. Figure 3.40a illustrates the optical design step for an undeformed lens, and Fig. 3.40b shows the process of designing a lens deformation. The membrane was modified to create a tilted beam. This design was entered into a mechanical simulation program, and a suitable membrane structure was determined by running an optimization loop between an FEM simulation of the elasticity and curvature and the structural deformation of the membrane. After the optimization phase, the design is printed and evaluated experimentally, and the results are fed back into the design.

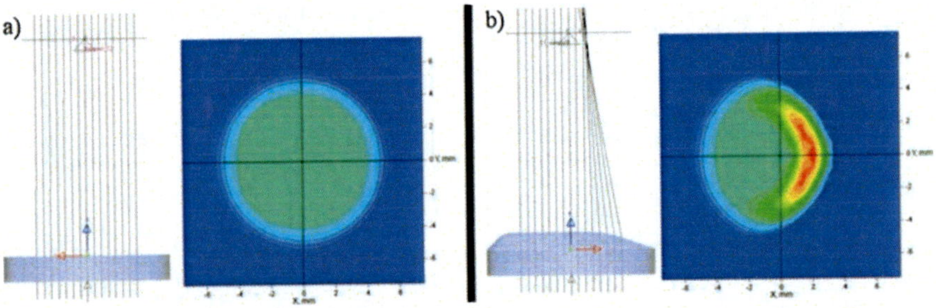

Fig. 3.40 (**a**) Default shape of the liquid lens in its undeformed state with a parallel beam; (**b**) deformation of the membrane of the liquid lens to produce a partially tilted beam

3.4.4 Freeform Lenses

3.4.4.1 Definition of Freeform Optics

The degrees of manufacturing freedom provided by additive manufacturing technologies make them especially suitable for producing components with high geometrical requirements. Freeform optics are especially relevant to the additive manufacturing of optical elements, since realizing these types of surface geometries tends to be challenging for classical optical manufacturing processes.

To differentiate freeform optics from conventional optics, a definition of the concept is required. The concept of freeform optics has previously been defined in several ways [23–25]. One reasonable definition of a freeform can be stated in terms of invariance about the axes of rotation [23]. Aspherical lenses are a special case of freeform optics with an invariant axis of rotation due to their rotational symmetry.

Due to the sheer variety of possible surface geometries, it is difficult to represent the surfaces of freeform optics with any single type of function. Aspherical lenses can be represented with the aspheric equation and a conic constant that describes their conic geometry. For surfaces that deviate from conic geometries, linear combinations of basis functions such as Zernike polynomials [26, 27] can be used to represent freeform surfaces. Even more complex continuous surfaces can be described using piecewise polynomials based on B-splines or NURBS [28]. In addition to continuous freeforms, freeform optics can also be designed to be discontinuous, like Fresnel lenses. In this case, each facet is defined on a restricted domain of definition.

In addition to a description in terms of functions, freeforms can also be represented discretely using a triangular mesh. To do this, the surface is first represented as a point cloud in a three-dimensional coordinate system. Adjacent points are then identified within the point cloud, and triples of adjacent points form a triangle on the surface. Most points are the vertices of more than one triangle. Since they discretized into flat surface elements, freeform surfaces usually have to be modeled at a higher resolution than when using the control points of B-splines or NURBS, for example. The large number of points and triangles therefore needs to be sorted efficiently and saved in a standardized file format. The Standard Triangulation Language (STL) format is currently the standard used by most 3D printers on the market.

3.4.4.2 Applications of Freeform Optics

The large variety of surface geometries allowed by freeform optics can be used to increase the optical performance. For example, adjusting the optics can allow aberrations to be corrected while simultaneously expanding the image field. Simplifying an optical system from a system of multiple lenses to a single optical component reduces the manufacturing costs, the size, and losses from scattered light. Combining multiple functional optical surfaces into a single component also reduces the costs associated with system integration and assembly.

An example of an additively manufactured freeform optical component that demonstrates all of these advantages was discussed earlier in Sect. 3.4.2 [29, 30]. In this system, a 360° panorama was mapped onto a ring shape on the image plane. The optical system was designed as a single optical element with multiple reflective and transmitting surfaces.

When designing optical systems consisting of multiple spherical lenses, it is common practice to replace one or several surfaces with an aspherical lens. Aberrations can be corrected by choosing the conic constant accordingly. In addition to spherical aberration, all other types of Seidel aberration can usually be reduced. In the panoramic objective discussed above, one of the surfaces was dimensioned as an aspherical lens to minimize a large part of the imaging errors.

Freeform optics also support applications in the fields of imaging and lighting. They are, for example, used in imaging to increase the resolution of telescopes (e.g., so-called Schmidt plates in Schmidt-Cassegrain telescopes) or to maximize the performance of head-up displays in terms of size, resolution, depth of field, and efficiency [31].

In lighting, optical elements are adapted to have specific ray characteristics. Catadioptric lenses allow LED rays to be collimated. Light pipes are used to guide light to certain target points by total internal reflections [32].

In the automotive sector, freeform optics are particularly advantageous for headlights. High beams and low beams each need to illuminate certain parts of the street. Accordingly, headlights are fitted with freeform optics to achieve the desired illumination. Freeform optics can also be used to project patterns. The possible applications include lighting effects on streets and wall surfaces by cars and streetlamps, but also projected patterns for metrology applications.

3.4.4.3 Computation of Freeform Lenses

Optical systems with aspherical lenses can be designed by parameter optimization in ray optics simulations. The optimization parameters are typically limited to the lens radii, the lens positions, and the conic constants.

To dimension a freeform optical element in accordance with its definition in terms of invariance about the axes of rotation, suitable algorithms are required to compute the freeform surface. The possible methods include multiparameter optimization of control points on the surface of the freeform, or alternatively direct computation approaches. These direct computational methods usually yield a precise solution but require certain framework conditions in order to be able to compute the freeform. An overview of computational methods for freeform optics is given in Sect. 3.4.6.

Software solutions often describe the freeform with partial differential equations that can then be solved numerically [33]. However, this method of computation is only suitable for punctiform or collimated light sources. The spatial dimensions of real sources produce aberrations that manifest as blurring.

The freeform is computed by mapping the ray characteristics of the source onto a corresponding target intensity distribution in the imaging plane as a function of the angle. The freeform surface is represented by control points on a surface interpolated by B-splines. Ray optics simulations additionally allow an idealized point source to be replaced by an LED that extends over space. This allows the effects of the nontrivial spatial dimensions of the source to be recorded and the setup to be optimized accordingly.

This method was used to calculate lens attachments for laser collimators, which are presented below as an example. The beam quality of a laser collimator eliminates any effects caused by assuming that the sources are punctiform, as well as any aberration in the collimation. Starting with the collimated rays, an optical element with a refractive effect (see Fig. 3.41) was used to generate an image in a target plane. Refraction at the freeform surface of the approximately flat optical element allowed the ray characteristics to be manipulated in such a way as to generate target figures in the image plane.

A target intensity distribution was defined in the target plane at a distance of 2 meters using the software. The intensity distribution was specified in the software using binary-encoded image files as shown in Fig. 3.42. In this example, a 3×3 checkerboard pattern and a ring structure were tested for the projected patterns.

The surface of the initially flat lens attachment was deformed by the software until the desired projected pattern was achieved. The deformation was observed by tracking the position of the control points in the freeform surface and the progression of the splines in the surface (see Fig. 3.43, example of a checkerboard pattern). Note that the spline density does not show the 3D shape of the surface but makes a statement about the local curvature. A color-coded 3D topology of the freeform is therefore also shown as an illustration. The control points in the freeform surface correspond to intersections of the splines shown in the image.

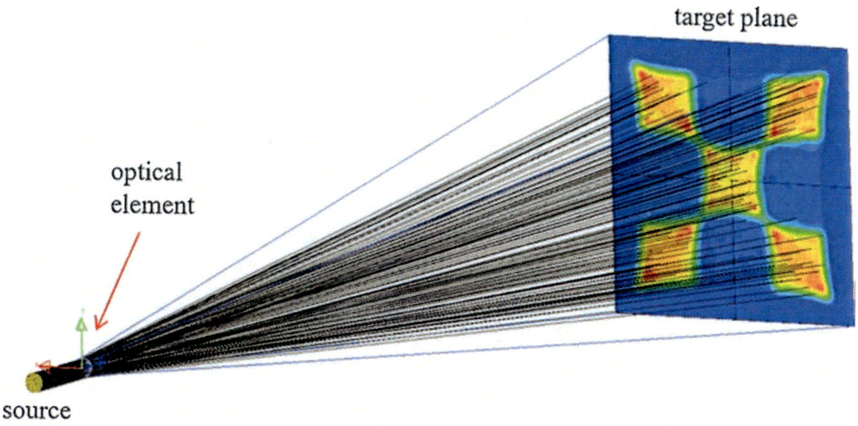

Fig. 3.41 Simulation (LightTools) of a pattern projected by a laser collimator. Starting with a collimated source, an intensity distribution is reproduced in the target plane using a transmissive optical element with a refractive effect. The intensity distribution is color-coded in the target plane in the simulation. The red and blue areas correspond to high and zero irradiance, respectively

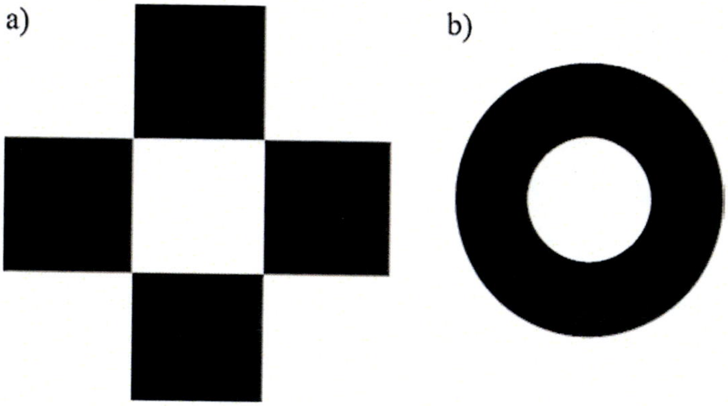

Fig. 3.42 Binary image files defining the intensity distribution in the target plane within an optical simulation of an optical attachment for a laser collimator. The specified binary pattern redistributes optical power to the brighter areas. Intensity distributions were defined for a checkerboard pattern (**a**) and an inverted ring (**b**)

The refractive effect of the lens attachment was investigated by a ray optics simulation. The Monte Carlo ray tracing software was used for the simulation. Figure 3.44 shows the simulated irradiance in the image plane as a 3D topology. For comparison, Fig. 3.47 shows the intensity distributions that were determined experimentally after the lens was additively manufactured.

Since the calculation method used by Freeform Designer is ultimately based on a numerical solution, residual background noise is also visible with this method. The quality of the projected pattern depends on the number of control points on the surface of the freeform. Modeling the surface with too few control points can result in blurred images.

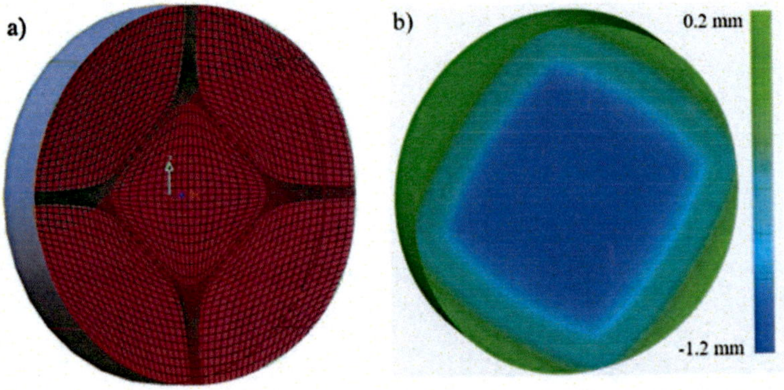

Fig. 3.43 Simulation model of a lens attachment (LightTools Freeform Designer) for a laser collimator to produce a refractive projected pattern. The spline density in (**a**) shows the value of the surface curvature. The height topology can be visualized by color-coding the height relative to a reference plane (**b**). The lens shown here was designed to reproduce the intensity distribution from Fig. 3.42a

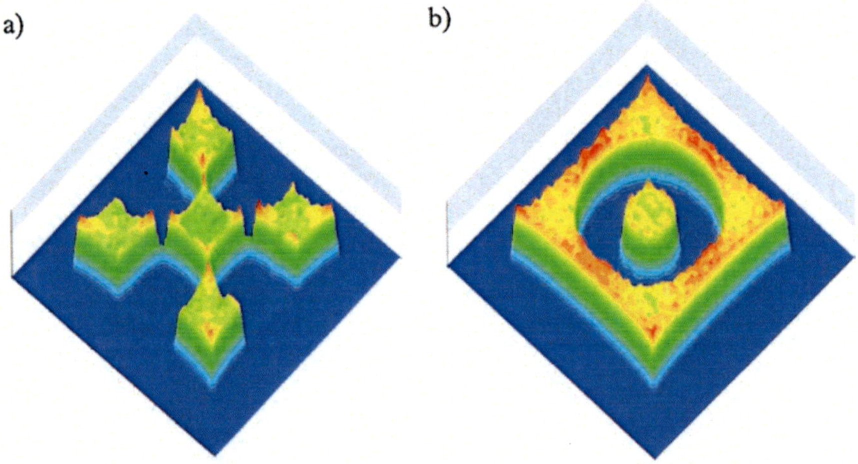

Fig. 3.44 Simulated intensity distribution of projected patterns created by lens attachments for a laser collimator. The lens attachments used for the simulation results shown here were designed to produce the patterns defined in Fig. 3.42. Here, the height topology is shown for intensity distributions corresponding to a checkerboard pattern (**a**) and an inverted ring structure (**b**)

3.4.4.4 Additive Manufacturing of Transmissive Freeform Elements

Since producing freeform optics is particularly challenging with classical manufacturing processes, additive manufacturing technologies offer a reasonable alternative for such optical components [32, 34, 35]. As discussed in Sect. 3.3.2, the surfaces are discretized into a layered structure. This creates a wave-shaped surface structure

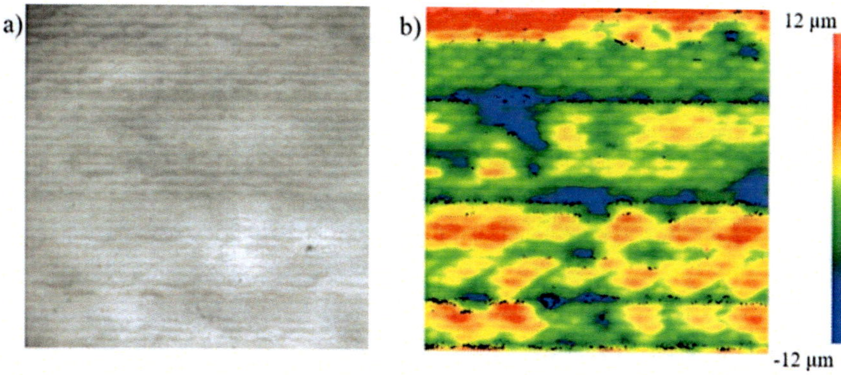

Fig. 3.45 Representation of the surface of a sample printed using the Keyence Agilista 3100 printer. (**a**) Surface of the sample under reflected light; (**b**) Height topology recorded by a white light interferometer. The color-coding reveals the wave structure caused by the printed layers, as well as longer-wavelength structures caused by a beat effect from superimposed spatial frequencies. The recorded image corresponds to an area of 3 mm by 3 mm. The surface was inclined at an angle of 22.5° relative to the print bed

(Fig. 3.45) with amplitudes in the micron range. The wave structure causes statistical scattering of rays at the surface of printed optical components, which prevents the originally intended optical functionality of the components from being achieved. Accordingly, printed optical components require post-processing.

Possible post-processing methods include polishing and coating. With coating, the surface is smoothed by surface tension. Epoxy resins or clear lacquers can be used as the material, applied by dip, spin, or spray coating (see also Sect. 3.3.3). Epoxy resins usually have a high viscosity that affects the dimensional accuracy. Furthermore, they need to be subsequently cured in another manufacturing step. One advantage of epoxy resins is that they allow the same material to be used for both 3D printing and coating. This avoids any layer boundaries between materials with different refractive indices. Clear lacquers can usually be applied in thinner layers. The solvents in the lacquer may however attack and erode the surface of printed optical objects, which can once again expose the layered structure if an overly thin layer is applied, as can, for example, occur with the spin coating procedure (see Fig. 3.56).

Figure 3.46 shows examples of printed optical components that were subsequently coated with an epoxy resin. The objects are lighting elements designed to homogeneously illuminate paintings [36, 37]. After printing, they were wetted with epoxy resin by being dipped into the printing material, which was then cured with UV radiation.

The example shows that the transmission through the surface was significantly improved. However, after applying material to the surface, high dimensional accuracy can no longer be guaranteed. This method is therefore most suitable for lighting optics that have lower requirements in terms of dimensional accuracy.

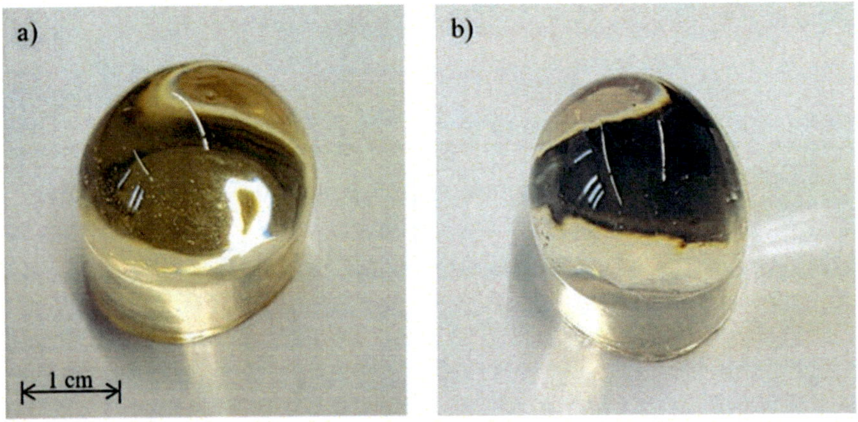

Fig. 3.46 Photographs of additively manufactured transmissive freeform optics. (**a**) Freeform optical element manufactured using the multijet modeling process (Keyence Agilista 3100 printer). A slight yellowing is visible. (**b**) Optical element manufactured by stereolithography (Autodesk Ember printer) with the Autodesk PR48 material. Both optics were coated with printing material using a dip coating process and UV-cured after printing

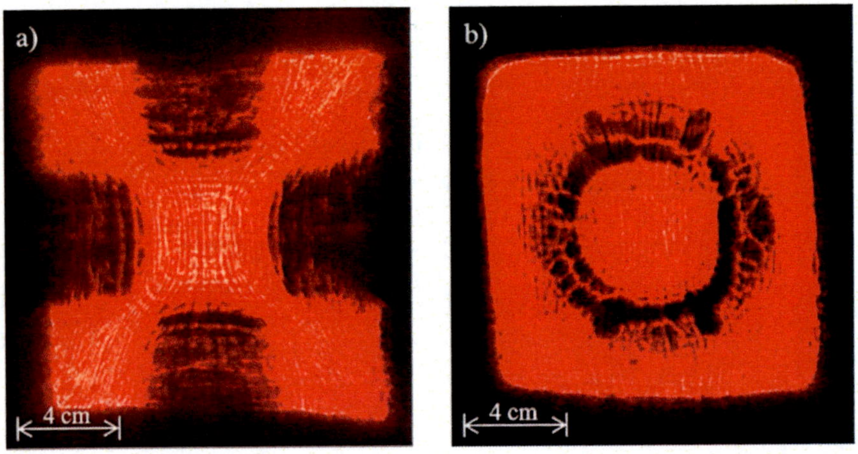

Fig. 3.47 Intensity distributions in the target plane generated by refractive lens attachments for a laser collimator. The dimensions of each pattern are 20 cm by 20 cm. A checkerboard pattern is shown in (**a**) and an inverted ring structure is shown in (**b**). Periodic defects can be observed, caused by residual surface structures from the Additive manufacturing process

Polishing is an alternative approach to coating. Automatic robot polishing is one possible method. However, generating the paths for this process is more complex than the additive manufacturing process itself. The benefits of additive manufacturing, namely, short cycle times and the low cost of adapting the machine to new components, are lost. Accordingly, hand polishing after printing is more viable. With manual processes, fluctuations in the surface quality cannot be excluded.

However, unlike the coating processes, the dimensional accuracy tends to be preserved.

The MJM process was also used to print the lens attachments simulated in Fig. 3.44 for a laser collimator. These optical components have high requirements in terms of dimensional accuracy. The surfaces of the printed lenses were therefore polished to optical quality by hand. The process used a wet polish with sandpaper with a grit size starting at 1000 and increasing up to 6000. Finally, the surface was further refined with an abrasive polishing agent. This polishing regime can achieve a roughness whose arithmetic mean lies in the two-to-three-digit nanometer range. The polished optical elements allowed the simulated intensity distributions shown in Fig. 3.47 to be achieved with the laser collimator.

The intensity distributions reproduce the simulated patterns. However, there are periodic defects that can be reliably reproduced across different samples. These defects are presumably attributable to residual surface structure from the layered character of the additive manufacturing process. In areas that are difficult to reach, manual polishing was not able to remove sufficient material to mitigate the ripples left behind by additive manufacturing without changing the shape of the freeform element. This limits the optical quality.

3.4.5 *Volumetric Displays Using Additive Manufacturing Processes*

3.4.5.1 Technologies for 3D Visualization

High-resolution digital displays are now part of our everyday lives. However, such displays can only represent images as two-dimensional arrays, typically on a flat surface. To render digital data sets in three dimensions, femtosecond lasers can, for example, be used to generate luminous voxels by locally ionizing the ambient air [38]. But this technology requires an elaborate setup and is currently limited to volumes of a cubic millimeter. Another strategy is to introduce targeted scattering centers into a transparent medium that can then be selectively illuminated by a projector unit [39]. These passive scattering centers can be generated as a matrix by frequency-doubled Nd:YAG lasers within a glass substrate.

3.4.5.2 Embedding of Scattering Centers in Transparent Substrates

With additive manufacturing, an alternative concept is possible. 3D printers based on the multijet modeling process often use different materials for the build material and the support material. If the build material is chosen to be transparent, for example, and the support material is chosen to be diffusely scattering, scattering voxels can be specifically embedded into a geometry by simultaneously printing the transparent and diffusely remitting media together.

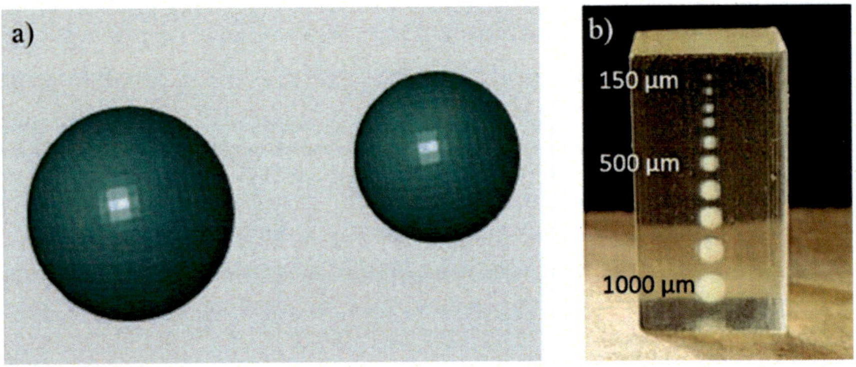

Fig. 3.48 Representation of voxels formed by embedded scattering centers in transparent substrates. Two spherical scattering centers defined discretely in a printing model using triangular meshes are shown in (**a**). An additively manufactured substrate (MJM process, Keyence Agilista 3100 printer) is shown in (**b**), indicating the values of the radii

Potential structure sizes for scattering centers must first be determined before embedding a voxel into a substrate. In general, a high volume resolution with small voxel sizes is desirable. However, if the scattering centers are embedded by means of an additive manufacturing process, the voxel size is limited by the printing resolution of the 3D printer. This is demonstrated below. Scattering centers were initially defined spherically and then discretized with a triangular mesh (Fig. 3.48a). Triangular meshes are advantageous because the models are typically transferred to the 3D printer in the STL format, which is defined in terms of sorted triangles. The printing system used for this example (inkjet) was capable of printing spherical scattering centers with a minimum radius of 150 µm (Fig. 3.48b). At smaller radii, no material was embedded in the substrate.

3.4.5.3 Design of a Volumetric Display

First, 3D data records are required for the object that will be displayed. These can, for example, be generated with 3D scanners (e.g., strip projection). The scan data are then converted into a geometry based on passive scattering centers and embedded into a cube.

Figure 3.49a–c shows an example of three-dimensional representation within a volumetric display. The scanned object (an automobile/Fig. 3.49a) was captured three-dimensionally by a pico projector (Celluon PicoBit) using the binary code method (for the scanning process, see Fig. 3.49b) and then printed as a point cloud within a cube-shaped solid (volumetric display). The distribution of the scattering centers was adapted to match the shape of the scan template.

The printed volumetric display was integrated into a setup consisting of a projector to illuminate the display and a mechanical bracket. When the lighting is switched

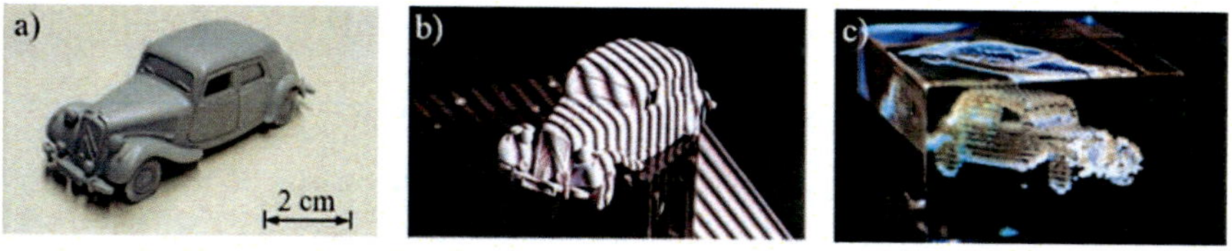

Fig. 3.49 (**a**) Model of an automobile with a length of 6 cm; (**b**) scanning process according to the binary code method with a Celluon PicoBit pico projector; (**c**) volumetric display with passive scattering centers in the shape of the scan template

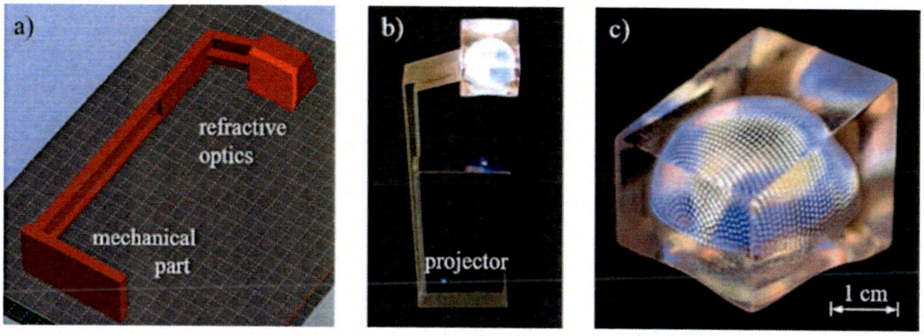

Fig. 3.50 (**a**) Model of a volumetric display with optomechanical elements; (**b**) additively manufactured object for demonstration (Keyence Agilista 3100 printer) combined with the Celluon PicoBit pico projector; (**c**) projection of a rotating globe using the Celluon PicoBit pico projector with diffuse scattering at scattering centers within the volumetric display

on, light enters the volumetric display and is scattered at the scattering centers within it, allowing the shape of the original object to be recognized (see Fig. 3.49c).

This setup also allows volumetric displays characterized by a homogeneous point distribution and selective illumination to be realized. These displays can show different objects with the same shape but different appearances. The planets of the solar system are presented as an example below (see Fig. 3.50). In the example, a volumetric display with a spherical point distribution was realized. The points were distributed equidistantly over a hemisphere. The volumetric display containing the hemisphere was then combined with a focus-free pico projector (Celluon PicoBit), so that the volumetric display serves as a three-dimensional screen. To optimally illuminate the hemisphere, a curved substrate surface with a diverging optical effect was integrated into the display, as well as a wedge that breaks any remaining transmitted ray components away from the viewer to guarantee safety when viewing with the naked eye.

In addition to the substrate of the volumetric display, the mechanical elements of setup were also incorporated into the model. Thus, the entire setup, consisting of the display substrate, the optical elements, and the mechanical elements, was manufactured in a single production step.

The complete setup was printed in a single run (Fig. 3.50a). The system thus realized is shown in Fig. 3.50b. In this case, the display was illuminated with textured lighting instead of homogeneous white light. To visualize the effect, a rotating earth was projected onto the volumetric display (Fig. 3.50c). Cubes with a side length of 400 μm served as scattering centers to represent the globe.

3.4.5.4 Evaluation of Additively Manufactured Volumetric Displays

Figure 3.50 shows that 3D printing using the MJM process enables diffuse remission at scattering centers embedded within the object. The scattering coefficient of the material chosen to print the scattering centers allows the display to be viewed from a 360° field.

In the displays realized to date, however, the scattering elements within the volume have only been arranged over a surface. The full potential of volumetric displays would be unlocked by combining a three-dimensional distribution of scattering centers in the volume with a selective illumination mechanism [39, 40]. The pixels of the projector should be mapped to the voxels that they illuminate. The challenge is to arrange the voxels within the substrate in such a way that none of them are in the shadow of the others. The scattering center must be distributed in such a way that the entire volume is filled with unobscured voxels. To do this, the voxels can be systematically arranged into suitable arrays.

3.4.5.5 Alternative Manufacturing Technologies for Volumetric Displays

An alternative to the 3D printing of scattering centers is given by 3D glass engraving. With this method, the sample is illuminated with two frequency-doubled Nd:YAG lasers. At the intersection of the rays, the material evaporates locally, and a scattering defect occurs within the material. Unlike the additive manufacturing of volumetric displays based on scattering centers, this type of display element cannot be realized as a monolithic optomechanical element (mechanical adapter for a projection unit connected to a refractive optical element). Furthermore, the advantages of 3D printing, such as immediate production of modified variants, cannot be exploited by the 3D engraving process.

3.4.6 Additively Manufactured Mirror Elements

3.4.6.1 Applications of Freeform Mirrors

Freeform mirrors are often used in the field of lighting optics. One of their advantages is that, unlike freeform lenses, freeform mirrors only have a single surface to consider. Furthermore, dispersion is not observed in mirrors, meaning that chromatic aberration does not need to be corrected. Parabolic mirror designs can be used to collimate the rays emitted from an approximately punctiform source.

One application of mirrors and freeform mirrors is in the area of so-called street lighting. The goal is to illuminate part of a street with a lighting system. Vehicle headlights or stationary systems such as information-transmitting streetlamps or architectonic lighting are also among the conceivable applications [41]. A parabolic reflector shape is sufficient to achieve homogeneous illumination. However, special ray characteristics or projected patterns are often required.

In particular, for automobile headlights, very precise ray characteristics are needed to produce high and low beams. Projected patterns that can convey information to the driver, e.g., for purposes of collision control, are also becoming increasingly relevant. For this purpose, selectively activated LED arrays or DMDs (digital micromirror devices) can be incorporated into the headlights [42]. Diffractive optics and holographic systems are other alternative approaches [43, 44]. These approaches take advantage of adaptive optical systems. But static optical elements can also be used to define the illumination of high-beam and low-beam headlights, among other things.

Concentrators are another application. This case corresponds to the inverse of lighting applications. The mirror of a concentrator is used to project parallel rays from the sun onto an absorber in order to generate energy. The spatial dimensions of the absorber and the requirement that the rays must be homogeneous when they enter the absorber means that the parabolic contour needs to be adapted into a freeform shape [45].

Freeform mirrors are also used for beam shaping. The intensity distribution of a laser beam cross-section is usually Gaussian. Some industrial and metrological applications may however require something different. Accordingly, mirror systems consisting of one or two freeform mirrors can be used to modify the intensity distribution of the beam cross-section, for example, into a round or rectangular shape.

3.4.6.2 Algorithms for Computing Freeform Mirrors

When dimensioning freeform optics, the objective is to map the source onto a target intensity distribution in the image plane. Possible methods for computing freeform optics include multiparameter optimization of control points on the surface of the freeform, or alternatively direct computational approaches. The direct computational methods typically construct a more precise solution. However, certain framework conditions must be satisfied in order to calculate the freeform.

Multiparameter optimization begins by describing the freeform in terms of piecewise polynomials using B-splines [46]. Manipulating the control points of the splines allows the shape of the surface to be manipulated. Between these control points, the surface is interpolated as a continuously differentiable polynomial. The initialization model of the mirror and a model of the source are then integrated into a ray optics simulation. A quality function is defined for the mirror image. Finally, the simulation is combined with an optimization algorithm, and the 3D positions of the control points of splines are released as optimization parameters. To deform the mirror, several hundred or even thousand control points are required on its surface.

Consequently, a rapidly converging optimization algorithm and a runtime-optimized ray tracer are required.

An alternative optimization method is to sort the facets of a mirror that is initially assumed to be discontinuous [47]. With this method, the projected pattern is described in terms of pixels in the image plane. For each pixel of the image, the normal vector corresponding to reflection in the direction of the pixel is calculated at the relevant position of the mirror. This defines a small, flat mirror element for each illuminated pixel. These elements are arranged into an array as the facets of a mirror, forming a discontinuous surface. Finally, the mirror facets are sorted with an optimization algorithm to make the mirror as continuous as possible.

In addition to optimization approaches, there are also mathematical computational approaches to generating freeforms. For less complex freeforms, a direct computational approach can be used to calculate a pointwise mapping, as shown in Fig. 3.51. This method approximates the source as a point source and maps it onto the image plane using surface elements. The slope or surface normal of each element is calculated in such a way that the reflected ray arrives at the corresponding target pixel. The intensity distribution is assembled from these specifically positioned images of the source. The surface elements are then continuously joined together. Thus, in principle, this approach is an inverse ray tracing method [45] that determines the normal vectors of the surface elements at the reflector surface instead of the direction of the reflected ray. The disadvantage of this method is that an exact solution can initially only be determined for a single cross-section. In the three-dimensional case, the number of degrees of freedom is too small to consistently assemble a reflector surface.

To compute a continuous freeform surface in three dimensions, the freeform is described with partial differential equations. The design of the optical lighting element can be computed exactly by numerically solving a differential equation [23,

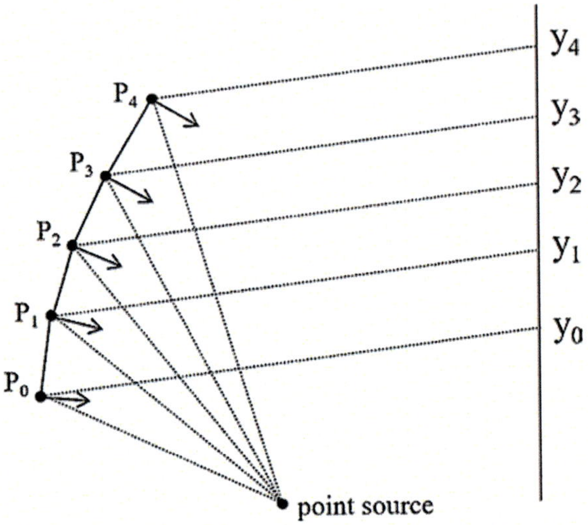

Fig. 3.51 Mapping of a point source to target points (y_n) using a freeform surface. The directions of the surface normals (P_n) are defined in such a way that the source is reflected towards the corresponding target point in each case

48–50]. However, some idealized framework assumptions are required. For example, the source is assumed to be much smaller than the distance from the reflector surface. Furthermore, only small sources guarantee a clear reflection on the freeform. The angle of the freeform surface relative to the local surface normal determines where the ray hits the target surface, and the curvature of the freeform determines the irradiance. These assumptions are used to establish the differential equation and find a numerical solution.

The so-called method of supporting ellipsoids offers an alternative to formulating a description in terms of differential equations [51–55]. This method begins by describing the freeform surface of the reflector piece by piece in terms of ellipsoids. Each ellipsoid maps the point source to a corresponding point in the target plane. The source and the target point both lie at the focal points of the ellipsoid (see Fig. 3.52). Initially, this produces a faceted reflector where each facet is represented by a section of the surface of an ellipsoid.

To generate a continuous freeform surface using the method of supporting ellipsoids, a numerical algorithm [56] is required to calculate the arrangement of the ellipsoids.

3.4.6.3 Additive Manufacturing of Freeform Mirrors with Metallic Substrates

In the context of the additive manufacturing of mirrors, for example, when realizing vehicle headlights [57–59], additive metal manufacturing processes such as selective laser melting (SLM) and selective laser sintering (SLS) can be used. These methods first apply the material layer by layer in powder form and then locally melt

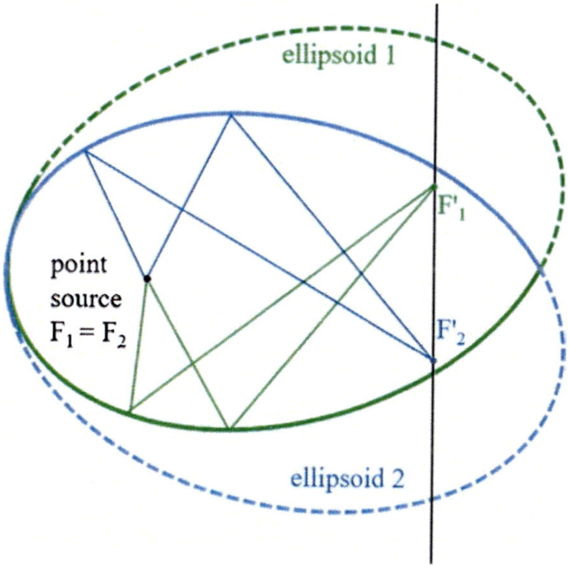

Fig. 3.52 Concept of a method for calculating freeform reflectors using supporting ellipsoids. Mapping of a point light source to two points in the target plane by truncated ellipsoids with focal points F_n and F'_n

or sinter the powder with a laser. Another method for additively processing metals is binder jetting, where the material is first locally extruded and then globally sintered at a later stage [60].

All of the manufacturing processes for additively processing metals mentioned above result in rough surfaces with roughness values in the micron range (approx. 15 μm with the SLM process [61]). For applications involving mirrors, post-processing steps are required. The possibilities include sandblasting and polishing [58]. Laser polishing and electropolishing can also be performed as an alternative to classical polishing.

Laser polishing involves remelting the surface of the optical component by locally exposing it to a laser after the additive manufacturing process is complete. The melting process allows the surface tension of the material to smooth out any unevenness while the material is in a liquid aggregation state [61]. The principal disadvantage of laser polishing is that it leaves behind traces of oxidation on the surface.

Electropolishing is another option. This method involves galvanically removing the surface of the additively manufactured optical element with an electrolyte, which produces a visible smoothing effect. As an example, this process was tested on a reflector element that was additively manufactured by the SLM process. The material AlSi10Mg was used, a standard material for additive manufacturing with the SLM process. The experiment demonstrated that the electrolysis removes metallic material. However, silicon residues remain on the surface, which significantly limit the reflectivity of the component. Instead of the standard material cited above, there are also silicon-free alloys that can be processed additively. Testing showed that electropolishing was able to achieve a pronounced improvement in the reflectivity of surfaces made from such silicon-free alloys.

3.4.6.4 Coating of Additively Manufactured Plastic Substrates

Another approach to the additive manufacturing of reflectors that does not require metal processing is to additively manufacture a plastic substrate and then apply a coating. The coating adds a series of very thin metallic layers to polished surfaces to achieve the desired reflectivity.

Before a polished substrate can be coated, the component must be cleaned, as coating processes are highly sensitive to contamination and the coated surface might otherwise appear dull or discolored. However, organic solvents have a corrosive effect on printed epoxy resins. This increases the roughness of the polished surfaces and reduces the directional reflectivity. The surface appears duller after the coating is applied.

Besides polishing, it is therefore advisable to apply a lacquer to plastic substrates. This involves polishing the additively manufactured substrate by hand in order to break the layered shape of the surface. A lacquer coating is then applied to the substrate. Spray coating is one example of a coating method. With this method, the surface tension of the layer of liquid lacquer smooths out the surface. The lacquered surface does not require cleaning.

After the lacquer process, the optical element is coated with a metallic material such as silver or aluminum. Physical vapor deposition (PVD) is one way to do this. With this technique, the metal is evaporated, forming a precipitate on the substrate. The process works in largely the same way with a wide range of different metals, such as aluminum, copper, and gold. With silver, chemical deposition processes are an alternative.

When silver is chemically deposited, it is applied to the optical surface as a solid. This forms a thin layer that is sometimes only a few atoms thick. The process uses silver nitrate to precipitate silver from a redox reaction with an ammonia solution. The exact chemical composition is critical for adhesion to the optical surface; with the wrong composition, the silver will precipitate within the liquid. Chemical kits known as "spray-on chrome" have been developed to optimize the adhesion of silver to surfaces. This method was tested with additively manufactured parabolic mirrors designed to collimate the rays of an LED, as shown in Fig. 3.53.

To achieve optimum adhesion of the silver to the surface, the surface tension of the mirror surface is first reduced by applying a suitable solution (known as a wetting agent) to improve the wetting. The silver nitrate solution and the ammonia solution are then applied simultaneously by spray coating. These coatings are however sensitive to mechanical loads, which cause the silver-coated surfaces to oxidize and discolor. Accordingly, it makes sense to apply another protective lacquer coating.

When applying the protective lacquer, it is important to remember that the reflective metallic coating is only a few atoms thick and is not chemically insulating. Consequently, acrylic lacquers will dissolve the undercoat under the metal layer and corrode the metal coating. To preserve the reflectivity, it is therefore essential to use a coating such as a urethane-based clear lacquer that cannot reliquefy when a solvent is used after curing.

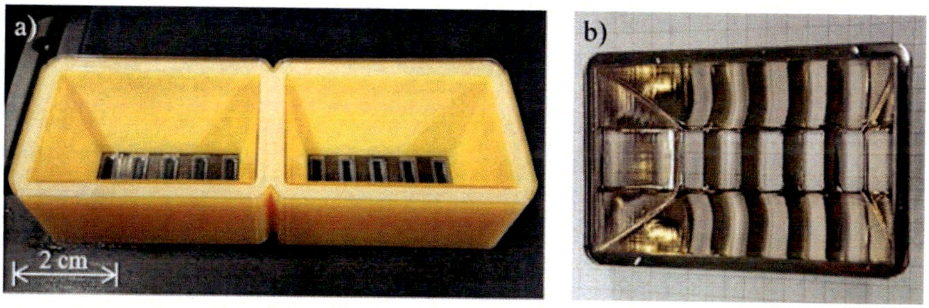

Fig. 3.53 (**a**) Two parabolically shaped reflectors designed to collimate the rays of an LED, currently surrounded by yellow support material after printing with the MJM process; (**b**) one of the reflectors from (**a**) after removing the support material, cleaning, and subsequent silver coating with the chemical deposition process

If the undercoat is a solvent-based lacquer, it has been observed that corrosive solvents may reexpose the layered structure. This happens because different degrees of curing are reached during the additive manufacturing process, which implies that the chemical resistance of the material may vary within the same layer. If the layered structure is exposed, ripples may form on the surface of the component. These ripples cause deviating surface normals that influence the ray characteristics and target light distribution. An example is shown in Fig. 3.54. A droplet of primer was deposited on a polished additively manufactured substrate. The droplet was then distributed over the sample by spinning (2000 rpm) while still in a liquid aggregation state. This created a very thin layer of lacquer through which a corroded surface exposing the additively manufactured layer structure could be observed. Ripples with a PV value of 1 micron are visible.

To prevent the formation of ripples, it makes sense to avoid applying any undercoat too thinly. This is why the spray or dip coating processes are preferable to spin coating. However, a thick lacquer coating reduces the dimensional accuracy. The spin coating method therefore tends to be more suitable for lighting optics than for imaging optics.

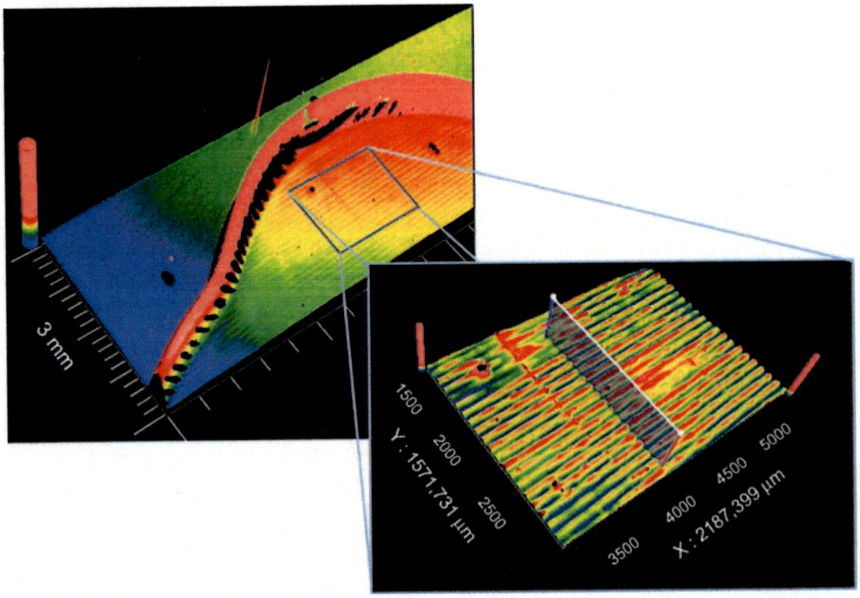

Fig. 3.54 A droplet of urethane-based lacquer on a polished additively manufactured substrate (AR-M2 material, Keyence). The droplet was distributed over the sample by a spinning process (2000 rpm) while still in a liquid aggregation state. This reexposed the corroded layer structure left behind by the Additive manufacturing process

3.4.6.5 Examples of Applications of Additively Manufactured Freeform Reflectors

The simplest approach to computing projected patterns with a freeform mirror is to consider a faceted reflector. First, the intensity distribution is discretized into pixels in the target plane. Then, for each pixel, the corresponding facet of the reflector is defined. Starting from a point light source, the inclination of the facet is adjusted by varying the surface normal as shown in Fig. 3.55 until the reflection is directed towards the corresponding pixel. The facets are arranged as an array, and the reflector is defined as a volumetric model using a triangular mesh. This reflector model can then be printed with very little effort using an additive manufacturing process.

The model from Fig. 3.55 was manufactured as a plastic substrate using the MJM process and then lacquered and chemically coated with silver (Fig. 3.56). In this

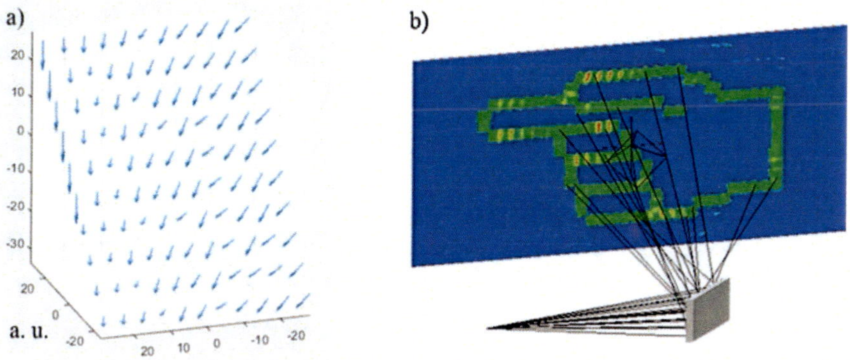

Fig. 3.55 Schematic representation of a faceted freeform reflector with 110 flat facets. (**a**) Representation of the facets in terms of the vector field of surface normals; (**b**) representation of the reflector in a ray optics simulation showing the simulated intensity distribution in the target plane

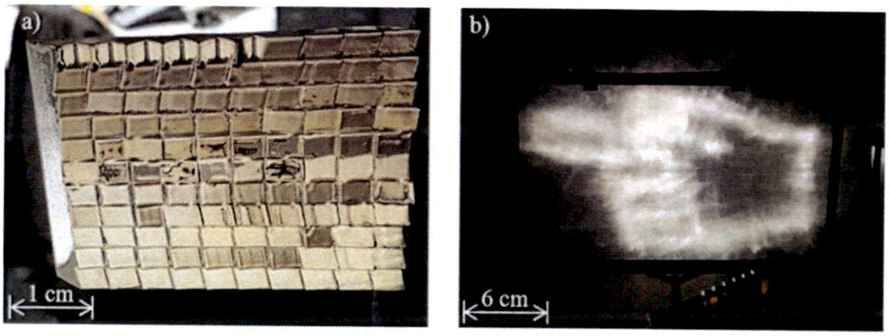

Fig. 3.56 (**a**) Photograph of a reflector with 110 flat facets that projects a pattern onto the target plane; (**b**) image obtained with illumination from an approximately punctiform source. The reflector has a size of 5 cm × 4 cm and is made from an additively manufactured plastic substrate (Keyence AR-M2) with a silver-coated surface

process, the resolution and contrast of the projected pattern are limited by the dimensional accuracy of the additive manufacturing process and the rounded edges that it leaves behind. In addition to the printing accuracy, the size of the reflector can also be adjusted. Making the reflector bigger while also increasing its distance from the source reduces the requirements of the reflector in terms of dimensional accuracy and the size of the point source.

The method shown here only uses flat facets that are arbitrarily arranged into an array. This limits the quality of the projected pattern. The arbitrary arrangement creates a strongly discontinuous surface that is susceptible to edge rounding from the manufacturing process. The flat facets generate trapezoidal images in the target plane.

To improve the quality of the projected pattern, the surface of each facet can be first be generated as a freeform. Figure 3.57 shows an example of one such faceted reflector. The image produced by each facet appears as a square in the target plane after the shape of the facet is adapted.

Adapting the shape of the facets already improves the resolution and contrast of the projected pattern. For further improvements, the reflectors need to be designed continuously. There are various calculation methods (see above) that allow such freeform surfaces to be computed (e.g., LightTools Freeform Designer module: method based on numerical solutions of partial differential equations). For a punctiform light source, the software allows the shape of a parabolic reflector to be modified to produce a projected pattern, as shown in Fig. 3.58. In this example, a shape is projected onto the target plane by a freeform reflector and a point source. Possible applications include the field of so-called street lighting.

The reflector design shown in Fig. 3.58 was manufactured as a prototype using the MJM process. First, the freeform was created as a plastic substrate, and then its surface was chemically coated with silver (see Fig. 3.59). The resolution and

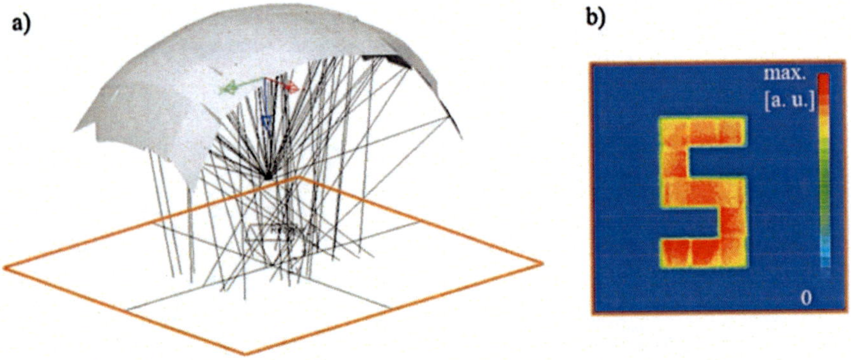

Fig. 3.57 Simulation of a faceted reflector (LightTools, Macro Focal module). (**a**) Representation of the reflector model in the ray optics simulation; (**b**) intensity distribution in the target plane. The adapted shapes of the facets produce square pixels in the target plane

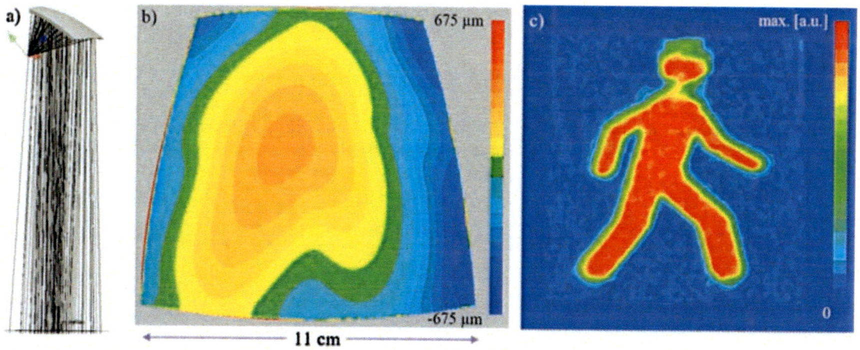

Fig. 3.58 (**a**) Parabolic mirror whose shape has been modified to produce a projected pattern in the ray optics simulation; (**b**) representation of the freeform mirror as a color-coded shape comparison relative to a pure parabola; (**c**) color-coded intensity distribution in the target plane calculated by the simulation

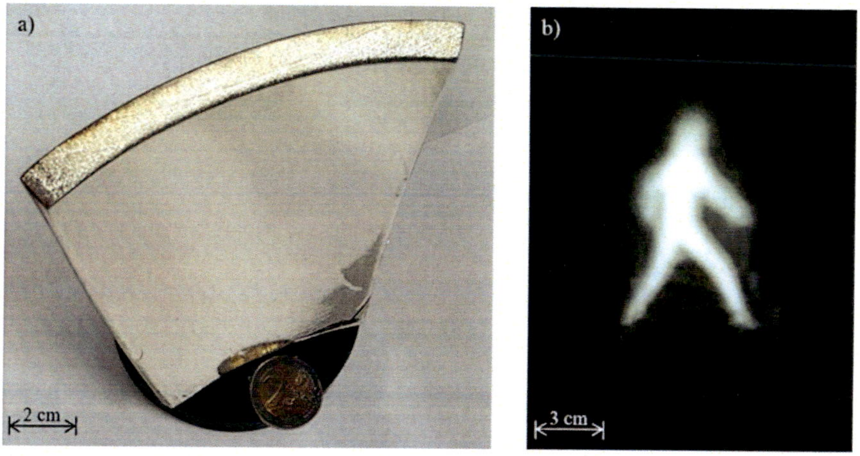

Fig. 3.59 (**a**) Photograph of an additively manufactured freeform reflector. The reflector was manufactured as a plastic substrate using the MJM process (Keyence Agilista 3100 printer) and then chemically coated with silver. (**b**) Intensity distribution generated by the reflector from a point source

contrast of the pattern projected by this prototype were limited by the dimensional accuracy of the additive manufacturing process. To achieve a sufficient resolution, a relatively large reflector with an edge length of 11 cm was manufactured. This improves the tolerance of the freeform surface to shape deviations.

The additive manufacturing and coating processes limit the resolution of the projected pattern achieved by this method. As an alternative to plastic substrates, reflectors can be additively manufactured from metal. In this case, the freeform is, for example, manufactured from aluminum or cobalt-chrome using the SLM process and then polished to optical quality.

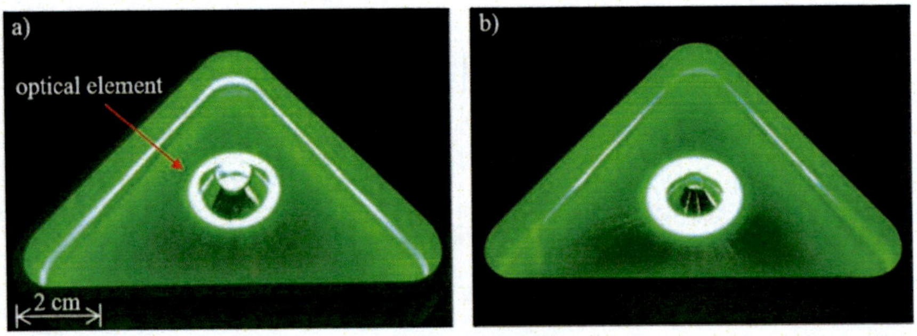

Fig. 3.60 Photographs of freeform reflectors designed to illuminate lines within a cavity; (**a**) the incident rays are focused into lines by adjusting their intensity. Only slight intensity fluctuations are visible along each laser line; (**b**) comparison: reflector with focusing function but without adjusting the intensity. In (**b**), significant fluctuations remain visible in the intensity

Figure 3.60 shows an example of a reflector that was additively manufactured using the SLM process. The application is an approximately conic reflector that projects collimated rays onto the surface of a triangular cavity. The additively manufactured freeform projects the rays onto the surface as lines while simultaneously budgeting the optical power. The power budgeting ensures that the irradiance remains constant throughout the laser lines. This optical component was used in a metrological application to record the interior of cavities in three dimensions by laser triangulation.

The reflector design was initially calculated with a numerical algorithm based on the faceted approach, as shown in Fig. 3.51. The facets were arranged continuously, and the reflector surface was discretized at a sufficiently high resolution that each facet was smaller than the resolution threshold of the 3D printer. The reflector model was then additively manufactured, and the initially rough surface was polished manually to optical quality. The evaluation demonstrated that the adapted freeform succeeded in reducing the intensity fluctuations to a fraction of their original values. A rotationally symmetric model with line focusing but without intensity budgeting resulted in intensity fluctuations with an amplitude of 60% of the mean signal power (see Fig. 3.60b). Adapting the freeform reduced the amplitude of the fluctuations to 5% of the mean signal power.

3.5 Additively Manufactured Microlenses

Microlenses have a wide range of applications in optics. For example, they can be used to collimate light beams [62]. Other fields of application include optimized light collection, e.g., in solar cells or cameras [63]. They are also used in optical sensors like Shack-Hartmann sensors: an array of microlenses integrated into the sensor allows the optical wavefront to be measured [64]. Further examples are found in microscopy and lithography. For these applications, realizing microlenses with a freeform or nonspherical surface is desirable to reduce optical aberration.

Technologies based on lithographic or printing and embossing processes have been examined as candidates for the realization of freeform-based microoptics [65–68]. However, these methods are generally not suitable for small series production, which is an obstacle, especially for individualization. The method of laser-induced forward transfer is one possible way of producing microlenses in small series [64].

But additive manufacturing also holds great potential for realizing microlenses. A major disadvantage of the additively manufactured macroscopic optical elements discussed above is that they experience strong light scattering as a result of both the reduced surface quality and the layered structure (see Sect. 3.3). This is not a problem with additively manufactured microlenses.

The additive manufacturing of spherical lenses is discussed below. A method that enables aspherical microlenses to be additively manufactured is also presented.

3.5.1 Additive Manufacturing of Spherical Microlenses

For the additive manufacturing of microlenses, dispenser or inkjet systems are a suitable choice [69, 70]. As shown in Fig. 3.61a, the advantage of these types of system is that individually defined droplets of printing material can be applied to a

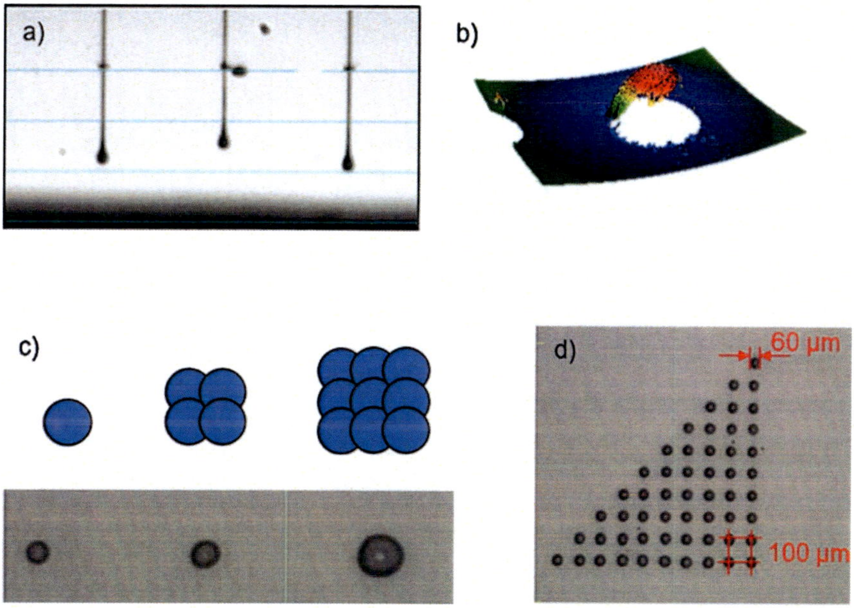

Fig. 3.61 Additive manufacturing of microlenses using an inkjet system. (**a**) Emission of material droplets from an inkjet print head; (**b**) spherical microlens surface measured with a white light interferometer; (**c**) increase in volume as the number of droplets increases; (**d**) printing of a microlens array

specific location. Each printing nozzle can also be controlled individually, allowing droplets with volumes in the pL range to be deposited locally. The droplet forms a spherical cap on the surface, which produces a spherical microlens. The liquid droplet can then be cured using UV radiation. Figure 3.61b shows the measured shape (white light interferometer) of one such microlens; a spherical lens whose shape is determined by the original surface tension of the droplet can be seen in the figure. Unlike macroscopic 3D-printed lenses, there is no layered structure. Furthermore, thanks to the surface tension, the surfaces of microlenses do not require post-processing. The final radius of curvature and volume are determined by the wetting behavior of the droplet, i.e., the interaction between the droplet, the substrate, and the surrounding medium (air) [71].

To modify the radius of curvature and hence the focal length of the lens, multiple droplets of printing material can be deposited at the same location. This increases the volume and modifies the radius of curvature accordingly (see Fig. 3.61c). This type of spherical lens element can also be manufactured into lens arrays, as is, for example, shown in Fig. 3.61d. In the example, droplets (10 pL) were deposited 100 μm apart and then cured. The diameter of each droplet was approx. 60 μm. The positioning accuracy of the printing system and the purity of the substrate surface are the decisive factors in the quality of such microlens arrays.

Instead of coating flat substrates, a coating can, for example, also be applied to optical fiber ends. Figure 3.62a shows the deposition of an out-coupling optical element directly onto an optical fiber. A total of 40 droplets were deposited onto the 125 μm optical fiber.

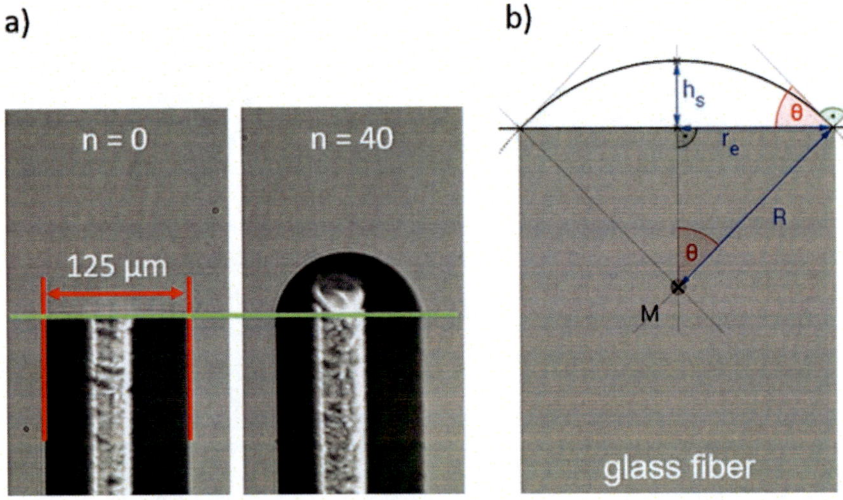

Fig. 3.62 Additive manufacturing of microlenses as out-coupling optical elements for an optical fiber. (**a**) Optical fiber before and after the deposition of the microlens; (**b**) model describing the droplet volume and radius of curvature

As before, the radius of curvature depends on the number of droplets deposited. This is described by a model. First, consider the volume of a spherical cap (see Fig. 3.62b):

$$V = \frac{\pi}{6} h_s \left(3r_e^2 + h_s^2\right) \tag{3.5.1}$$

This formula can be rewritten as a cubic equation:

$$h_s^3 + 3r_e^2 * h_s - \frac{6}{\pi} V = 0 \tag{3.5.2}$$

The real solution of this equation describes the height of a microlens as a function of its volume V and contact radius r_e:

$$h_s = \frac{\sqrt[3]{\sqrt{\pi^2 r_e^6 + 9V^2} + 3V}}{\sqrt[3]{\pi}} - \frac{\sqrt[3]{\pi} r_e^2}{\sqrt[3]{\sqrt{\pi^2 r_e^6 + 9V^2} + 3V}} \tag{3.5.3}$$

When printing onto an optical fiber, the contact radius can be taken to be constant. The volume of the microlens may be deduced from the number of printed droplets and the volume of each droplet. Finally, the radius of curvature may be expressed as follows by substituting the constant contact radius and the calculated height:

$$R = \frac{h_s^2 + r_e^2}{2h_s} \tag{3.5.4}$$

The radius of curvature of the microlens now only depends on the number of printed droplets and can be modified as required.

When measuring the geometry of a printed microlens, the volume is calculated directly from the geometric data as follows:

$$V = \frac{\pi}{3} h_s^2 \left(3R - h_s\right) \tag{3.5.5}$$

or

$$V = \frac{\pi}{6} h_s \left(3r_e^2 - h_s^2\right) \tag{3.5.6}$$

The measured relationship between the volume of a microlens on a 125 μm optical fiber and the number of droplets is shown in Fig. 3.63. The volume of a single droplet may be determined from the line of best fit. In this case, the droplet volume is 8.3 pL. This result is close to the volume of 8 pL measured on a glass slide once the droplets were fully cured.

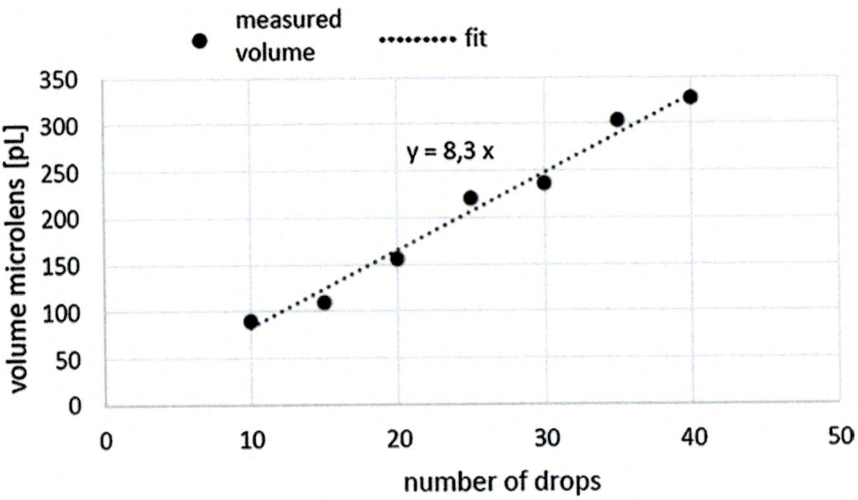

Fig. 3.63 Determining the volume of a single droplet printed onto a 125 μm optical fiber

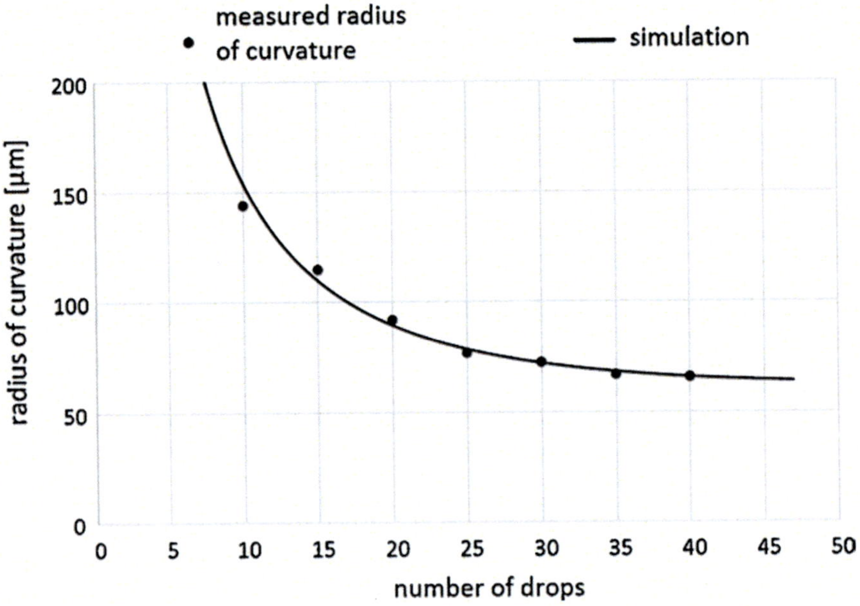

Fig. 3.64 Measured and simulated radius of curvature of a microlens as a function of the number of droplets printed onto a 125 μm optical fiber

The radius of curvature can now be simulated for any number of droplets by assuming a volume per droplet of 8.3 pL and a contact radius of 62.5 μm. The measured results and the simulated results derived from the above equations are shown in Fig. 3.64. There is good agreement between the model and the experimentally determined values.

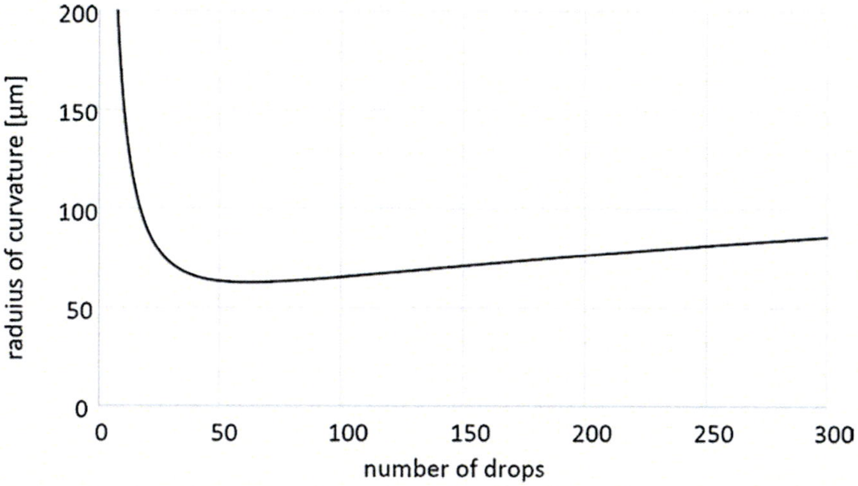

Fig. 3.65 Simulated radius of curvature of microlenses on a 125 μm optical fiber

The radius of curvature can therefore be predicted for a large range of volumes or droplets. Figure 3.65 shows the results up to a droplet count of 300. The smallest radius of curvature that the lens can achieve is 62.5 μm. This value is the radius of the optical fiber and is reached after printing approximately 60 droplets. At this point, the microlens is hemispherical in shape. The radius of curvature then increases again if the number of droplets is further increased.

3.5.2 *Individualized Microlenses*

To correct optical errors such as spherical aberration, nonspherical microlens shapes may be desirable. Since the polymers used to additively manufacture microlenses are dielectric, this can be achieved with electric fields. This principle is illustrated in Fig. 3.66. In Fig. 3.66a, a droplet of liquid (polymer) is shown on a substrate. The dipoles in the liquid are arranged disorderly within the spherical cap. As described above, the droplet forms a spherical shape. But if an electric field is now applied to the droplet, the dipoles align themselves parallel to the field lines of the electric field, i.e., the lines of action of the electric force. This induces a deformation in the original spherical cap. This effect can be used to deform any printed polymer droplets that have not yet been cured to achieve a predefined freeform surface that can then be cured with UV light.

The experimental setup is shown in Fig. 3.67. A metal tip (2), which serves as an anode, is placed inside an evacuable glass cylinder (1). This tip can be used to apply a voltage of up to 2.5 kV from a high-voltage source. The microlens, still in a liquid state, is on top of a substrate placed on the cathode (3). Once the sample is in position, the glass cylinder is evacuated and flooded with SF6 to avoid the risk of

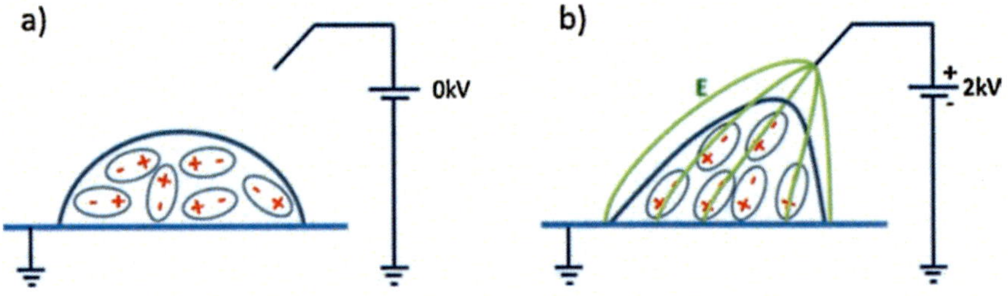

Fig. 3.66 Deformation of a liquid microlens in an electrical field. (**a**) Schematic representation of the printed droplet, which consists of dipoles on top of the substrate; (**b**) alignment of the dipoles under an electrical field and the resulting change in the shape of the droplet

Fig. 3.67 Experimental setup to induce a deformation of a liquid microlens within an electric field. (**a**) Side view; (**b**) top view of the central area of the experimental setup; (**c**) deformation of a droplet that has not yet been cured under an electric field of 0 and 1.9 kV

flashover between the anode and the cathode. The exact positioning of the microlens below the anode is adjusted with an x-y-z kinematics mechanism (not shown in the image) that is connected to the setup by a flexible joint (4). To allow precise positioning (accuracy ~5 μm), the microlens silhouette is observed from two directions with two cameras (5) in collimated lighting (not shown) (see top view of setup in Fig. 3.67b). The setup also contains UV LEDs to cure the deformed lens (not shown).

A typical result is shown in Fig. 3.67c (silhouette of the microlens taken using a camera). The anode/metal tip can be seen in the upper part of the image. The liquid microlens is approx. 320 μm below it (diameter of the lens, approx. 200 μm). When no voltage is applied, the lens has a spherical shape as expected. If a voltage such as 1.9 kV is now applied, the lens surface experiences a drastic aspherical deformation. The magnitude and the nature of this deformation depends on the droplet size, the field strength, the anode shape, and the position of the anode relative to the lens. This allows individually adapted lens shapes with adapted and optimized optical functionality to be realized.

The general mathematical description of lenses can be stated in terms of Zernike polynomials. These polynomials allow the shape of the lens or the wavefront that it generates to be described. Zernike polynomials are decomposed into an azimuthal component and a radial component ρ.

The Zernike polynomial series expansion can be stated in terms of the following rotationally symmetric equations:

$$Z_k(\rho,\phi) = Z_n^m(\rho,\phi) = R_n^m(p)\begin{cases}\cos(m\phi) : m > 0 \\ \sin(m\phi) : m < 0 \\ 1 : m = 0\end{cases} \tag{3.5.7}$$

The radial component satisfies

$$R_n^m(\rho) = \sum_{\frac{n-m}{2}}^{l=0} \frac{(-1)^l (n-l)!}{l!\left[\frac{1}{2}(n+m)-l\right]!\left[\frac{1}{2}(n-m)-l\right]!}\rho^{n-2l} \tag{3.5.8}$$

Equation (3.5.10) defines the Zernike polynomial of n-th radial and m-th azimuthal order. The term l denotes the starting point of the Zernike expansion.

Based on this representation, the general Zernike polynomial of n-th order is now given a weighting factor. The weight is called the n-th order Zernike coefficient c_n. A description of the wavefront is obtained by summing over all Zernike components:

$$W(\rho,\phi) = \sum_{\infty}^{k=1} c_k Z_{k(\rho,\phi)} \tag{3.5.9}$$

In general, including a greater number of Zernike polynomials yields a better approximation of the wavefront or lens geometry.

A description of the surface of the deformed microlens can now be established in terms of these Zernike polynomials. This can then be imported into an optical design software program to simulate the effect of the microlens, for example.

The systematic procedure followed to establish this description is shown in Fig. 3.68 for an anode positioned directly above the vertex of the droplet. The arrows indicate the sequence of process steps. First, the camera records an image of the

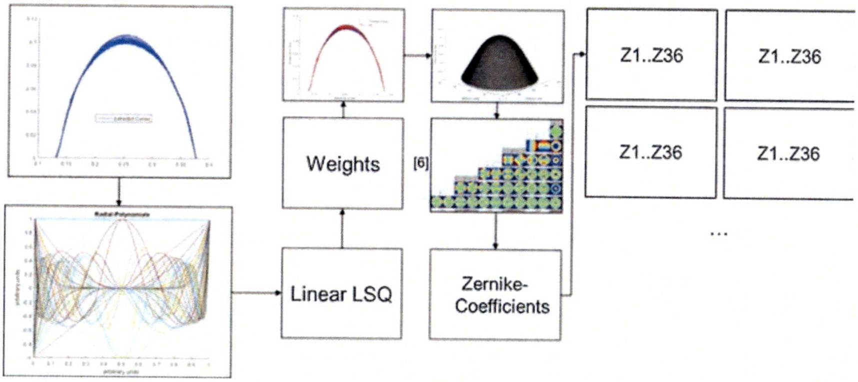

Fig. 3.68 Systematic representation of the process chain to express the surface shape of the deformed droplet in terms of Zernike polynomials

silhouette of the deformed droplet at a specific voltage (not shown in the figure/equivalent to Fig. 3.67c). The contour of the droplet (= cross-section through the vertex) can be extracted from this image using image processing algorithms. This is shown for different voltages by the many blue lines in the top left of Fig. 3.68. In the next step, the radial components of the first 36 Zernike polynomials are generated (see bottom left of Fig. 3.68) and fitted to the contour lines deduced from the droplet based on a least-squares approach that takes into account a weighting of each polynomial. Thus, the fit curves shown in red are computed for the measured contour lines shown in blue. Since the deformation of the droplet is assumed to be symmetrical, the fit curves are then rotated to model the "3D shape" of the deformed droplet. The first 36 complete Zernike polynomials are then generated and fitted to the 3D model thus obtained by varying the Zernike coefficients accordingly. For each voltage, this yields a set of Z1-Z36 Zernike coefficients that describe the surface of the droplet under this voltage.

With the symmetrical anode positioning described above, changing the voltage induces a symmetrical change in the curvature of the lens and hence produces optical defocusing. This is, for example, described by the Zernike polynomial Z4. The dependency of the Z4 coefficient on the voltage is shown in Fig. 3.69. Finally, a function is fitted to these measurement points using least-squares fitting based on the Levenberg-Marquardt method. This is performed not only for the Z4 coefficient but for every relevant Zernike polynomial, which allows a definitive model of the surface deformation to be established in terms of the Zernike polynomials Z1-Z36.

The surface shape of the droplets can now, for example, be transferred to an optical simulation software program based on the Zernike polynomial model, and the expected light distribution can be simulated as a function of the deformation/voltage.

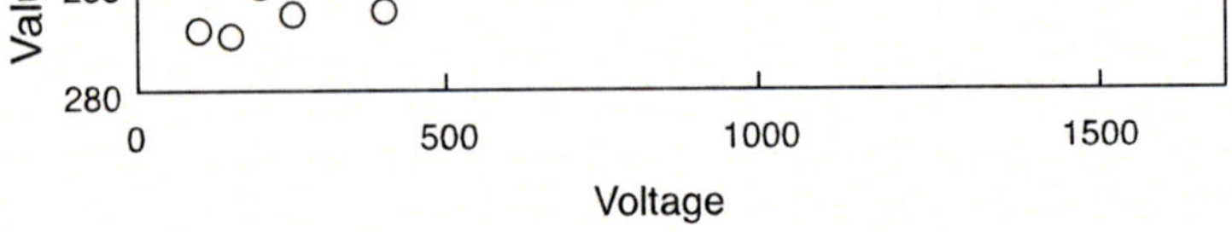

Fig. 3.69 Dependence of the Zernike coefficient Z4 (defocusing) on the applied voltage

3.6 Additively Manufactured Light Sources

The benefits of additively manufactured optical elements extend further than simply realizing complex geometries. They also enable very different materials to be combined. In this way, additional functionality can be incorporated into the optical components or realized alongside the printed optics.

One attractive approach to achieving a greater functionalization of additively manufactured optics is the integration of light-emitting materials. This topic is discussed below.

3.6.1 Organic LEDs

Integrating light sources into the additive manufacturing process is another way to functionalize 3D-printed optics. Organic light-emitting diodes (OLEDs) are a promising alternative to conventional light sources based on semiconductors. OLEDs open up new possibilities, especially during manufacturing. Inkjet printing can be used to realize individually shaped polymer-based light sources with low material consumption thanks to the drop-on-demand technique. Organic solar cells can already be produced in this way from layers manufactured exclusively by inkjet printing [72]. Furthermore, the inkjet process is not limited to flat surfaces (see Sect. 3.7). By the end of the process chain, a complete lighting element with housings and brackets but also optics and light sources can be manufactured entirely additively.

Using jettable materials with low solids content to additively manufacture semiconductors poses several challenges relating to the accuracy of the printing system, the surface roughness of the substrate, and the layer design. A layer stack consisting of inks with dissolved semiconducting polymers was therefore tested to investigate functionality and compatibility on conventional substrates (ITO anode).

However, the approach described here used transparent 3D printing materials as the basis of the substrate on which the OLED was built. To compensate for the absence of the conventionally used glass substrates with an indium tin oxide (ITO) electrode, a dissolved semiconducting polymer was also used as an ITO replacement [73].

For the additively manufactured OLED, a four-layer OLED on a polymer anode and a 3D printing substrate were used. Figure 3.70 shows a relative band diagram highlighting the functionality of each layer with its lowest unoccupied molecular orbital (LUMO) and highest occupied molecular orbital (HOMO). If a positive voltage is applied from the anode to the cathode, charge carriers are injected from each electrode into the layer stack, recombining within the emissive layer (EL) while emitting visible light. The hole injection layer (HIL) transports positive charge carriers (holes, white points) from the anode to the EL. Similarly, the negative charge carriers (electrons, black points) enter the layer stack through the electron transport layer (ETL). An optional hole blocking layer (HBL) prevents positive charge

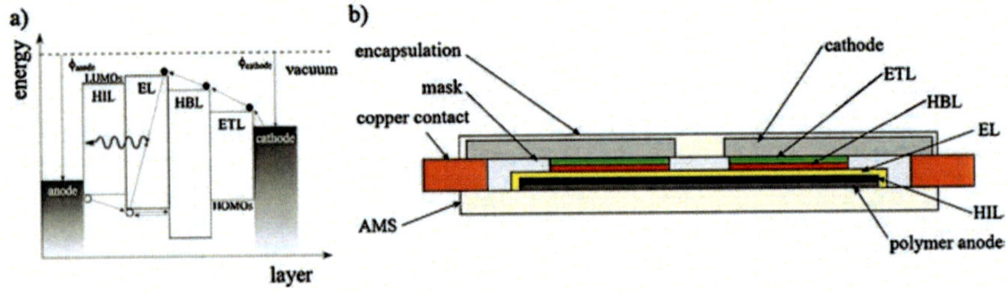

Fig. 3.70 (**a**) Schematic energy diagram of the four-layer OLED presented here; (**b**) cross-section sketch of the OLED design for Additive manufacturing by inkjet (excluding Galinstan). Layer thickness not to scale

carriers from drifting to the electron transport layer instead of recombining within the EL to emit light [74]. The structure of an additively manufactured OLED is shown schematically on the right-hand side of Fig. 3.70. In practice, the thickness of each layer can sometimes be as low as 100 nm. In the design shown in the figure, the polymer anode, the HIL, and the EL (EML) span a large area on an additively manufactured substrate (AMS). A mask is then applied to divide the large area into smaller active areas where the HBL (not shown) and the ETL are locally printed. The rear contact is established by locally dispensing a liquid metal alloy that is then encapsulated in the same material as the substrate. The copper plates used for the electrical connection of both the anode and the cathode are not shown.

A commercial inkjet system (Fujifilm DMP-2850) was used to realize the inkjet OLEDs by integrating them onto additively manufactured optical components. This system allows the various inks to be filled and the parameters to be configured in such a way that the print head releases droplets of approx. 10 pL. The inkjet printer was used to produce the substrate and the mask, for sealing, and to print the organic layers themselves. The materials can easily be exchanged thanks to a cartridge system. The strongly solvent-based inks also require intensive drying to produce homogeneous layers. For an optimized process, multiple layers of printing must therefore be combined with drying cycles.

Figure 3.71a shows an inhomogeneous layer after a printing-drying cycle. Although the layer initially covered the entire surface in its liquid state, patches formed during intensive vacuum drying, meaning that planar conductivity was not achieved for the polymer anode or the HIL. Nevertheless, since water-soluble layers are sometimes used but other layers may be degraded by the presence of even the slightest amounts of water in their environment, drying is a greatly important part of achieving a functional layer stack. Figure 3.71b shows multiple layers of the polymer anode (arrows) printed with an offset, as well as the HIL on the right. Although low-viscosity inks were used and a fully covering layer was created during printing, the printed surfaces exhibit a characteristic line structure caused by the sequential line writing process, as can be seen in Fig. 3.71b and c. Up to the line artifacts visible at the edge, the layer is fully covering in this example. In Fig. 3.71c, the masking of the EL can also be seen as a thick bulge to the top and to the left. The mask is

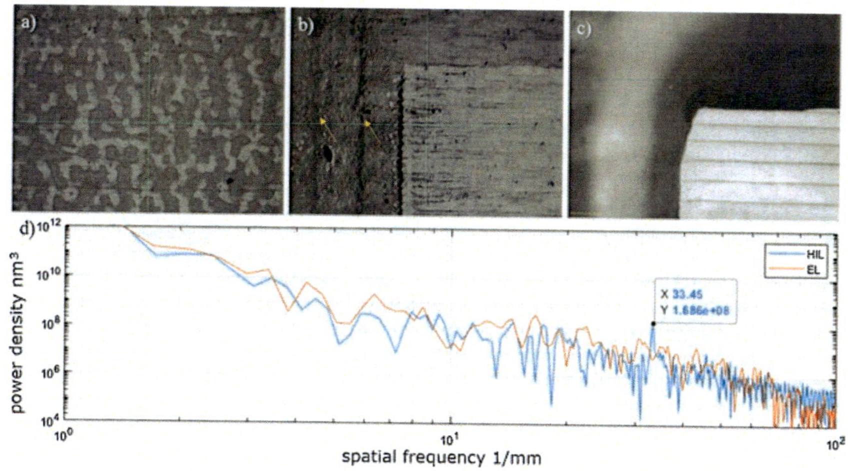

Fig. 3.71 (**a**) Inhomogeneity in a single layer after vacuum drying; (**b**) multilayered printing of the anode and the HIL; (**c**) EL with the edge of the mask; (**d**) PSD analysis of different printed layers

greatly important, since any contact between the HBL and the ETL or the cathode and the HIL would create a short circuit, and the leakage currents would significantly impair the efficiency of the system [75].

The mask severely limits the usable area of the EL. However, it prevents a potential defect at the edge of a layer from causing the entire OLED to fail. Since the mask is made from the same material as the substrate material, the UV dose is a critical aspect of curing, since the EL is exposed and unprotected at this stage and will degrade if subjected to excessive radiation [76]. The same material is also used to encapsulate the OLED. The encapsulation on one side and the substrate on the other protect the semiconductor stack from contact with atmospheric oxygen, which would also cause the OLED to degrade [77].

The small layer thicknesses mentioned above can lead to problems, especially when a layer is applied to a rough substrate surface or on top of rough sublayers. A topology measurement obtained with a white light interferometer after each processing step shows that the surface roughness of each layer is approx. $S_a = 1$ μm but can reach values (S_z) of more than 10 μm wherever contaminants are present. These roughness values are the greatest challenge to integrating OLEDs into 3D-printed optics. Figure 3.71d shows a more in-depth analysis of the surfaces in terms of the power spectral density (PSD). Although the material fully covered the surface when it was applied, the line spacing of the printer can still be identified once the layer has dried. The indicated measurement shows an increase in the amplitude at 33.45 1/mm for the HIL, which matches the line spacing of 30 μm of the printing process. However, this spatial frequency is smoothed out when the EL is subsequently applied [78].

Since the emitted light is coupled out through the HIL, the anode, and the substrate, the efficiency of the overall stack depends on the transmission of these layers. As the number of layers increases, the surface resistance decreases the transmission. According to measurements, the OLED emits light in the spectral range from 525 to 640 nm (FWHM). Transmission analysis of the anode using a spectrometer based on the printed substrate as a reference found average values of 99% for the anode layer, 97% for two layers, and 92% for four layers. Since the refractive index of the substrate and the PEDOT:PSS used as a replacement for ITO are almost identical, the losses due to Fresnel reflection are negligible. A compromise must therefore be found between conductivity and transmission. Since a single conductive layer is susceptible to failure, two layers were used for the anodes of the additively manufactured OLEDs presented here.

The electrical characteristics were determined using a constant current source producing currents of a few nanoamps. The voltage drop across the OLED and the forward current were recorded in parallel. The results achieved by the realized OLEDs correspond to the expected behavior of a typical semiconductor [78]. Figure 3.72 shows a completely additively manufactured OLED combined with a 3D-printed substrate as an example. The view through the anode in Fig. 3.72a shows four OLEDs with grid-like contact surfaces around the edges. The yellowish EL is visible through the dark blue HIL. The top and bottom copper plates are in contact with the planar anode. Figure 3.72b shows the rear side of the OLED, where the liquid metal contacts with the left and right copper plates can be seen as separated cathodes. Figure 3.72c shows one of the four OLEDs in its active state.

3.6.2 *Additively Manufactured Optical Converter and Random Laser*

This section investigates whether an optical converter or laser can be realized additively. For this purpose, a design similar to a fiber laser and another design based on the principle of random lasers are discussed.

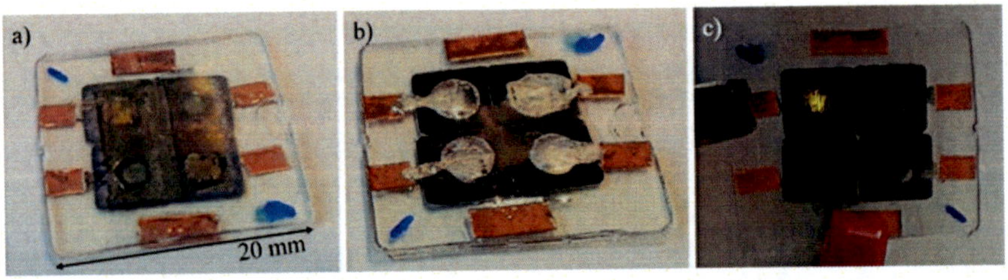

Fig. 3.72 Additively manufactured OLED on a 3D-printed substrate; (**a**) view of anode with active surfaces; (**b**) view of cathodes with liquid metal contacts; (**c**) illumination from the OLED

3.6.2.1 Optical Converter

This idea is based on the concept of a fiber laser. Fiber lasers have a fiber core that contains laser-active dopants that are excited to produce laser activity using an external pump light source. The laser resonator is, for example, realized with fiber Bragg gratings in the core. The total reflection induced by the difference in the refractive index between the core and the cladding allows the light propagation to be directed within the core.

Based on this principle, the concept shown in Fig. 3.73a was created for a converter. A 3D-printed rod doped with Nd:YAG nanoparticles is placed between two mirrors. When this rod is pumped from an external source, the Nd:YAG nanoparticles are excited, causing them to emit light at 1064 nm. Ideally, the light should be guided in a preferential direction, like in an optical fiber. To achieve this, like in an optical fiber, the refractive index must vary along the direction perpendicular to the rod. This can be accomplished with additive manufacturing. If the rod shown in Fig. 3.73a is printed by stereolithography and a DLP projector is used for curing, the pixel structure of the projector will induce a column-shaped curing pattern on the material. This is shown in Fig. 3.73b. A 3D-printed rod is shown from the side. Each pixel or DLP micromirror of the projector represents a column. The formation of these columns is caused by the "dead area" between two pixels (DLP mirrors). In these areas, the material is cured less when the pixels are projected onto the resin. Since the refractive index of the resin used in this application depends on the curing of the material, the refractive index is typically 0.01 lower in these transition areas (see Sect. 3.3.4). Each 3D-printed column therefore acts like a separate optical fiber. This is also illustrated by an image of the surface topology taken by white light interferometer (top view/see Fig. 3.73c). This figure also shows typical measurements of the refractive index.

This principle was used to 3D print an optical converter. For this purpose, cuboids of 5 × 5 × 12 mm³ (see Fig. 3.74a) were realized with a standard SLA/DLP printer. However, the printer was modified to support multimaterial printing (see Sect. 3.7.3). This allowed the printer to alternate between layers with doped and undoped resin. Nd:YAG particles (diameter 40 nm) were added to the doped resin. The layer

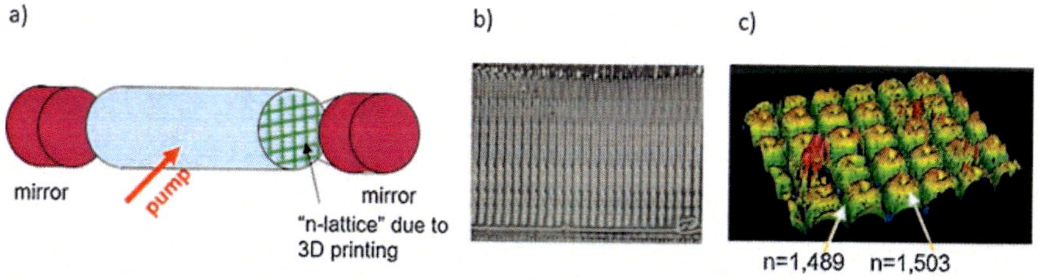

Fig. 3.73 (**a**) Principle of a 3D-printed converter; (**b**) side view of the 3D-printed rod: column structures are visible. Each column represents a pixel of the DLP projector; (**c**) top view (white light interferometer) of the rod. At the transition between columns, the refractive index is lower, as the curing occurred less completely

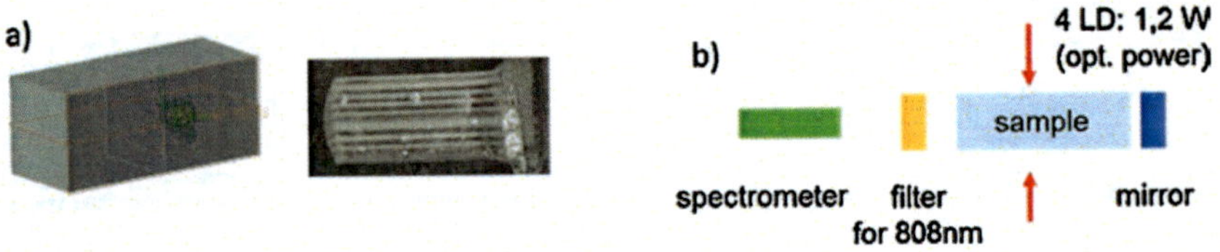

Fig. 3.74 (**a**) 3D-printed rod; (**b**) experimental setup

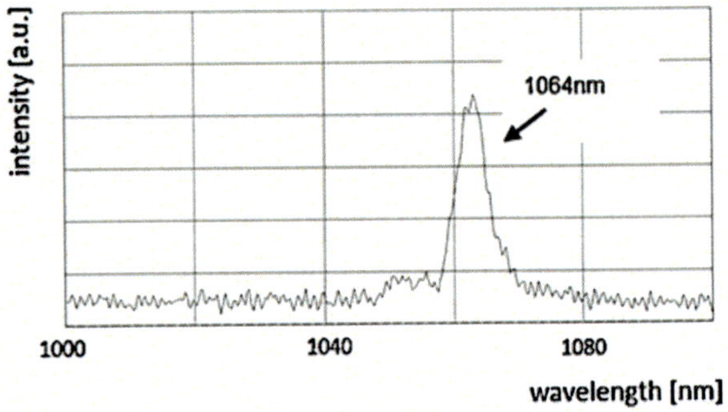

Fig. 3.75 Measured emission spectrum with excitation at 808 nm

thicknesses were between 50 and 150 μm. As described in Sect. 3.3.4, the refractive index was also reduced at the interface between two layers, meaning that each layer helped to facilitate the preferential horizontal light transmission. After the rod was printed as described above, it was installed into an experimental setup like the one depicted schematically in Fig. 3.74b. The pump sources were four laser diodes (808 nm, optical power per diode, 1.2 W). The emission signal was recorded with a spectrometer. An added filter at 808 nm ensured that none of the signal from the pump light source entered the spectrometer.

Figure 3.75 shows a typical result: there is a clear signal indicating spontaneous emission by the Nd:YAG particles at 1064 nm induced by the excitation at 808 nm.

These 3D-printed samples demonstrate that additive manufacturing can be used to successfully implement an optical amplifier or converter. Thanks to the degrees of freedom allowed by 3D printing, these components can be made in any desired shape (e.g., even curved shapes). 3D printing can also print directly onto existing optical components and thus directly couple the emitted light into another optical element.

To realize a laser on the basis of the above, a resonator is required. A series of ongoing studies are therefore seeking to apply reflective layers to the end surfaces by inkjet printing in order to develop a fully 3D-printed laser. However, this is a challenging problem due to the high quality required by the mirrors. Consequently, a second approach is being investigated in parallel, as described below.

3.6.2.2 Random Lasers

A conventional laser consists of a medium that allows optical amplification through stimulated emission. The optical medium is contained within an optical cavity that partially traps the light. The system starts to lase if the total gain is larger than the total losses.

Random lasers work differently. Instead, nanoparticles are, for example, mixed into a gain medium (see Fig. 3.76). The gain medium is pumped by an external source, emitting light by spontaneous emission. This light scatters elastically on the nanoparticles, forming modes. Thus, a random laser is a material where light is multiply scattered by the randomness of the scattering partners incorporated into the material (e.g., nanoparticles) and then amplified by stimulated emission. As in conventional lasers, there is a threshold value above which the total gain is higher than the total loss [79].

This section reports a study of whether a random laser can be realized using additive manufacturing. First, rhodamine 6G was dissolved in ethanol as the optically active material, and Nd:YAG nanoparticles (40 nm diameter) were dissolved in isopropyl alcohol. The 3D printing material (Luxaprint) was heated to 120 °C, and both the dissolved rhodamine and the nanoparticles were added to it and agitated with a magnetic stirrer. Samples with a wide range of geometries were then printed by stereolithography, and their laser activity was investigated. The experimental setup is shown in Fig. 3.77. A pulsed Nd:YAG laser system (532 nm, 10 Hz, 10 ns) was used as the pump light source. The light from the pump laser was first passed through a bandpass filter (532 nm). Part of the laser light (4%) was then coupled out to determine its energy or directed onto the sample via a beam expander. At the position of the sample, the pump laser produced a spot of approx. 2 mm in diameter. The spectrum emitted by the sample was recorded using a spectrometer.

Figure 3.78 shows a few typical measurement results. Figure 3.78a shows a spectrum measured when the sample was excited below the laser threshold. The signal

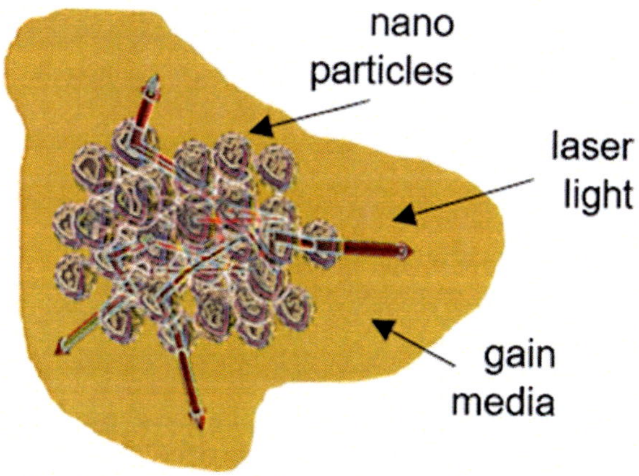

Fig. 3.76 Principle of a random laser

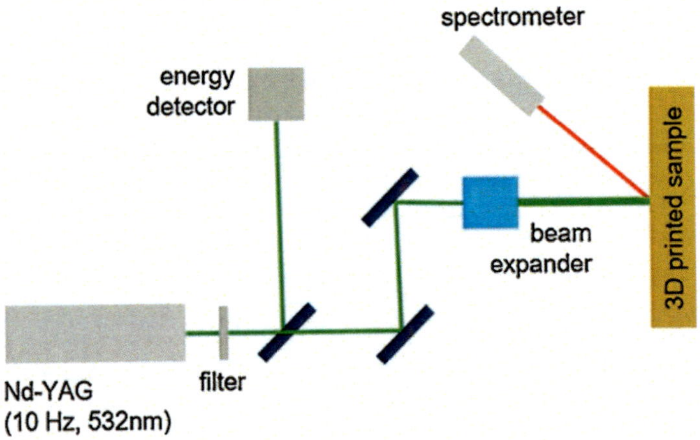

Fig. 3.77 Experimental setup to investigate the random laser activity of 3D-printed samples

from the pump light source is visible at 532 nm. In the range from 540 to 650 nm, the spontaneous emission of the rhodamine excited by the pump light is visible. Once a specific energy density is exceeded at the position of the sample, another prominent peak appears at 575 nm (Fig. 3.78b). This demonstrates that the 3D-printed sample exhibits the properties of a random laser.

Figure 3.79 illustrates the same results again. Here, the spectrometer signal measured at 575 nm is plotted as a function of the pump energy input. Above a value of 1.2 mJ (approx. 10 mJ/cm^2), the laser threshold is exceeded, and the total loss in the random laser is smaller than the total gain. As the energy density of the pump laser increases, the random laser activity of the 3D-printed laser also increases. It therefore exhibits comparable behavior to conventional lasers.

Since the random laser effect is based on a multiple scattering process, the concentration of the scattering particles (nanoparticles) must also influence the modes that are created. To verify this hypothesis, samples were printed with different concentrations of Nd:YAG particles. Figure 3.80 shows the dependence of the wavelength of the emitted laser light on the Nd:YAG concentration. When the concentration of nanoparticles is increased by a factor of 15, the emission wavelength decreases by 4 nm from 575 to 571 nm.

Finally, Fig. 3.81 shows a possible application for a 3D-printed random laser. They could be used as light sources for integrated photonics. The laser would be 3D printed onto a silicon chip in combination with printed waveguides, optical isolators, and so forth.

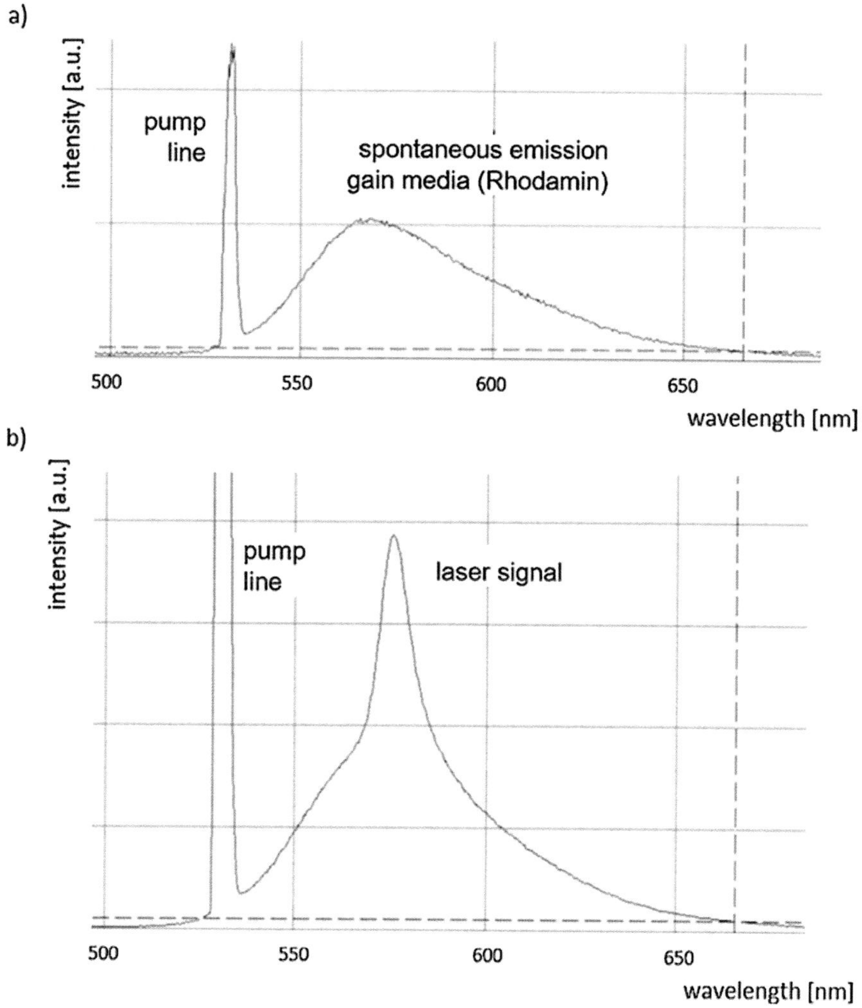

Fig. 3.78 Emission spectrum measured (**a**) below and (**b**) above the laser threshold

3.6.3 Additive Manufacturing of Photoluminescent Optics

3.6.3.1 Functionality in Optical Elements

The use of 3D printers in the fabrication industry is on its way to becoming an indispensable fabrication tool. This is indicated by the variety of materials and process methodologies being invented and modified to enhance and rectify the commonly used systems. The amount of freedom in developing and modifying the designs onsite and immediately implementing changes into the production line provided by additive manufacturing is exceptional when compared to conventional fabrication

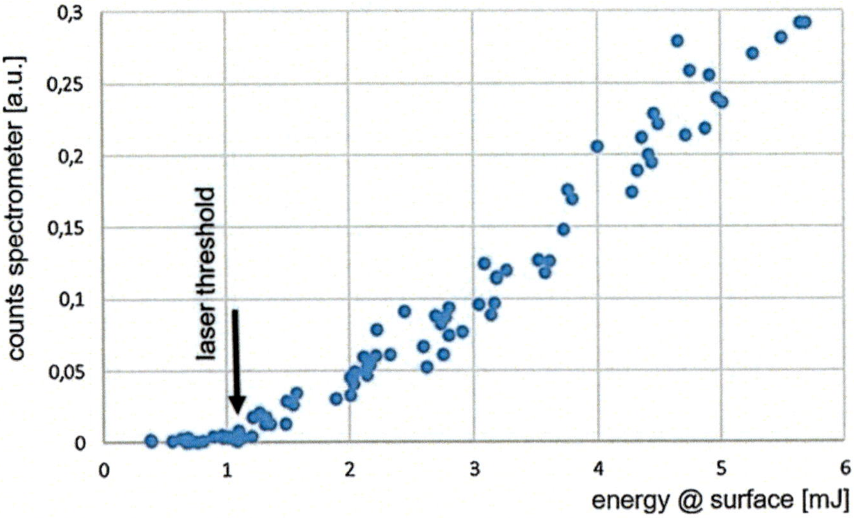

Fig. 3.79 Dependence of the measured intensity of the emitted laser line on the pump energy input

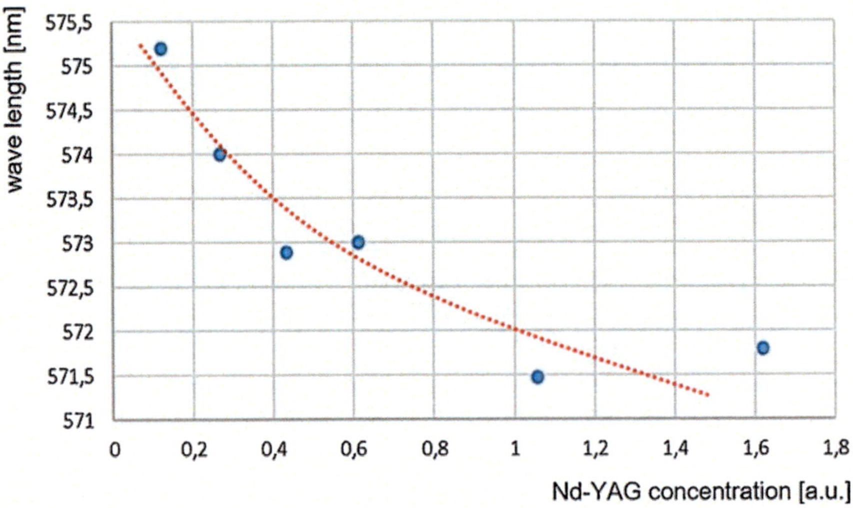

Fig. 3.80 Dependence of the emitted laser line on the nanoparticle concentration

processes. In addition to variations in the geometrical aspects of the product, additive manufacturing preserves the ability to introduce specific traits into fabricated parts by manipulating the raw material. One way to do this is to experiment with the composition of the existing palette of materials to vary existing properties. Certain new additives could also be added to the resin composition to implant the properties of the additive into the printed component. In this section, one such experiment is discussed where the desired property is the photoluminescence of quantum dots.

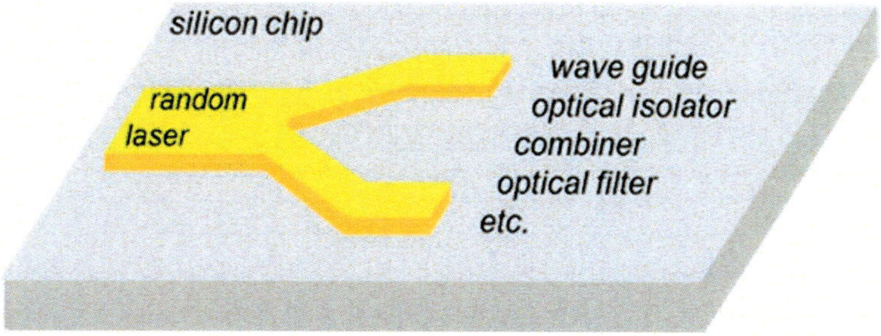

Fig. 3.81 Possible application for 3D-printed random lasers

3.6.3.2 Introducing Functionality by Altering the Raw Materials in Additive Manufacturing

Quantum dots (QDs) can be defined as organized clusters of molecules that are physically sized anywhere from 1 to 100 nm. Conventionally, this mesoscopic size restriction of quantum dots leads to the phenomenon of photoluminescence with an exceptionally narrow emission spectrum. This is explained by the theory of quantum confinement, according to which the size restriction confines the free electrons within the molecules of the quantum dot to very specific energy levels of excitation, as compared to their bulk counterparts, where the abundance of such molecules within a three-dimensional volume widens the energy band of excitation and emission. In terms of reactivity to the environment, the sensitivity of quantum dots to the presence or absence of certain chemical molecules is much higher than their bulk counterparts. This is due to the exceptional surface-to-volume ratio of the molecular arrangement in the quantum dot nanoparticles [80].

Conventional production of quantum dots involves reactants that react under a controlled environment to form product molecules that accumulate and grow into small clusters in an organized manner under a continuously monitored reaction environment [81]. This growth is stopped when these molecule clusters reach the nanometer size range, thereby yielding the quantum dots. Such individual nanometric clusters are engineered to alter the material band gap with precision by comparison with the material's bulk counterparts. As a rule of thumb, conventional quantum dots with an identical composition show a strong correlation between the dot size and the emission wavelength, in accordance with quantum confinement [80].

In the current scenario, a multitude of fabrication methods for conventional quantum dots (CdSe/ZnS; core-shell quantum dots) were established in order to integrate quantum dots into additive manufacturing materials. Additionally, an array of alloyed quantum dots (ZnCdSeS; low-cadmium quantum dots) was used. The molecule composition of such alloyed quantum dots is modified to fine-tune the emission wavelength of the alloyed quantum dots. In other words, the emission wavelength of alloyed quantum dots has a stronger dependence on the chemical composition, rather than just the quantum size effect [82]. The quantum dot size of the alloyed quantum dots used in the experiments outlined in this section was in the range of 6 nm in dot diameter. The reasoning behind the choice to use alloyed quantum dots instead of conventional quantum dots was the superior surface modification of the alloyed QDs, being coated with ligands and therefore better suited for stable interaction in a chemical environment like a conventional 3D printing polymer.

3.6.3.3 Preparation of 3D Printing Materials Enriched with QDs

The first step towards creating photoluminescent optical components is to synthesize the material required for 3D printing. The preparation step of this QD-enriched raw material is shown in Fig. 3.82. First, the quantum dots in their customized ligand-protected format (for instance, L-type ligands like 1-Hexadecylamine) were blended into a suitable colloidal solvent (e.g., THF). The presence of colloidal solvents when preparing the printing material leads to a better homogenized spread of the quantum dots across the resin matrix compared to incorporating quantum dots in their powdered form. The choice of the colloidal solvent depended upon its compatibility with the quantum dot ligands and the photoreactive resin chosen as the base matrix. Commercially available photoreactive resins used for earmold fabrications, as well as resins based on organically modified ceramics, were tested as base matrices for preparing the QD-doped resin mixtures.

After mixing the dissolved QD colloid with the resin, vaporization of the colloidal solvent from the mixture was actively encouraged during the mixing process to minimize the traces of the solvent in the nanocomposite, for several reasons. One reason is that, with this printer setup, a PDMS window is needed for the printing

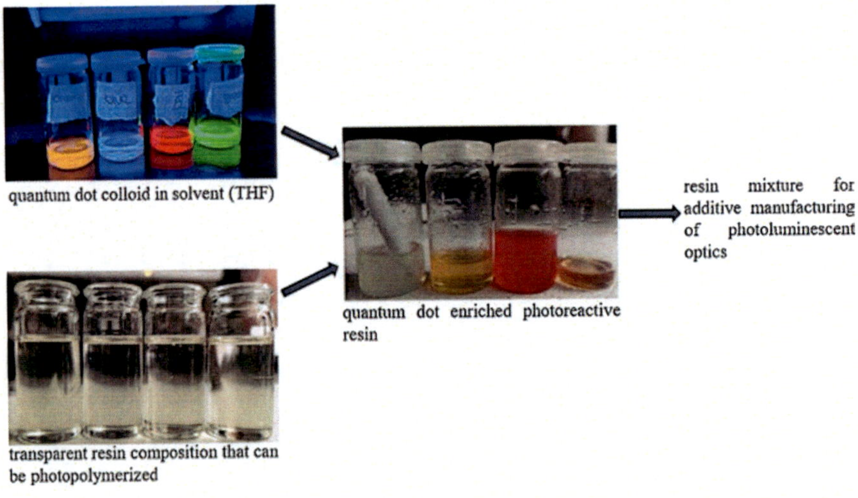

Fig. 3.82 Pictorial flow diagram depicting the stepwise procedure for obtaining the QD-enriched printing material from the ligand-protected quantum dots, the colloidal solvent (tetrahydrofuran in this case) and the photoreactive resin transparent to visible light

process (see Sect. 3.7.3). Solvents that can be used for quantum dot colloids swell the PDMS layer [83], thereby causing clouding and a decrease in transparency. A decrease in transparency causes erroneous variations in the energy dosage that reaches and polymerizes the printing material liquid layer. Also, while traces of solvents are often incorporated into certain resin compositions, presence of the solvent in large proportions within the printing material can decrease the polymer network density of the printed component and hence adversely affect product properties like brittleness.

The quantum dots are very sensitive to their surrounding chemical environment, leading to the development of quantum dot clusters inside the printing material. Such localization of the QD clusters can cause statistical variations in the photoluminescence intensity depending upon the spatial point of measurement within the sample. The mixed materials are hence consumed for additive manufacturing as soon as possible to avoid an increase in cluster formation within the composite.

3.6.3.4 Fabrication of Photoluminescent Optics Via Multimaterial Printing

The synthesis of the liquid QD-enriched printing material is followed by the fabrication of components with the mixture. In the experiments discussed in this section, a stereolithography printer (Ember from Autodesk) was used for additive manufacturing with nanocomposites. It is mandatory to have separate resin vats for different mixtures in order to account for the possibility that the nanoparticles may diffuse into the PDMS layer of the printing setup as a result of continued print jobs with nanocomposites. Experiments to investigate the idea of using multiple quantum dot mixtures to create a single component were performed on the same printing apparatus by making certain modifications to the printer, as explained in Sect. 3.7.3. Printing with multiple nanocomposites in this context should be interpreted as synthesizing printing materials with quantum dots of identical chemical composition and a similar absorption wavelength range but which emit at different wavelengths in the visible spectrum.

3.6.3.5 Results and Discussion

Detection of Fluorescence from Printing Materials Other than the Quantum Dots

Photoreactive resins were incorporated with quantum dot nanoparticles with traces of colloidal solvent to investigate the idea of imparting the optical property of photoluminescence to components fabricated with these mixtures. Attention must therefore be paid to any autofluorescence of the polymers used as the 3D printing material and the solvents used to dissolve the QDs; this autofluorescence is mainly present in the deep blue region [84]. Weak peaks were detected in the region of

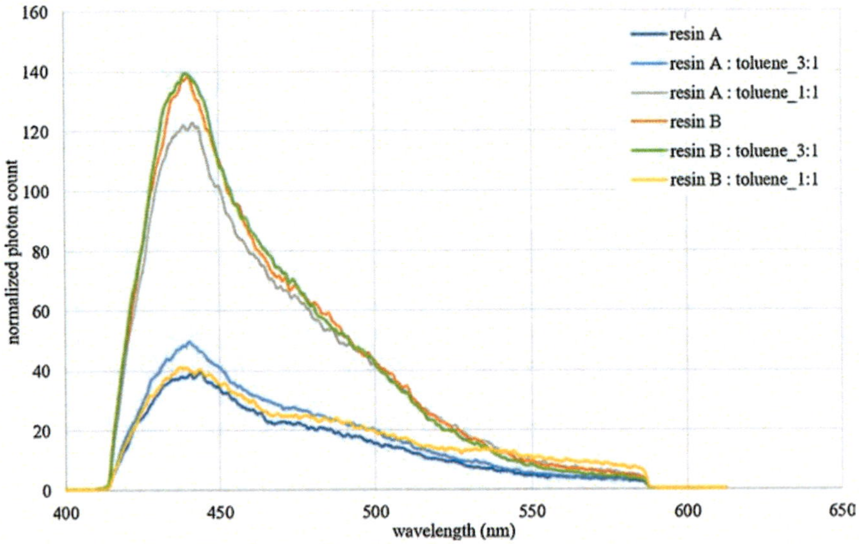

Fig. 3.83 Emission spectra of the two least autofluorescent photoreactive resins that are commercially available and variation in the resin autofluorescence observed when the resin is diluted with a solvent (toluene in this case) in different proportions

expected quantum dot emission. In some cases, these unexpected fluorescent peaks were so strong that the QD peaks were barely detected or completely absent. This initiated a search for alternative photoreactive resins and colloidal solvents that nullify or minimize the autofluorescence from the raw material. A range of commercial resins including the open source resin with the strong blue fluorescent peak (PR48) were tested and compared for autofluorescence. It was found that all commercial resins showed inherent fluorescence to varying degrees in the deep blue region. Out of the various choices investigated, two of the least fluorescent resins were tested along with the solvent toluene, as shown in Fig. 3.83.

Polymers with aromatic or heterocyclic units are known to be capable of photon absorption in 300–500 nm range for π–π^* transitions. These chromophore excitons can either non-radiatively relax to the ground state or choose the radiative decay path resulting in photoluminescence [85]. Since the compositions of commercial resins are mostly secret, it is difficult to pinpoint the cause of this resin fluorescence to a precise origin for most commercial resins. On the other hand, examination of an open source resin formulation like PR48 found that the formulation consisted of a thiophene-based UV blocker component used to control the curing depth of the photoreactive resin (see Sect. 3.2). Thiophene chemical groups are known to be blue light-emitting fluorophores and can, for example, be used in polymer LED fabrication [85]. This may account for the bright fluorescence of PR48 when exposed to ultraviolet radiation. The literature mentions the effect of polymerization on the structural arrangement of certain polymer chains that lead to chromophore stacking. This results in excimer emission and the development of fluorescent peaks. Such a possibility also cannot be ruled out with the polymer networks created from the photoreactive resin compositions.

Apart from the origin of the inherent fluorescence from the resins, another interesting observation is the effect of the solvent and its proportion in the resin-solvent mixture, which affects the intensity of autofluorescence from the resin. As can be observed in Fig. 3.83, the two resins show different responses to the same colloidal solvent toluene. The autofluorescence peak is strengthened in intensity by the presence of toluene in the resin A-toluene mixture and shows a threefold increase when the resin composition is 50% diluted in the solvent. The autofluorescence peak is weakened in intensity by the presence of toluene in resin B-toluene mixture and is reduced to one third of its strength when the solvent concentration in the mixture increases to 50% by volume. Although the literature mentions the capability of a polar solvent to cause bathochromic shifts in the emission wavelength of fluorescence from polymers due to stabilization of the excited or ground states of the chromophores, almost no wavelength shift was seen in the spectra of both mixtures being discussed here [86]. This could be linked to the nonpolar nature of the solvent toluene, but there still is no precise explanation for the mechanism of interaction between the toluene molecules and the resin composition or possible routes to emission due to internal charge transfer states [86] that could lead to the strong changes in the intensity of photoluminescence. But the theory of emission due to internal charge transfer, among other criteria, would require the presence of electron-donating and electron-accepting properties of groups within or attached to the fluorophore, which brings back the conundrum of an ambiguous unknown composition.

Both resin A and resin B compositions were considered for photoluminescent 3D printing material preparations. The first trials were done with resin B, since the presence of toluene seemed to weaken the autofluorescence, thereby giving an edge to the photoluminescent signals from the quantum dots embedded in the polymer matrix. But this decision was met with challenges in the fabrication step, as the solvent corroded the transmission window in the printing setup much faster than expected. The presence of the solvent in the printed components could also alter the mechanical properties of the 3D-printed component. This reinforced the idea that the presence of the solvent in the final nanocomposite raw material should be limited as much as possible. In this situation, resin A provided an edge with lower autofluorescence and a lesser concentration of the solvent.

Dependence of the Photoluminescence on the QD Concentration

The effect of increasing the concentration of the QDs in nanocomposites was observed in single quantum dot [108]. Figure 3.84 shows one such effect, where PL can be observed visually from samples fabricated from 0.6% concentration by weight (Fig. 3.84a, right, top without UV light illumination; bottom, under UV light illumination) as compared to 0.03% by weight in QD nanocomposite samples (Fig. 3.84a, left), where the PL signals are too weak to be detected without

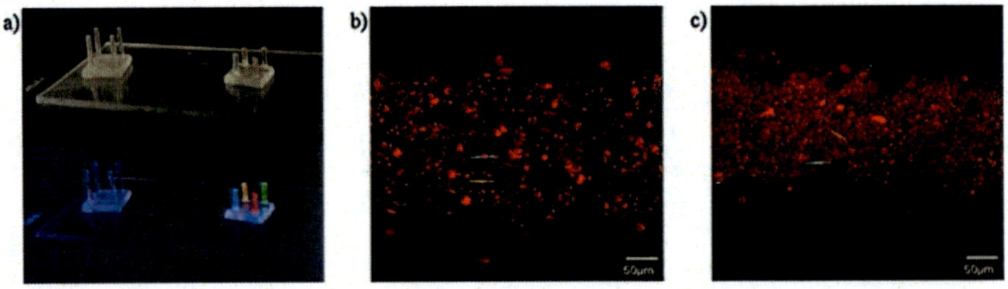

Fig. 3.84 (**a**) Increase in concentration, increasing the photoluminescence from the quantum dots, illustrated over four different emission wavelengths. The sample to the left was fabricated with 0.03% QD nanocomposite, whereas the sample to the right was fabricated with 0.06% QD nanocomposites; (**b**, **c**) an LSM image comparison from QD nanocomposites of different QD concentrations: 0.01% (**b**) and 0.08% (**c**)

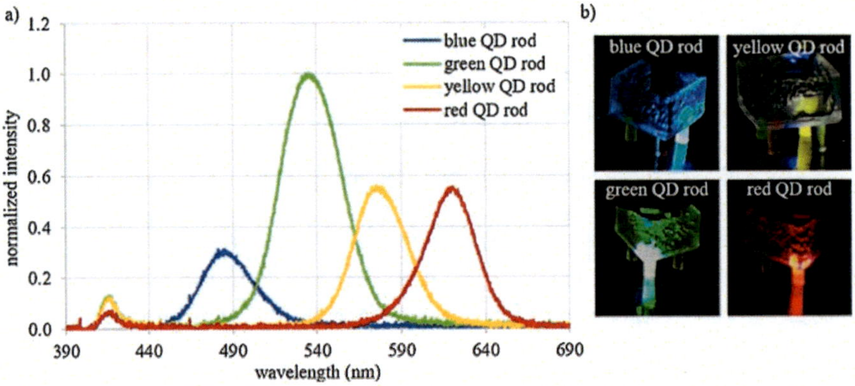

Fig. 3.85 (**a**) The PL spectra measured from a 3D-printed "chip" capable of emitting multiple-wavelength PL from different spatial locations; (**b**) the spatial locations of measurement are shown

fluorescence spectroscopy. The sample shown in Fig. 3.84a demonstrates the use of multiple material printing techniques to print a single "chip" that is capable of emitting different wavelengths from different spatial locations within the chip. The PL spectra of similarly printed chips are demonstrated in Figs. 3.85 and 3.86.

Samples with two different QD concentrations were also observed with light sheet microscopy. Light sheet microscopy is a microscopic technique with true potential for imaging fluorescent biological samples. In light sheet microscopy, a light sheet of micrometer thickness is generated in the sample, thereby illuminating and hence exciting a thin slice within the sample thickness [87]. A section of the thick sample is accurately imaged in the form of a stack of images, where each image in the stack is generated by moving the "light sheet" up or down within the sample. When quantum dot-mixed 3D-printed samples were observed under light sheet microscopy, a difference in QD cluster patterns could be seen. To demonstrate this, the resin mixtures with two different quantum dot concentrations were used to fabricate two 3D-printed samples. These samples were mounted onto a light sheet

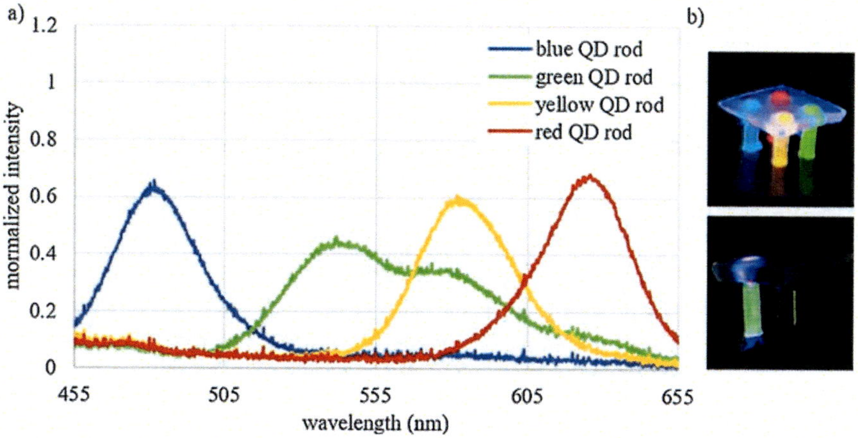

Fig. 3.86 (**a**) Barrier diffusion-induced variation in the photoluminescence seen in the green spectrum, indicated by a dip in the expected PL peak intensity (at 550 nm) as well as the appearance of a new shoulder at 576 nm, which is around the PL peak for the adjacently printed quantum dot rod in the yellow spectrum; (**b**) top, the measured sample printed with multiple quantum dots capable of photoluminescence in different wavelengths; bottom, the green rod shows an inflex of yellow QD material in the PL spectra

microscopy setup, and a stack of images was taken. A thin illuminated slice from the sample with lower QD concentration (Fig. 3.84b) reveals bigger clusters that are more localized, forming voids in between, as compared to the high concentration sample (Fig. 3.84c), where the clusters are smaller in size and spread out over the 40 μm sheet of the sample. Lower concentrations of QD in the mixed material made it difficult to detect the weak photoluminescent signals from the 3D-printed samples made from this mixture (for instance, see the 0.03% concentration sample in Fig. 3.84a, left). Samples fabricated from a higher QD concentration (e.g., 0.06% concentration) registered strong photoluminescent signals. One such example is shown in Fig. 3.85.

The sample shown in Fig. 3.85 was 3D printed using the multiple material technique, where five variations of the base resin were printed in a single print job. The resulting sample exhibits photoluminescence from different wavelengths in the visible region, depending on the spatial location of measurement. The base measurement was performed in the middle of the chip and therefore only captures the response of the undoped printed resin. It can be observed that the PL intensities of the different wavelength-emitting sections are not similar, even though all the resin mixtures used for the fabrication were of similar QD concentrations. Another fabrication-oriented variation in the photoluminescence emitted by the additively manufactured samples is shown in Fig. 3.86. The spectrum measurement of the 530 nm emission QD nanocomposite displayed a strong shoulder around 570 nm, contributed by the neighboring 570 nm QD composite rod, indicating barrier diffusion between the green and yellow QD composites. In other words, the liquid resin mixtures pose a risk of mixing between the materials during the print, and this diffusion is visible in the PL spectra measured from the green rod.

Dependence of the Photoluminescence on Prolonged UV Irradiation

Apart from the quantum dot concentration in the nanocomposites, the input excitation also had a profound effect on the variation of the photoluminescence from the 3D-printed samples. The extent of variation depended upon the energy dosage in the excitation. In other words, the longer the sample was subjected to ultraviolet radiation, the stronger the intensity of the photoluminescence from the sample. An example is shown in Fig. 3.87. A 2-mm-thick cuboidal sample was printed from an alloyed quantum dot nanocomposite with a colloidal emission wavelength of 630 nm. From the measurements, it is safe to assume that the photoluminescence emission from the quantum dots increases as the sample is subjected to longer periods of excitation. Over the span of 2 h of continuous ultraviolet irradiation on the sample at 398 nm, the photoluminescence intensity from the quantum dots increased by four times the value of a fresh 3D-printed sample. No significant shifts in the emission wavelength were detected. The inherent fluorescence from the resin was also strengthened by the continuous UV irradiation, increasing in intensity by a factor of 3 by the end of 3 h. It is interesting to note that the peak wavelength of this autofluorescence is blueshifted by 5 nm. An analogous example for the inherent photoluminescence from the resins can be derived from the light-emitting polymers, as explained above. There exist measurement records of alternating copolymers exhibiting photoluminescence at 422 nm for 360 nm excitation that undergoes blue shift when the solvent is switched from THF to hexane [85], or when certain side

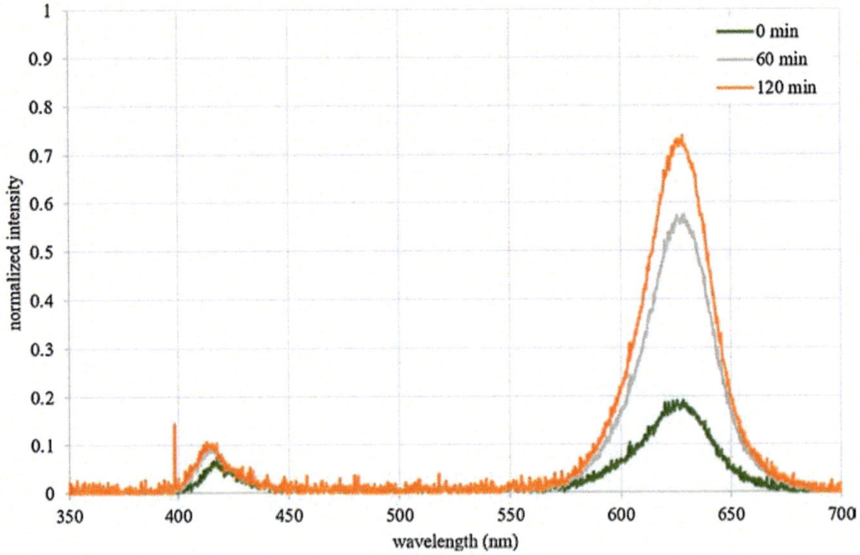

Fig. 3.87 Steady increase in photoluminescence observed as the exposure of the sample to ultraviolet radiation is increased over time (indicated in the legend). The sample being inspected was printed with a 0.6% QD nanocomposite, colloidal emission wavelength 630 nm

groups are exchanged on the polymer backbone [85]. The process of polymerization and cross-linking may also contribute to the observed intensity of the autofluorescence from the resin. For instance, cross-linking between PVA and PPV polymers increases the polymer's photoluminescence intensity [85].

There are several hypotheses to explain the variation in the emission of the quantum dots. One possibility is Förster resonance energy transfer, according to which there exists a route of non-radiative energy transfer from the fluorophores of a wider emission band gap (here, the autofluorescence) to a narrower emission band gap fluorophore (the quantum dots), thereby intensifying the PL peak from the quantum dots. The occurrence of this possibility however is slim in this scenario, as the autofluorescence of the resin increases under continuous ultraviolet radiation. This is contradictory to the reasoning that the non-radiative energy transfer would weaken the autofluorescence from the resin. Another possibility is reduced quenching of the excitons that form as a result of quantum dot excitation. The reason for this may be the isolation of the excitons within the polymer network (trapping of the formed excitons within the polymer network), thereby reducing the chances for the excitons to migrate into quenching sites that could result in photoluminescence quenching. It is known that 100% conversion of the acrylic resin to polymer network cannot be expected from a fresh 3D-printed sample under atmospheric conditions. This is one of the reasons why post-processing is such an important step in additive manufacturing. Exposing a 3D-printed sample to prolonged ultraviolet excitation immediately after the print may be increasing the degree of cure within the polymer network, thereby increasing the chances of the "exciton trapping" mentioned above. References in the literature report that a light-emitting polymer may show enhanced photoluminescence in the polymer blend when diluted in an inert polymer like PMMA [85].

3.7 New Approaches to the Additive Manufacturing of Optics

3.7.1 Robot-Based Additive Manufacturing

As presented in earlier sections, the additive manufacturing of optical elements presents various challenges relating to the choice of materials and the printing process itself. To avoid the disadvantages of conventional processes, new technological approaches are needed.

The layered structure is one of the key causes of reduced optical quality in macroscopic components. This structure produces two negative effects: a stepping structure forms on the surface of curved objects, and the refractive index can vary across layers (see Sect. 3.3). In addition to issues associated with the layered structure, many conventional processes encounter problems when realizing overhanging

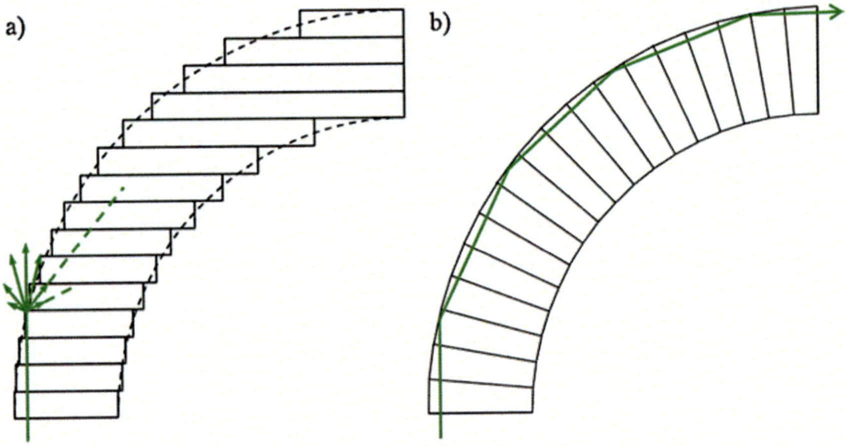

Fig. 3.88 Schematic representation of light transmission through a curved light guide (**a**) with unadapted layer orientations and (**b**) with adapted layer orientations

structures. The support structures or materials required for overhanging parts often significantly reduce the surface quality of the component.

With most conventional printing processes, the primary problem of the layered structure cannot be eliminated. The orientation of the layers relative to the direction of light propagation plays a critical role. Therefore scattering increases significantly when the light propagates through bended structures (see Fig. 3.88a) due to the step-like structure. However, if the layers are oriented perpendicular to the light propagation, it can potentially be guided, like in an optical fiber, thanks to the variation in the refractive index. Accordingly, the problem of light scattering can be significantly reduced (see Fig. 3.88b).

To adapt the layer orientations to the light propagation, the number of degrees of freedom of the printing process needs to be increased from three to six. The goal is to allow the printing unit or build platform to rotate about three axes in addition to translational motion (x, y, z). If the new layer orientation can be achieved by rotating the build platform, there will also no longer be any need for support structures. The layer orientation can then be used to specifically guide the rays as desired.

The concept of one such additive manufacturing platform is shown in Fig. 3.89. The printing unit is attached to a 6-axis robot. This allows the rough positioning to be flexibly adjusted. The build platform is represented as a hexapod with a repeat accuracy of 100 nm. This allows the component to be precisely handled to match the desired layer orientation at any given point. A second robot is also used, equipped with various contactless analysis tools for in situ analysis during the printing process. The various printing and analysis units can be exchanged as needed thanks to the standardized interface. Each unit of this platform is presented in more detail below.

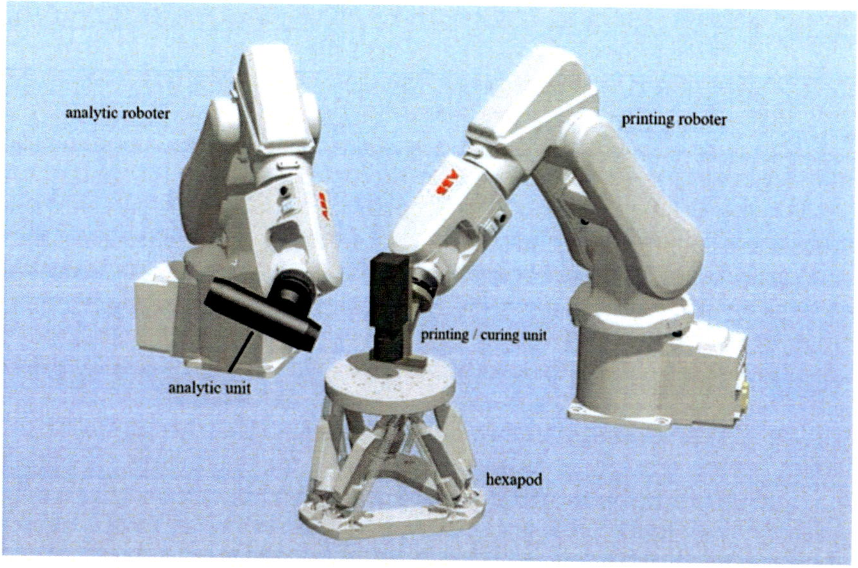

Fig. 3.89 Overview of the components of an Additive manufacturing platform designed to allow printing with six degrees of freedom

3.7.1.1 Printing Unit

The material can be applied either by inkjet printing or with a dispenser. Both technologies have different advantages and areas of applications that are described in more detail over the following sections.

Inkjet

Inkjet technology is a flexible and very versatile printing technique. The earliest functional systems were manufactured by IBM as early as the 1970s [88]. Since then, applications have been further developed in the private sector, most notably industrial applications. Inkjet technology can be used as a noncontact printing technique on a wide variety of substrates. Examples of typical industrial applications include printing onto textiles, ceramic tiles, or wood veneers [88].

UV-curable materials also allow inkjet technology to be used in additive manufacturing. The decisive factor is the development of suitable materials with the right viscosity range for inkjet technology. The viscosity can be further reduced by using solvents or increasing the temperature of the material to make it jettable. When doing this, the process parameters need to be precisely adapted to the print head and the material. Developing a suitable waveform to control the print head plays a key role. One way to develop these waveforms is presented below.

Strictly speaking, the concept of inkjet encompasses multiple different technologies. A distinction can be made between thermal, continuous, and piezoelectric inkjet techniques [88, 89]. However, the most common variant is the piezoelectric inkjet technique, which is the approach considered below. The technique involves carefully controlling the deposition of a droplet of liquid and its positioning on the substrate. This is often described as "drop on demand."

A piezoelectric inkjet print head usually consists of several nozzles that can eject droplets independently from one another. Studies and simulations of the droplet generation process have been performed since the very beginning of inkjet technology [90]. The wide range of effects exerted by the fluid and the pressure chamber have been described in extensive detail. As manufacturers developed their own print heads, many different variants of the droplet generation process were introduced. However, the underlying principle remains the same (see Fig. 3.90). The print head consists of a chamber with an inlet for the liquid. It is often also equipped with a filter. The other end of the chamber is closed except for a small opening—the nozzle. Depending on the print head, the diameter of the nozzle can be as small as a few microns. The droplet volumes are typically in the range of 5–50 pL. A diaphragm with a piezoactuator is fitted to the side of the chamber.

Before the printing process, the chamber is filled with material. It must be ensured that no air bubbles remain within the chamber. Air bubbles have a damping effect that may negatively impact the jetting behavior. Since the nozzle diameter is typically around a few microns, the material must also be filtered to prevent the nozzles from becoming obstructed. A slight negative pressure is also applied to the pump chamber to prevent the material from dripping out of the nozzle. This results in a slight negative meniscus on the surface of the liquid at the nozzle.

To generate a droplet, the first step is to squeeze the diaphragm by applying a voltage within the pump chamber. This forces material out of the nozzle. The diaphragm is then relaxed again, which causes part of the material that has already been ejected to be drawn back into the chamber. If the diaphragm stroke is large enough, and hence a sufficient amount of material has passed through the nozzle, this breaks the material thread outside of the nozzle, forming a droplet. If the diaphragm stroke is too small, all of the material will be pulled back through the nozzle. The relaxing motion of the diaphragm not only pulls back material through the nozzle; it also draws in new material through the liquid inlet to compensate for the

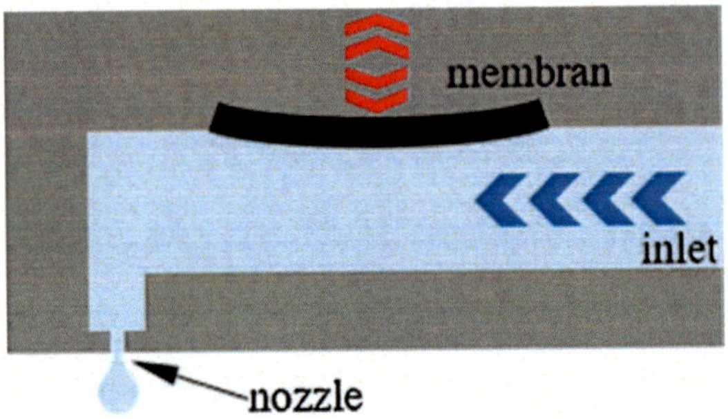

Fig. 3.90 Schematic drawing of an inkjet nozzle with a piezoelement

material lost to the newly formed droplet. This prepares the chamber for the next droplet.

The voltage curve applied to the piezoactuator to control the behavior of the diaphragm is called the waveform. This waveform determines the jetting behavior and must be adjusted as a function of both the print head and the material. Figure 3.91 a shows a simple waveform that produces the diaphragm behavior described above. The waveform is divided into different sections, each consisting of a voltage change followed by a hold time during which the new voltage is maintained. The waveform shown here contains the minimum number of segments necessary to create a droplet: apply a voltage to deform the diaphragm (pressure increase within the chamber), maintain the voltage (constant pressure within the chamber while the droplet forms), and reduce the voltage (pressure decrease within the chamber). Since the timing of the voltage applied by the piezoactuators can be specified very precisely, piezoelectric inkjet printing is significantly more precisely configurable than thermal techniques [91, 92]. The rise and fall times of the voltage represent optimizable waveform parameters, as well as the voltage hold time. The total duration of the waveform to create a droplet is approximately 20 microseconds. Droplet formation can be optimized by inserting additional segments such as the echo waveform shown in Fig. 3.91b. The follow-up negative pulse damps the vibration in the material induced by the diaphragm, which minimizes any influence on the next droplet. In principle, there are no limitations on the complexity of the waveform.

Analyzing the droplet formation with a dropwatcher allows multiple phases to be observed. An example of droplet formation is shown in Fig. 3.92a. First, the droplet is pushed out of the nozzle, pulling a thread of material known as the ligament behind it. If the droplet breaks off, the ligament initially remains intact before combining with the main droplet due to surface tension, forming a complete droplet. The length of the ligament depends on the waveform, as well as the viscosity and surface tension of the material. Under some conditions, the ligament may not reconnect with the main droplet, instead forming so-called satellite droplets (see Fig. 3.92b. The satellite droplets hit the substrate after a delay, which blurs the printed image. Their formation must therefore be avoided by configuring the waveform or material parameters accordingly.

The choice of materials is primarily limited by the viscosity. The optimal viscosity range is usually 1–10 mPas [91, 93]. At lower viscosities, the droplet formation

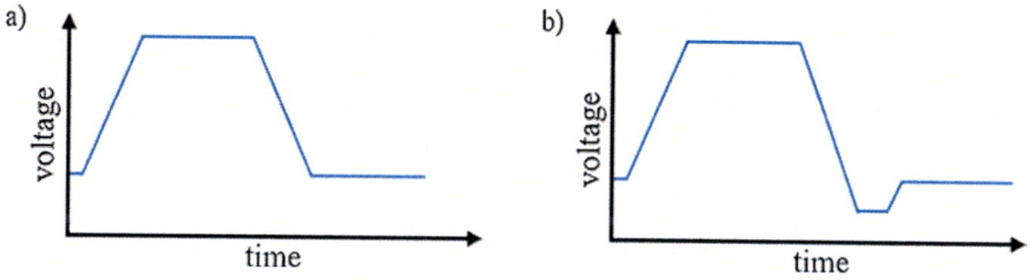

Fig. 3.91 Simple waveform for controlling the piezoelement of an inkjet nozzle (**a**). Waveform with an echo pulse to damp vibrations within the nozzle (**b**)

Fig. 3.92 (**a**) Formation of a droplet in an inkjet process. Initially, the droplet pulls a thread of material (ligament) behind it, until the ligament ultimately separates from the nozzle. Surface tension causes the ligament to recombine with the droplet after separating from the nozzle; (**b**) formation of satellite droplets

can be adjusted using the waveform. At higher viscosities, the material becomes too viscous to flow through the printing system. The diaphragm cannot generate enough force to create a droplet. In this case, the viscosity can be reduced by adding solvents or heating. If the material cannot be adapted accordingly, an alternative printing system is required. Microdispensers are an example of a system that can be used at high viscosities, as described in the following section.

Dispenser

Individual droplets of fluid can also be produced by a so-called microdispensing system. This is referred to as a dispenser below. There are various approaches to implementing such a system. For example, some dispensers operate in a manner similar to a syringe [94], and some dispensers eject material using ultrasound [95]. Here, a dispenser that works by delivering fluid with a piezoactuator is discussed [96]. The key advantage of this system is that it supports highly viscous materials with a viscosity of up to 2000 Pas [97].

Figure 3.93 shows the basic design of the dispenser. The valve body (1) contains the control electronics and the piezoactuator. The tappet (2) protrudes from the bottom of the valve body. The tappet is moved up and down by an actuator. The tappet seal (3) guides the tappet and seals off the chamber, which is filled with fluid. The pressurized medium container (7) supplies fluid through the chamber to the nozzle insert (4). The tappet is accelerated by the actuator and pressed into the nozzle

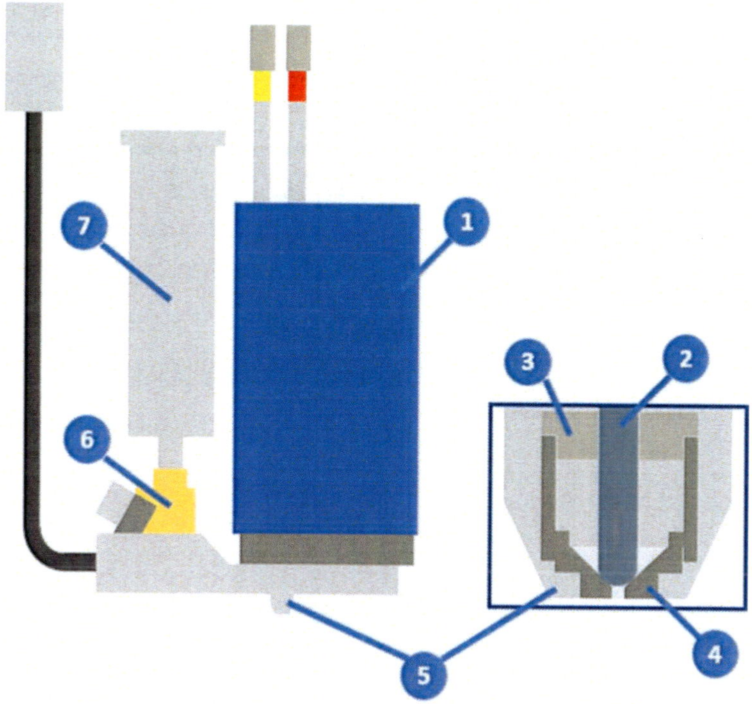

Fig. 3.93 Basic design of a dispenser

insert. This pushes the medium through the nozzle opening. The nozzle fixing nut (5) holds the nozzle insert in place within the fluid system. An additional heating element (6) is integrated into the fluid system and thermally decoupled from the actuator [98].

The dispensing process deposits a droplet of the dosed medium onto a substrate. This is triggered either by manually actuating the control unit or via an external signal such as a programmable logic controller or serial interface. A step-by-step diagram that explains the process is shown in Fig. 3.94a. The needle lift (NL) is shown in Fig. 3.94b as a function of time. In the initial state, the tappet is pressed into the conic nozzle opening of the nozzle insert, which seals off the opening. This prevents liquid from escaping from the nozzle opening. If the dosing process is now triggered, a variable response time t_{delay} elapses before the tappet moves up by an amount corresponding to the NL over a duration corresponding to the rising time (RI). The response time is determined by the computing time needed by the control unit to process the command. The tappet now remains at this NL in this position for a period corresponding to the open time (OT). During this interval, medium can flow into the chamber from the reservoir (2). Once the OT has elapsed, the tappet moves downwards by an amount corresponding to the NL over a duration corresponding to the falling time (FA). During this interval, the tappet displaces the medium before ultimately hitting the nozzle. The medium in the mouth of the

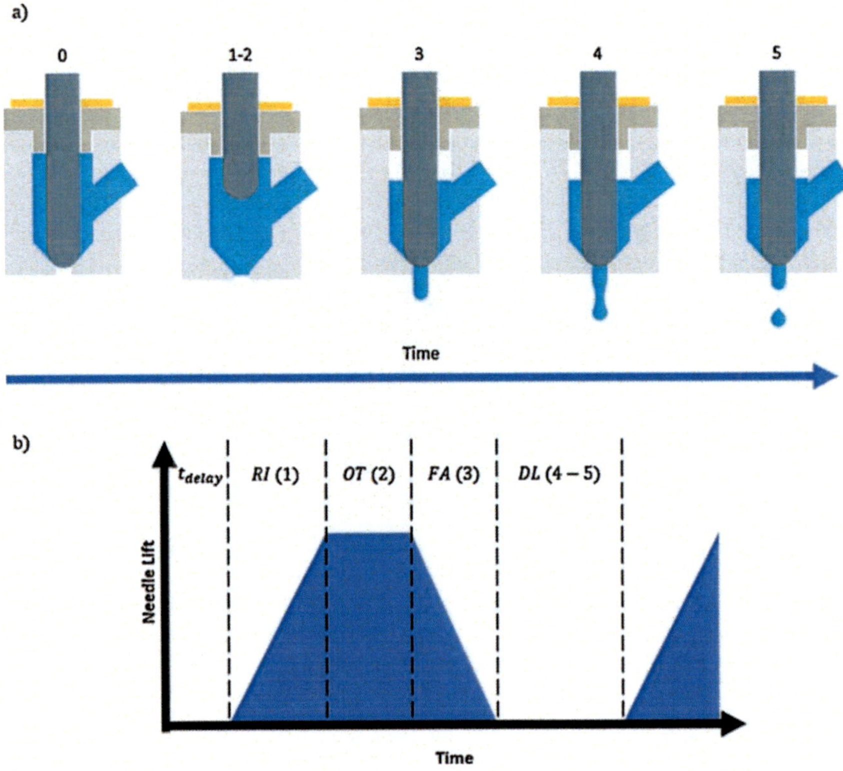

Fig. 3.94 (**a**) Dispensing procedure; (**b**) needle lift as a function of time during the dispensing process

nozzle is pushed out of the nozzle opening (3). The dispenser has now returned to its starting position. The droplet that forms at the opening falls away from the nozzle (4). After the droplet has fallen for a short interval, it constricts and breaks off (5). To dispense more than one droplet, a new dispensing process is initiated after a fixed delay time (DL) [98].

The process is influenced by various factors that can be manipulated either directly or indirectly (see Fig. 3.95). The output parameters are the droplet volume or the resulting droplet size on the substrate. In addition to the control parameters cited above, the number of dispensing pulses (NP) emitted after each dispensing signal is another direct factor of influence. This corresponds to the number of droplets dispensed.

The indirect influencing factor of the supply pressure regulates the pressure difference between the overpressure in the supply container and the printing chamber. It should be possible to adjust this pressure difference to within an accuracy of 0.1 bar. The fluid temperature and the related viscosity of the material also have a major influence on the process [99]. The temperature typically ranges from 30 to

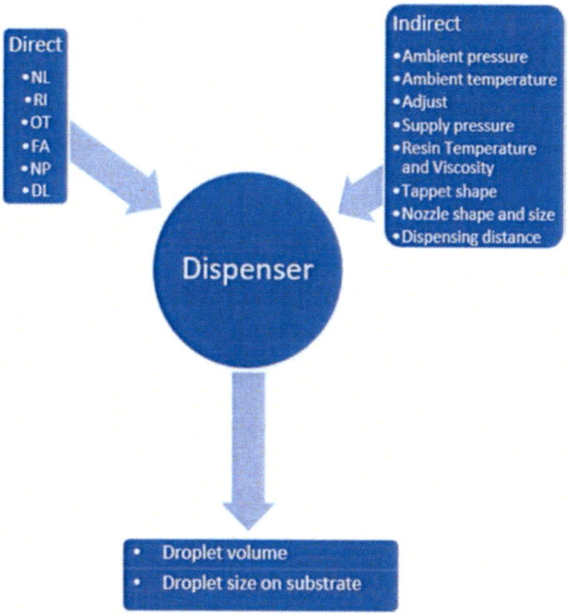

Fig. 3.95 Direct and indirect input parameters and resulting output parameters

70 °C. It is essential to ensure that the temperature and supply pressure remain constant throughout the entire printing process. The shape of the tappet and the nozzle insert also affect the dispensing behavior. The way that the resin is dispensed from the nozzle (speed profile within the nozzle insert) depends on how the tip of the tappet is shaped. Differences between nozzle inserts primarily concern the length of the nozzle outlet. This affects the flow rate during dispensing. The dispensing distance is the distance between the bottom edge of the nozzle and the substrate. The minimum distance is typically around 0.5 mm but can range up to several millimeters.

Like for the inkjet process, the biggest challenge is to find the right process parameters for each resin. These parameters can be determined using a so-called dropwatcher as discussed below.

3.7.1.2 Dropwatcher

A technique known as dropwatching is an important analysis tool for both inkjet and dispenser printing. A dropwatcher is a device for visualizing and analyzing small droplets of material as they travel through the air. This type of analysis is necessary to find the optimal process parameters for the printing unit when using a specific printing material. In the context of inkjet printing, these parameters specify the waveform. For dispenser printing, they are the control parameters. Key criteria include, for example, the behavior of the droplet ligament and the formation of satellite droplets. However, the droplet speed and size are also relevant. In addition to the wetting behavior and droplet density on the substrate, the droplet size plays a role in determining the thickness of the printed layer and must therefore be controlled precisely. Dropwatchers are also an important tool when printing with six degrees of freedom. For example, the flight behavior of droplets of material changes when the printing unit is tilted. The path computed for the printing unit needs to be corrected accordingly.

The key challenge to observing droplets of material is their high speed. In inkjet printing, droplets can leave the print head at a speed of up to 15 m/s (typical values range from 4 to 6 m/s) [100]. A high-speed camera would therefore be necessary for a precise analysis of droplet formation. However, for the two printing processes discussed above, an alternative approach can be adopted. The droplet generation process is very stable over time. The shape and speed of consecutively created droplets only differ marginally. Instead of considering the formation of a single droplet over time, multiple identical droplets can be analyzed in succession at different times. The observation period is shifted slightly for each new image, which allows the entire creation process to be visualized by considering multiple droplets over time. This does however require short exposure times. To achieve this with conventional industrial cameras, stroboscopic lighting is used. The measurement setup is based on recording the silhouettes. The printing unit is placed in the beam path of a collimated stroboscopic LED lighting system to allow the silhouette of the droplets to be captured by a camera with a telecentric objective.

The dropwatching system is synchronized with the printing unit using a control board. The control board sends a TTL signal whenever a droplet is about to be generated. This trigger signal is processed by the flash controller to further fine-tune the synchronization of the camera and the lighting. The flash controller performs multiple functions. Firstly, it allows the LED to be activated for very short times (approx. 0.5–10 μs). This means that the LED can be operated in an "overstrobing" mode with up to ten times more current. This produces nominally higher illumination and therefore a better signal-to-noise ratio in the image. A camera with an exposure time of 10 ms can therefore be used, since only the duration of the flash matters for image acquisition. Accordingly, inexpensive conventional industrial cameras can be used. The camera exposure is also activated by a trigger from the control board and is only deactivated once the flash has ended.

Another important function of the flash controller is setting a delay before the trigger signal. This delay is needed to observe how the droplet develops over time. The longer the delay, the more the droplet will have developed. An example is shown in Fig. 3.96. To visualize the entire droplet creation process, the delay is increased after taking each image. The camera exposure time, however, remains the same.

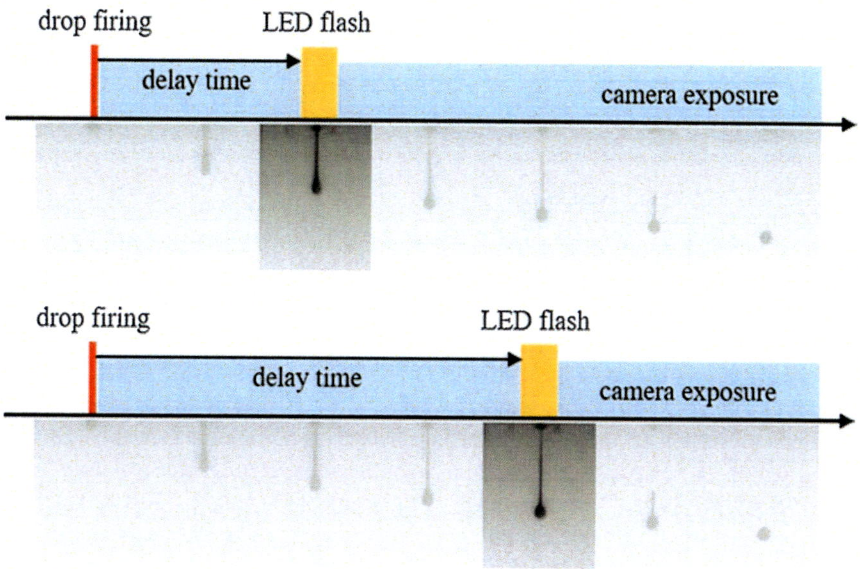

Fig. 3.96 Controlling the camera and LED in the dropwatcher. The delay time between droplet generation and the LED flash determines the position at which the droplet is captured and can be analyzed

3.7.1.3 Kinematics

The kinematics of conventional 3D printers generally consist of three linear axes for the X, Y, and Z directions. This is relatively convenient and straightforward to control. By contrast, robots have six degrees of freedom. This allows individual layers to be freely oriented, which provides significant benefits when printing components and realizing parts that span a certain volume.

The six degrees of freedom of the robot may be divided into three degrees of freedom for translation through space in the X, Y, and Z directions and three degrees of freedom for the orientation of the robot head about these same three axes. As a general rule, each joint of the robot gives it another degree of freedom.

The tool center point (TCP) is a key concept in this context. This is the point with respect to which the tool coordinate system is defined. The TCP and all other points can be described in terms of so-called quaternions. Quaternions allow a mathematically elegant description of three-dimensional Euclidean space and other spaces, especially those encountered in contexts involving rotations [101]. An alternative description can be established in terms of Euler angles. Euler angles define each position as a chain of three successive rotations about the axes of the reference system. The orientation is described by the angles of these three rotations. Accordingly, the TCP can be represented as the following vector [102, 103]:

$$\mathrm{TCP} = \begin{pmatrix} X \\ Y \\ Z \\ R_x \\ R_y \\ R_z \end{pmatrix} \tag{3.7.1}$$

where X, Y, and Z describe the position within Cartesian space and R_x, R_y, and R_z are the three angles of rotation. The TCP is defined with respect to a basis coordinate system.

The TCP must be chosen to coincide with the working point of the tool attached to the robot. In the example of the dispenser, this means that the TCP should be defined at the exact tip of the nozzle outlet. It is recommended to use a "zero-point clamping system" to attach the print heads to the robot. This makes it easier to replace print heads without needing to remeasure them after they are fitted.

Importantly, the kinematics must be integrated within a comprehensive software program that controls the printing process, the motion of the build platform (hexapod), and the motion of the robot. The path programming of the robot is especially important and is presented below.

In general, path generation is understood as a procedure that generates a path followed by the kinematics mechanism, in this case the robot, from a template such as a 3D model. The path can, for example, consist of a series of individual points to which the robot must travel in order. As with conventional 3D printing, a slicer that converts a given 3D model into a layered model can be used to generate paths for each layer. This produces a file written in so-called G-code. Since G-code is systematically built up from commands, it can be converted into a cloud of points with Cartesian coordinates relatively easily. These coordinates are then assigned a G-mode, which determines whether or not material should be dispensed. A tool number can also be specified for each point to identify which of the print heads should be used. Figure 3.97 shows the information flow from a CAD file to the dispenser software.

However, pure path generation is not sufficient for dispenser printing. To build up a predefined geometry with individual droplets, certain calculations must first be performed. These calculations are presented below.

The goal is to compute the correct parameters for the dispenser and the robot from the coordinates, the G-mode, and the tool number. As multiple dispenser print

Fig. 3.97 Information flow from the CAD file to the dispenser

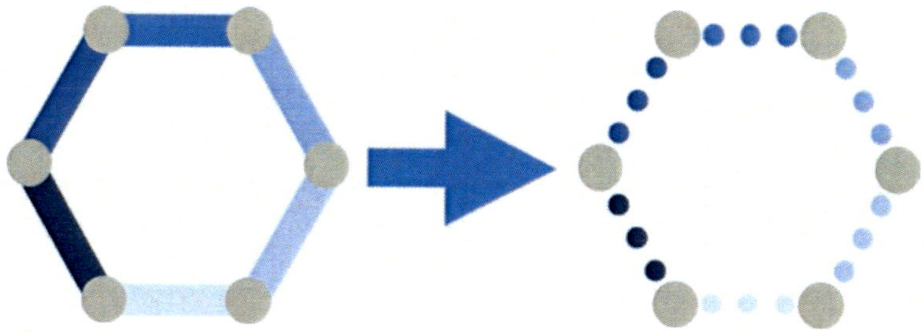

Fig. 3.98 G-code path to dispensed geometry

heads can be used in parallel in the context of multimaterial printing, the tool number specifies the correct print head.

The calculations seek to "translate" the input variables for the dispenser and the robot in such a way that the desired geometries can be printed. This is illustrated in Fig. 3.98. The left side shows the contour of a hexagon realized in standard slicing software. The slicer generates six points from the CAD file of a hollow hexagon and defines paths between these points. Along these paths, the G-mode would, for example, simply be set to extrude for conventional FDM printing. However, for robot-based dispenser printing, the G-code must perform another calculation to convert the original hexagon into the hexagon shown on the right of Fig. 3.98. The new hexagon consists of individual droplets that are dispensed with a certain spacing.

This calculation takes the coordinates, the G-mode, and the tool number as basic inputs. However, to obtain the geometry shown on the right of Fig. 3.98, other input parameters are required to describe the robot and the dispenser. These parameters are, namely, the robot acceleration a_{robot}, the robot speed v_{robot}, the preferred droplet spacing d_{space} for the given dispenser, and the dispenser parameters discussed above. The droplet spacing represents the preferred distance between two droplets in the translated code. This is shown in Fig. 3.99. Another parameter shown in the figure is the path d_{path} from point 1 to point 2. This path corresponds to the distance between P_1 and P_2.

Like the TCP, these two points are robot coordinates with six components:

$$P_1 = \begin{pmatrix} x_1 \\ y_1 \\ z_1 \\ R_x \\ R_y \\ R_z \end{pmatrix}; P_2 = \begin{pmatrix} x_2 \\ y_2 \\ z_2 \\ R_x \\ R_y \\ R_z \end{pmatrix} \tag{3.7.2}$$

The x-, y-, and z-coordinates are the Cartesian coordinates extracted from the G-code. The R-coordinates describe the orientation of the TCP of the robot, as described earlier. For the printing process described here, these coordinates are constant across all points. To calculate the distance d_{path} between the two points, two assumptions can therefore be made. Firstly, the R-coordinates can be ignored, since the orientation of the TCP never changes during the printing process. Secondly, since the G-code generated by the slicer consists of layers, and each layer has a fixed z-value, this z-value can also be ignored. This means that the x and y-coordinates are sufficient to define the two points. The same is true for all subsequent points in this calculation. The points can therefore be simplified to

$$P_1 = \begin{pmatrix} x_1 \\ y_1 \end{pmatrix}; P_2 = \begin{pmatrix} x_2 \\ y_2 \end{pmatrix} \tag{3.7.3}$$

To calculate the distance between two points in space, it suffices to calculate the magnitude or length of the vector $\overrightarrow{P_1P_2}$ [104]:

$$d_{\text{path}} = \left|\overrightarrow{P_1P_2}\right| = \sqrt{(x_1 - x_2)^2 + (y_1 - y_2)^2} \tag{3.7.4}$$

The number of droplets n_{drops} between P_1 and P_2 given a preferred droplet spacing of d_{space} can now be calculated from d_{path} and d_{space}:

$$n_{\text{drops}} = \frac{d_{\text{path}}}{d_{\text{space}}} \tag{3.7.5}$$

This formula needs to be extended by +1 for the droplet pattern to reproduce the one shown in Fig. 3.99. This ensures that there is both a first droplet and a last droplet:

$$n_{\text{drops}} = \frac{d_{\text{path}}}{d_{\text{space}}} + 1 \tag{3.7.6}$$

The number of droplets can only be an integer value. This means that n_{drops} needs to be rounded depending on its value after the decimal point [105]:

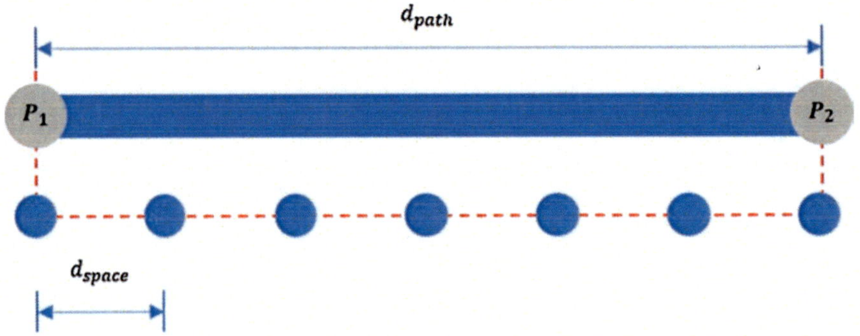

Fig. 3.99 Path from G-code and droplets

$$n'_{\text{drops}} = \{ \begin{matrix} n_{\text{drops}} f \overset{¤}{\mathrm{u}} r \left| n_{\text{drops}} - n'_{\text{drops}} \right| < 0.5 \\ n_{\text{drops}} f \overset{¤}{\mathrm{u}} r \left| n_{\text{drops}} - n'_{\text{drops}} \right| < 0.5 \end{matrix} \tag{3.7.7}$$

This gives a new length d_{path}:

$$d_{\text{path}} = n'_{\text{drops}} * d_{\text{space}} = \{ \begin{matrix} 14\text{mm}\ f \overset{¤}{\mathrm{u}} r d_{\text{drops}} = 3.5\text{mm} \\ 16\text{mm}\ f \overset{¤}{\mathrm{u}} r d_{\text{drops}} = 4\text{mm} \end{matrix} \tag{3.7.8}$$

Thus, the length of the distance between points P_1 and P_2 depends on the number of droplets. This would change the dimensions of the target geometry, which must be avoided. Since d_{path} is a constant value that depends on the points and n_{drops} was configured to specify the properties of the dispenser, only d_{space} can be modified without changing the overall geometry or producing an invalid value for the number of droplets. This gives

$$d'_{\text{space}} = \frac{d_{\text{path}}}{n'_{\text{drops}} - 1} \tag{3.7.9}$$

where d'_{space} is the actual droplet spacing. This leads to an inaccuracy between two droplets, since the actual droplet spacing d'_{space} and the preferred droplet spacing d_{space} do not coincide. The inaccuracy is given by

$$\Delta d_{\text{space}} = \left| d'_{\text{space}} - d_{\text{space}} \right| \tag{3.7.10}$$

where Δd_{space} is the deviation from the preferred droplet spacing. This inaccuracy is only relevant if it is larger than the repeat accuracy of the robot (e.g., 0.1 mm).

Another problem is the synchronization between the robot and the dispenser. There are several challenges. The most obvious is that the robot has a certain speed profile, as shown in Fig. 3.100a. The robot requires some time to reach full speed. For the example with droplets discussed above, this creates the problem that the first and last droplets are not deposited in the correct positions. This is visible in Fig. 3.100b as an offset in the droplets. The printed path becomes shorter as a result. The spacing between the first and second droplet is reduced to d_{start}, and the spacing between the last droplet and the last-but-one droplet is reduced to d_{stop}. This effect is caused by the acceleration and a_{robot} of the robot, which requires an acceleration time $t_{\text{acceleration}}$ before reaching v_{robot}. Figure 3.100c gives another visual summary of the situation.

To correct this error, delays must be introduced during the acceleration phase to ensure that the droplet spacings are identical.

A robot-based printing approach comparable to FDM printing is possible based on the techniques described above. A typical result is shown in Fig. 3.101.

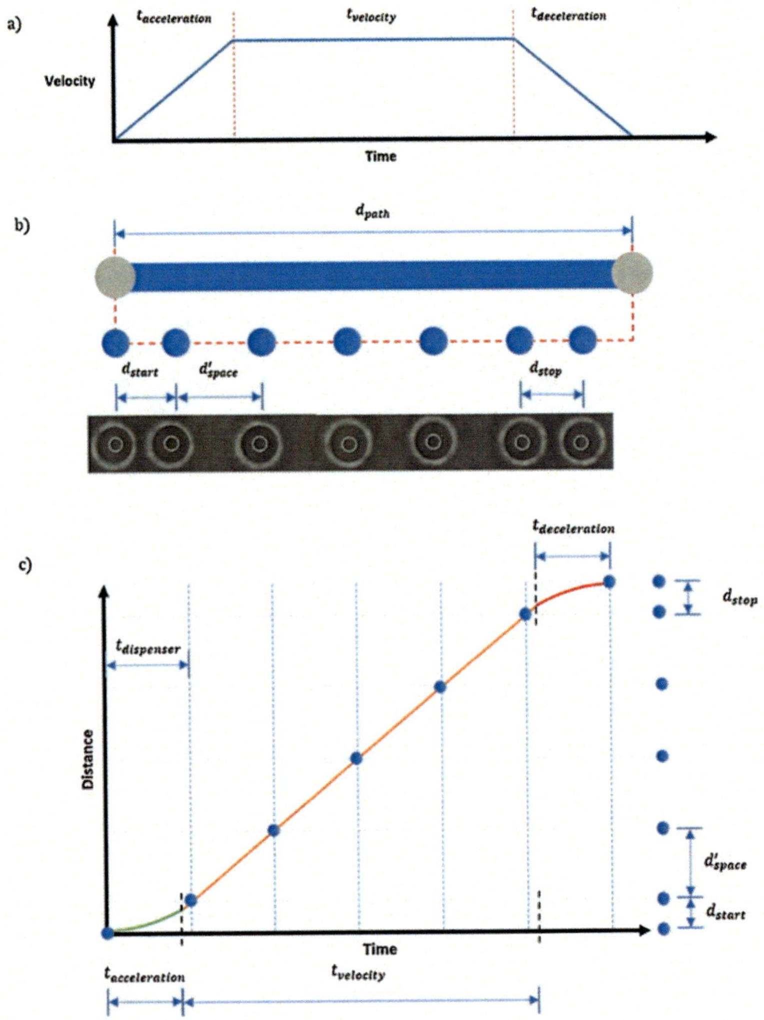

Fig. 3.100 (**a**) Droplet distribution; (**b**) velocity profile when the robot moves; (**c**) distance over time with droplets

Figure 3.101a shows a pyramid before printing, and Fig. 3.101b shows the realized component placed on a slide. Some shape deviations are visible, especially around the edges of the pyramid. This is attributable to the surface tension of the resin. Small bubbles can also be seen within the pyramid. These bubbles arise when the process control is too fast. Figure 3.101c shows printed letters. Problems relating to the wetting of the substrate by the resin can be seen around the edges.

In some applications such as gradient-index optics, the printing of multiple materials is desirable. The material or corresponding print head is selected using the "tool number" parameter. An example of the parallel printing of two materials is shown in Fig. 3.102. The different hatching patterns of the sliced model in

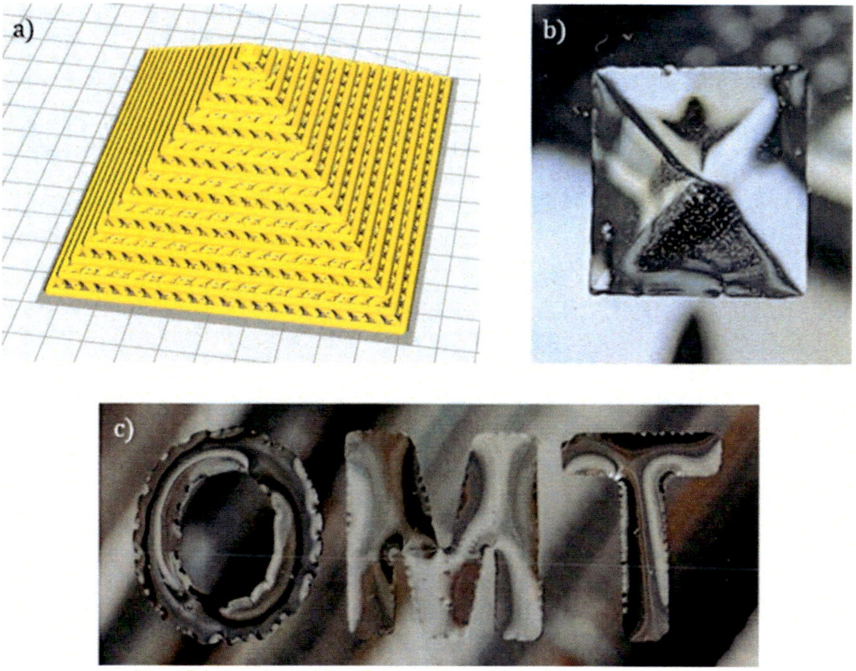

Fig. 3.101 (**a**) Schematic representation of the pyramid after the slice process; (**b**) printed pyramid on a slide; (**c**) printed text

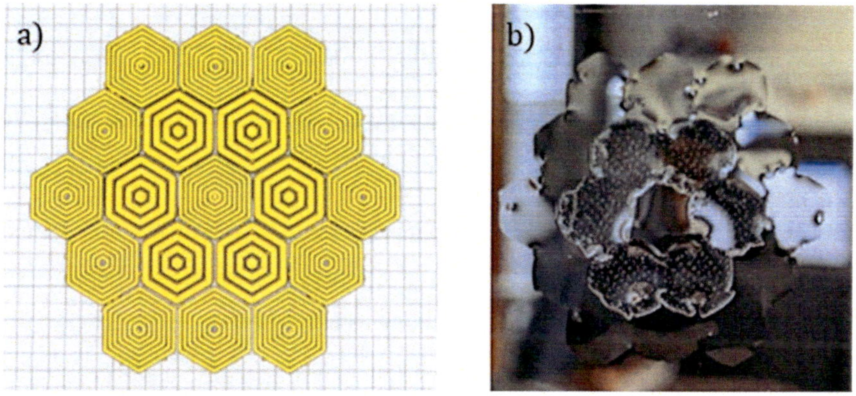

Fig. 3.102 (**a**) Schematic representation of two materials after the slice process; (**b**) structure realized on a slide

Fig. 3.102a correspond to different materials. Figure 3.102b shows the printed result. The material of the inner circle has significantly higher viscosity, which causes strong bubble formation in some parts of the material. In practice, this is undesirable; however, in this case, it allows the two materials to be distinguished more easily.

The path generation procedures considered so far are not yet capable of orienting the robot with six degrees of freedom. To do this, a robot path must be generated along the surface of the object while also defining a local coordinate system that uniquely specifies the orientation of the robot for each point on the surface.

The basic procedure is illustrated below using the pyramid shown in Fig. 3.103a as an example. First, the coordinates of the surface are extracted from the data generated by the slicer. For the object shown in the figure, only the corner points of the surface are defined in the STL file (import coordinates). To ensure that there are sufficiently many points for later processing, the next step is to interpolate between the points (interpolate points). Next, in a manner equivalent to FDM printing, the number of lines that need to be printed adjacent to one another on the component (number of lines) is calculated from the specified line width. Based on this and a tolerance value, a plane is extracted from the points, e.g., in the x-direction with a fixed y-value (assign points). Of these points, only the points that describe the surface are retained (clean slice). These points are linearly interpolated so that the

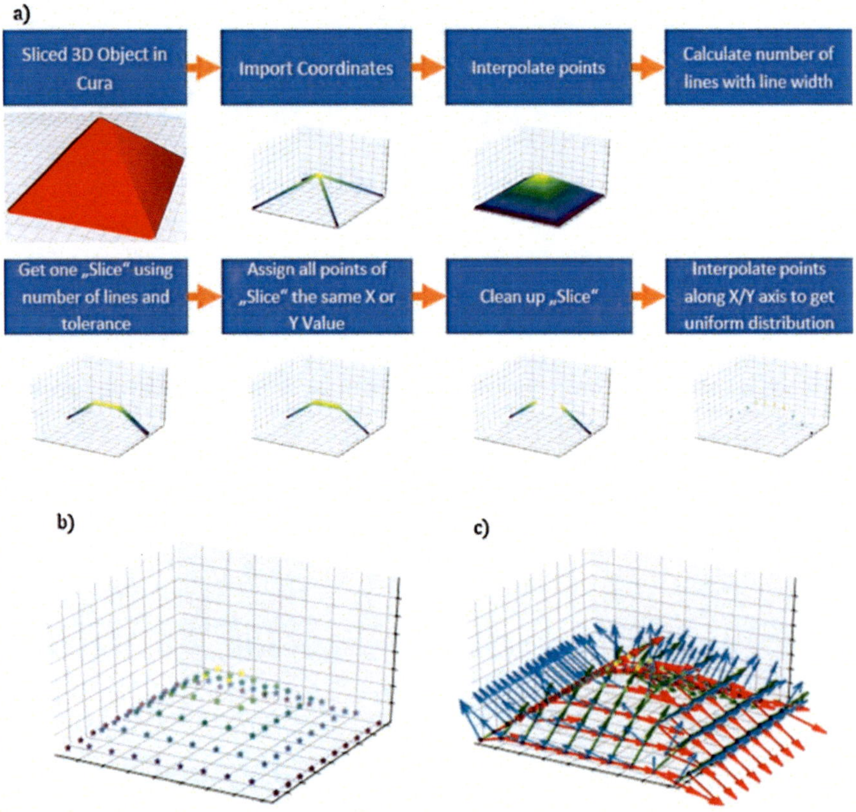

Fig. 3.103 (**a**) Basic procedure for generating paths with six degrees of freedom; (**b**) resulting point cloud for the generated paths; (**c**) local coordinate systems assigned to each point to specify the perpendicular orientation of the print head

spacing between the points is always constant (interpolate points). This is performed for each line, with meandering connections between the lines. The resulting point cloud is shown in Fig. 3.103b. Finally, the orientation of the robot is determined for each point. The guiding principle is that it should be oriented parallel to the local surface normal. The local coordinate system needed for this can be constructed using vector calculus from each point and its relation to its closest neighbors. The final point cloud and the corresponding local coordinate systems are shown in Fig. 3.102c. This specifies both the path of the robot and the orientation of the robot or print head.

3.7.1.4 Metrology Unit

As shown in Fig. 3.89, the build platform is a hexapod. A key aspect of the control system is that both the position of the robot or print head and the position of the hexapod must be known within the same coordinate system. This means that the position and orientation of both objects must be captured by a measurement system. A diagram of such an optical measurement system is shown in Fig. 3.104. Measuring plates with three ceramic spheres each are mounted on both the robot head and the hexapod (spheres A and spheres B). The positions of the ceramic spheres relative to one another are precisely determined using a coordinate measuring system. The positions of the spheres themselves with respect to TCP_{Robot} and $TCP_{Hexapod}$ are also known, described by the matrices ${}^{T}_{A}M$ and ${}^{B}_{P}M$. Furthermore, the radius of sphere

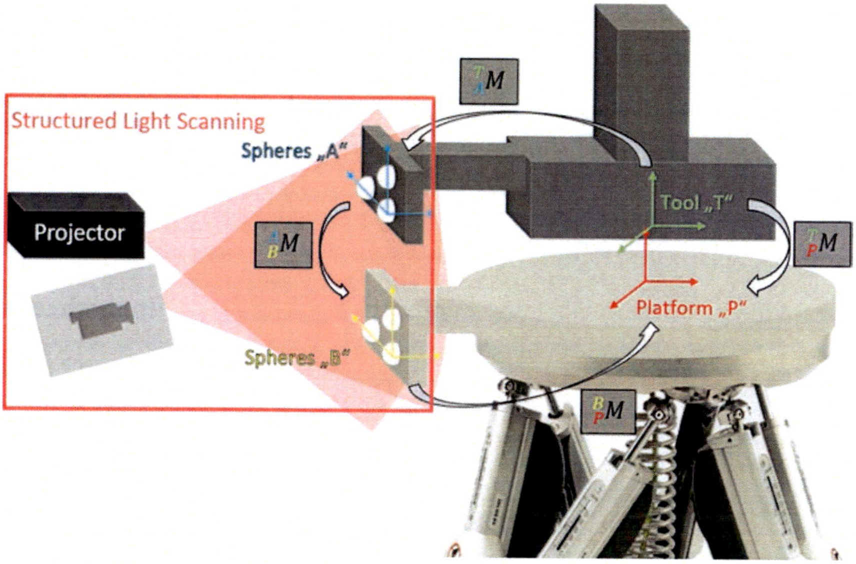

Fig. 3.104 Schematic representation of the optical measurement procedure to determine the position and orientation of the robot and the hexapod

A differs slightly from the radius of sphere B and is also precisely known. This is necessary to ensure that the spheres can be distinguished and correctly assigned during the optical measurement process.

A fringe projection system was used for optical measurement. This system projects a striped pattern onto the ceramic spheres. The deformation of the projected lines induced by the ceramic spheres is recorded by a camera system and converted into a point cloud (strip projection method). These spheres are identified within the point cloud, and, after calibrating the system, the position of the spheres relative to one another can be calculated to obtain the transformation matrix ${}_{P}^{B}M$. At this point, all required transformations ${}_{A}^{T}M$, ${}_{P}^{B}M$, and ${}_{P}^{B}M$ are known, and the transformation matrix ${}_{P}^{T}M$ establishing the relation between TCP_{Robot} and $TCP_{Hexapod}$ can be calculated by matrix multiplication.

Figure 3.105 shows an example. Figure 3.105a shows each tool plate with the ceramic spheres. To demonstrate the positioning, tips are mounted on each of the robot and the hexapod and positioned relative to one another. The figure also shows the striped pattern generated by the projector and captured by the camera. Once the positions of the spheres have been determined in space, the rough positioning is adjusted by moving the robot. A second measurement process is now performed. This adjusts the fine positioning by precisely moving the hexapod. The end result is shown in Fig. 3.105b. This method can achieve a positioning accuracy in the range of 10 µm. To the first order of approximation, the accuracy depends on the field of view of the camera, the camera resolution, and the width of the narrowest projected strip.

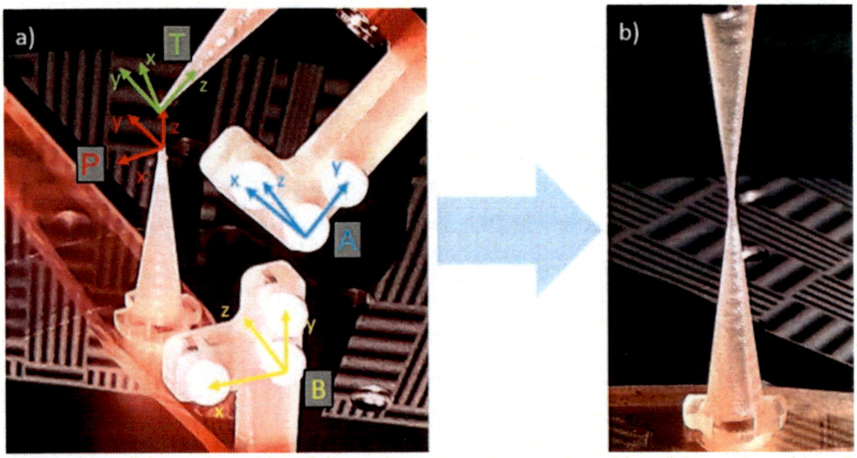

Fig. 3.105 Example of a positioning process: position and orientation of the robot and hexapod (**a**) before alignment and (**b**) after alignment based on the position determined for both objects in space

3.7.1.5 Curing Unit

Like the printing unit, multiple possible variants are possible for the curing unit. Curing units are also mounted on a second robot with a zero-point clamping system. One variant is a laser scanning system that guides a 405 nm UV laser beam over the resin using two galvo scanners to cure it. Another variant is a DMD-based process. In this variant, a UV LED is used to illuminate a DMD in such a way that curing is achieved by two-dimensional projection. This allows new processes to be integrated into the printing process. One example of this is a process based on the double patterning method. This process is described in more detail in the next section.

3.7.2 *DMD-Based Additive Manufacturing of Optical Components*

DMD projectors are ideal for curing an entire layer of a component with a single exposure. The minimum lateral structure size is determined by the size of the imaged pixel. However, especially in the context of optical applications, it is desirable to further reduce this structure size. One way to achieve this is with a subpixeling process called the double patterning process. This method allows large areas to be cured with a DLP projector while simultaneously producing small structures on these areas.

3.7.2.1 Curing at Subpixel Resolutions by Multiple Exposure

In standard DLP projectors, the edge length of the imaged pixels is around 40 μm. A circular area would therefore be discretized into steps of size 40 μm. Similarly, the smallest lateral structure size is 40 μm wide. To achieve a finer discretization and smaller structures with the same projector, the double patterning method known from lithography can be modified to work with a DLP-based process. To do this, the energy of the exposure mask is divided into two pulses. Before the second portion of energy is applied, the substrate is moved so that the two exposure masks overlap slightly, as shown in Fig. 3.106a. The energy of each pulse is configured to be

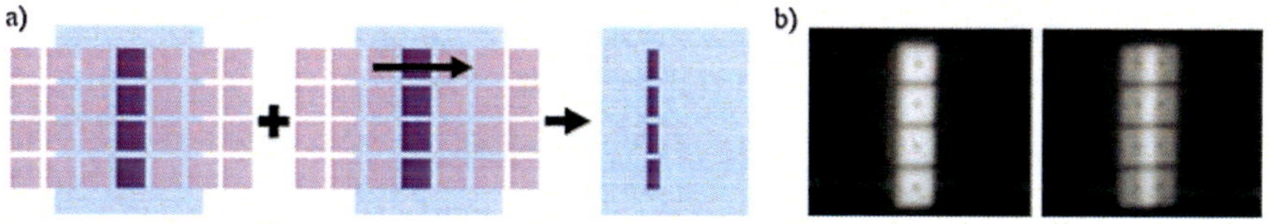

Fig. 3.106 Principle of double patterning. (**a**) Schematic representation of the superposition of pixel columns; (**b**) image of an overlapping pixel column. The overlapping area is approx. 10 μm (bright areas in the right-hand image)

sufficiently low that the material is not fully cured by a single exposure. The material is only cured in the overlapping areas that receive energy from both pulses. Thus, a row of pixels with a width of 40 μm can, for example, be used to expose a bar that is 10 μm wide (Fig. 3.106b). The minimum structure size is no longer limited by the pixel size but by the positioning accuracy of the translation unit and the minimum curable structure size of the resin. In the example discussed here, a hexapod with a positioning accuracy better than 100 nm is used (see Fig. 3.89).

The polymerization reaction can be divided into multiple parts. At the beginning of the process, the resin consists of mobile monomers and photoinitiators. UV radiation radicalizes the photoinitiators, which triggers a chain reaction. As the length of the resulting oligomers increases, the viscosity of the material increases, until the material is fully cured. Before the material transforms into a solid, it is in a gel-like state where it is dimensionally stable but can easily be deformed by mechanical action. This gel phase can be reverted, for example, using solvents. By contrast, solvents only affect the cured material to a very limited extent. Figure 3.107 shows a diagram of this process. The energy of the first exposure mask cures the material up to the gel phase. The substrate is then moved to the second position and subjected to another exposure. In the overlapping area, the material is fully cured. This cured area is surrounded by a layer containing gel-like material. The liquid and gel-like material is then removed with a solvent bath.

To prevent the material from being fully cured by the energy from the first exposure mask, the critical energy of the transition between the gel-like state and the cured state needs to be known. Determining these material properties is therefore essential, as discussed in Sect. 3.3.1.

There is a logarithmic relationship between the surface energy input and the curing thickness (see Formula 3.3.1). This relation can be used to determine the E_c and D_p values of the material. A standard procedure for doing this is to measure the curing thickness as a function of different surface energy inputs and plot it semilogarithmically as shown in Fig. 3.108a. The slope of the plotted line is the curing depth (D_p). The intersection with the x-axis is the critical surface energy.

When configuring the parameters of the double patterning process, only the area very close to the critical surface energy is important. This is shown in more detail in Fig. 3.108b. An approximately linear relationship can be seen between the curing thickness and the energy input around the intersection with the x-axis. Before determining the exact value of E_c, the value is first determined roughly using the method

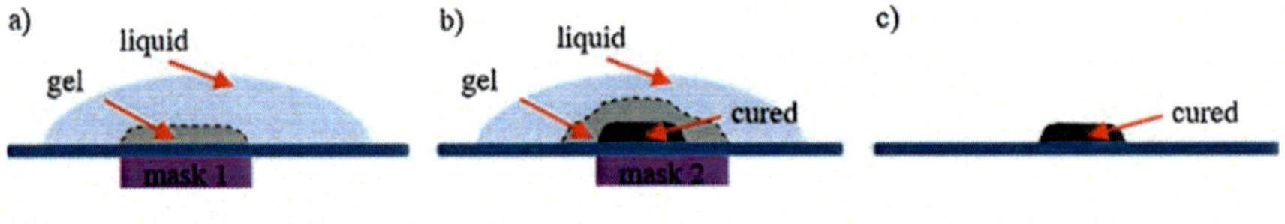

Fig. 3.107 Schematic representation of the double patterning method. (**a**) Exposure of the liquid resin droplet with the first exposure mask; (**b**) displacement of the substrate and exposure with the second exposure mask; (**c**) any material still in a liquid or gel-like state is removed. Cured structures only remain in the areas where the two exposure masks overlapped

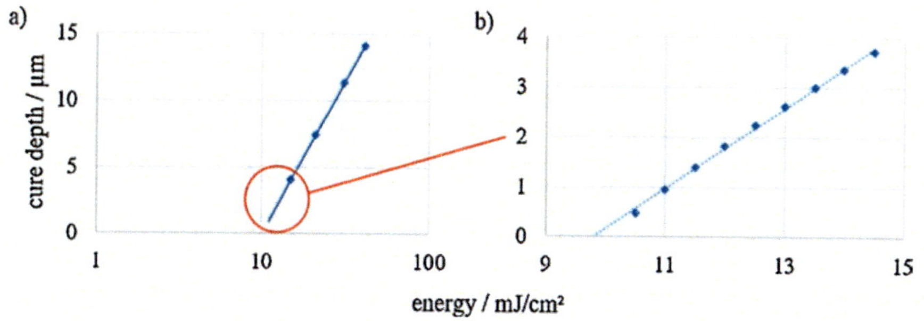

Fig. 3.108 (**a**) Logarithmic representation of the curing depth as a function of the energy input; (**b**) linear approximation of the curing depth close to the critical surface energy

described above. Several values of C_d (in this case, nine) are then determined near this energy. These nine values form an approximately straight line whose intersection with the x-axis specifies a new, more precise value of E_c.

The resin used in the example is based on acrylates whose curing process unfolds by radical polymerization. One known problem with this reaction is that polymerization can be inhibited by oxygen present within the surrounding atmosphere. This occurs when oxygen molecules bind to the free radicals of the polymer chains or photoinitiators, thereby stopping the formation of chains. Maximum conversion of monomers is therefore not achieved at the surface of the cured layer, resulting in the formation of a gel-like film.

To develop a curing technique based on the double patterning method for a given resin system, a series of basic studies of the behavior of the polymerization reaction as a function of the exposure parameters are required. Inhibition by oxygen is an undesirable effect that needs to be avoided. This can be achieved by performing the tests within an atmosphere of nitrogen or carbon dioxide.

To guarantee stable and known experimental conditions for a parameter study, the resin can also be placed on a slide. The resin is cured from behind through the glass slide (see Fig. 3.109). The excess layer of resin over the structure offers sufficient protection against the oxygen present in the environment.

After curing, the liquid and gel-like resins need to be removed from the cured structure to be examined. To do this, the substrate is immersed in a bath of isopropyl alcohol for 4 minutes and dried with a gentle stream of nitrogen. The cured structure is measured with a white light interferometer. The thickness of the sample is then determined from the measured topography.

3.7.2.2 Parameter Study

First, the dependence of the D_p and E_c values of the resin on the process parameters is investigated. The configured irradiance of the DLP projector, the size of the irradiated area, and the pause time between the two exposure pulses are taken into consideration.

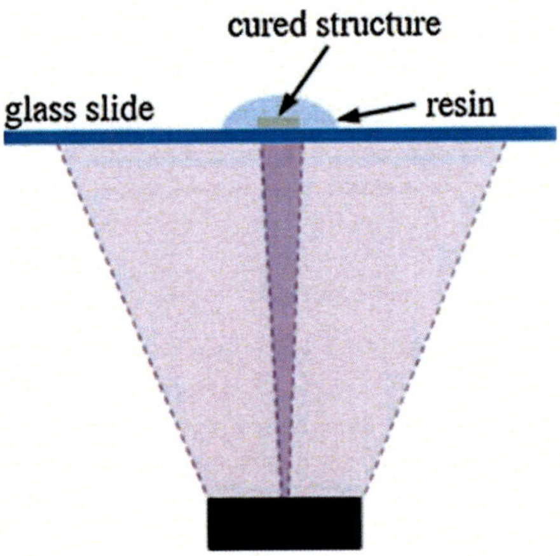

Fig. 3.109 Curing principle that avoids the problem of inhibition by oxygen. The DLP projector is positioned underneath the glass slide. The structure on top of the slide is cured from within the resin droplet

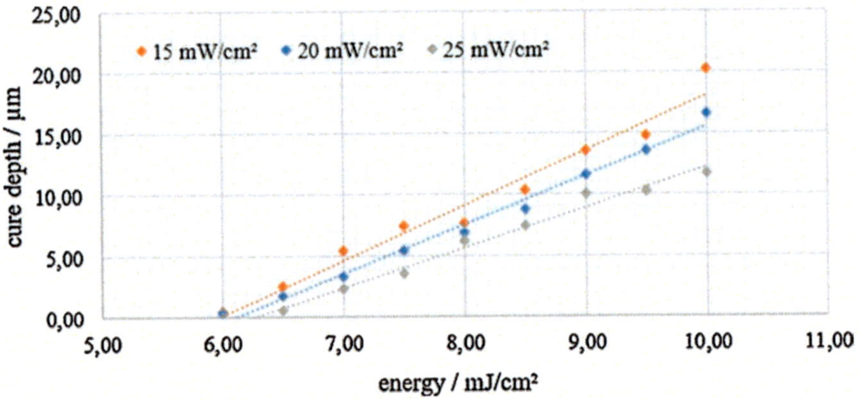

Fig. 3.110 Curing depth as a function of the irradiance

Figure 3.110 shows the thickness of several layers cured with the same energy input but different levels of irradiance. A linear relationship is visible. If the curing is performed at a lower irradiance, the curing thickness is higher, provided that the surface energy input is kept the same. Thus, if the photons are supplied over a longer period of time, the polymerization reaction unfolds more effectively. This also shifts the value of the critical surface energy. Accordingly, the D_p and E_c values need to be determined as a function of the irradiance.

When large areas are exposed (several mm^2), the surface area does not have a significant influence on the D_p and E_c values. However, since double patterning involves the exposure of very small areas, this factor still needs to be considered in more detail. To do this, the D_p and E_c values are determined using three different exposure masks. Figure 3.111 shows the cured thickness for exposure masks consisting of 4 × 4 pixels, 3 × 3 pixels, and 2 × 2 pixels, respectively. It can clearly be seen that both the intersection with the x-axis and the slope of the curve vary

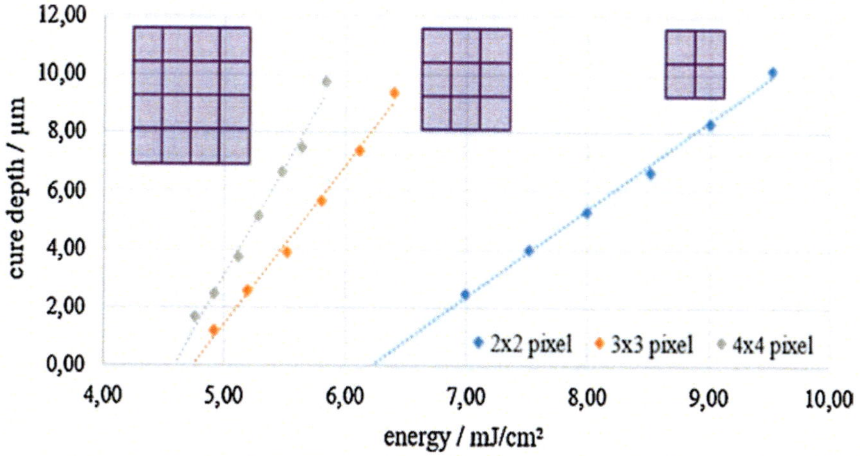

Fig. 3.111 Curing depth as a function of the exposed area

significantly. This means that the energy of the exposure pulses must be precisely adapted to the size and overlapping area of the exposure mask.

Figure 3.112 shows the E_c value as a function of the area of the exposure mask. The area is specified in units of pixels. Each imaged projector pixel corresponds to an edge length of 40 µm. The change in the E_c decreases as the size of the exposed area increases. The target structure sizes of just a few microns are consequently located in a strongly nonlinear range.

Since the substrate needs to be translated mechanically, the two exposure pulses cannot occur in immediate succession. Consequently, the supply of energy to the polymerization reaction is interrupted for a certain time. Figure 3.113 shows the dependence of the cured thickness on the pause time between two exposure pulses. As the pause time increases, the curing thickness decreases, even though the same energy is supplied in each case. This is related to the kinetics of the polymerization reaction.

Finally, two examples are presented to illustrate the curing of structures in the subpixel range (see Fig. 3.114). In this case, a single row of pixels is used as the exposure mask. The imaged pixels once again have an edge length of 40 µm. The substrate is shifted by 30 µm between exposures to create an overlapping area of 10 µm. Figure 3.114a shows the topology of the cured structure. The measured strip has a width of approx. 10 µm and a height of 1.3 µm.

For the second example, an absorption grid was cured. This was done with a period strip mask. Each stripe consists of a row of pixels. The stripes themselves are spaced by two rows of pixels. The mask was shifted by 25 µm between exposures to create an overlapping area of 15 µm. Measurements of the cured grating taken by white light interferometer are shown in Fig. 3.114b. The width of the strip corresponds to the target of 15 µm, and its height is 1.8 µm. The spacing between the strip is 80 µm. The optical behavior of the grating when transmitting a 632 nm laser is shown in Fig. 3.114c. The resulting diffraction pattern is as expected for this type of grating.

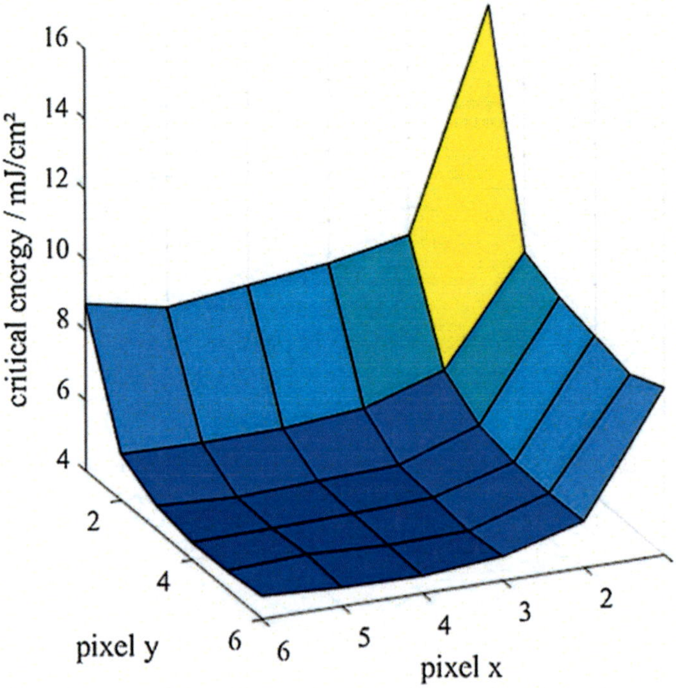

Fig. 3.112 Critical surface energy as a function of the exposed area. For exposed areas larger than 3 × 3 imaged projector pixels, the critical surface energy is approximately constant. The imaged pixel size is 40 μm in this example

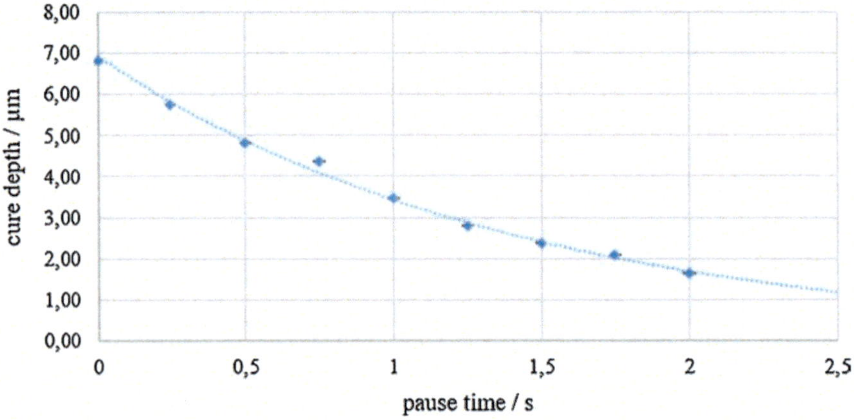

Fig. 3.113 Curing depth as a function of the pause time between mask exposures. The total energy input is 8 mJ/cm² for every cured sample

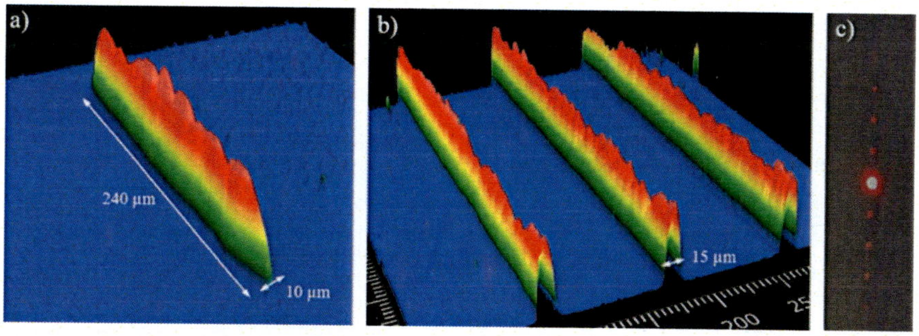

Fig. 3.114 (**a**) Strip with a width of 10 µm cured using the double patterning method; (**b**) optical absorption grating; (**c**) diffraction pattern generated by (**b**)

3.7.3 3D Printing of Multiple Materials

The final section considers multimaterial printing with a conventional STL printer. The goal is to use different materials in both the lateral and axial directions within a single printing process (see Sect. 3.6.3).

3.7.3.1 3D Printing with a Single Material

An indication of the extent of research done into additive manufacturing is seen through the abundance of materials and the corresponding manufacturing processes being developed. Depending on the application of the fabricated part and its geometrical complexity (precision required for its macro and micro features, for instance), a wide range of polymers, ceramics, and metals are available for fabrication with additive manufacturing [106]. The use of several materials within the same fabrication step is often required to enhance the performance of the manufactured part. In the context of stereolithography, where thin layers of the material are accumulated on top of each other according to the component's design, using multiple materials can either mean printing with different materials within the same layer or printing with different materials per set of layers. An experiment on the printing of multiple materials is discussed in this section after making minor changes to the fabrication process used by the Ember printer from Autodesk. Under normal circumstances, the printer works according to the principle of photopolymerization by irradiating single slices of a select photoreactive resin in liquid form with ultraviolet light. The UV dosage, which is predefined for a specific layer thickness, initiates polymerization of the liquid layer, solidifying it on the printing platform, as shown in Fig. 3.115.

Under normal printing conditions, one material is used to build a solid version of the computer-aided design (CAD) of the part. The resin vat is filled with 75 ml of photoreactive resin. One layer is printed on the printing platform with each completed printing cycle. Each printing cycle consists of four steps, as shown in

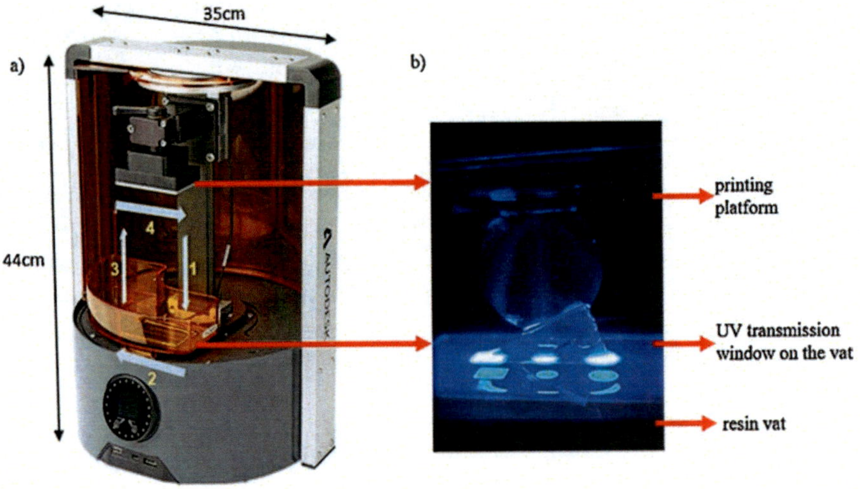

Fig. 3.115 (**a**) 3D printing machine used for multiple material printing; (**b**) normal operation of the printer, yielding single-material prints—a print job is shown during the exposure step of the print cycle. Each printing cycle accumulates a layer on the printing platform, thereby additively manufacturing the sample

Fig. 3.115a. The platform moves down and is immersed in the resin vat in the first step. At this position, only a thin layer of liquid resin remains in between the platform and the transparent window. Subsequently, this thin layer of the liquid photoreactive resin is exposed to a precalculated dosage of ultraviolet radiation in the shape of the slice of the part according to the design, as shown in Fig. 3.115b. Each slice corresponds to the cross-section of one layer of a specific thickness, created by slicing software from the CAD design of the part. Within the thin layer that is exposed to the radiation, there is a dead zone consisting of the resin directly in contact with the oxygen diffused into the PDMS layer in the transmission window that does not respond to the excitation (see Sect. 3.2). The liquid layer segment above this dead zone polymerizes onto the print platform. In the second step, the printed layer on the platform is separated from the transmission window via a sliding motion as the resin vat rotates, in order to compensate for the strong adhesion forces between the layer formed on the print build head (platform) and the liquid dead zone segment beneath it. In the third step, the platform moves up from the vat. Finally, in the last step, the resin vat rotates back to its initial position. With each cycle, the platform moves up by one layer thickness when it immerses into the resin to accommodate the next layer cured onto the platform.

3.7.3.2 3D Printing Multiple Materials with a Conventional STL Printer

Modification of the Fundamental Fabrication Process

When using multiple materials to 3D print a component using a printer designed for one material, the first task is to accommodate all the raw materials while maintaining the capability to polymerize each material onto the platform as per the design. This step was realized using a metal barrier that partitioned the transparent window into different sections so that one material is held in each section. The printing platform was built with projections that could be fully immersed into these sections to print the material from each section onto the corresponding projection of the print platform according to the design. It should be noted that the metal barrier/rim is held in position solely by the adhesion forces between the PDMS layer on the transmission window and the barrier. Therefore, steps 2 and 4 of the normal printing cycle, where the resin vat rotates, were removed to avoid the possibility of mechanical disturbances that might displace the barrier placement altogether. A shortcoming of this sectioning method is that the materials are at the risk of mixing after long prints (see Fig. 3.86).

Possible Methods of Multiple Material Printing

With the printing setup presented in the previous section, multiple materials can potentially be incorporated into the printed sample in two ways: by printing with different materials sequentially, as shown in Fig. 3.116, and/or by printing different materials simultaneously, as shown in Fig. 3.117 [108].

Printing with multiple materials sequentially involves printing with one material for a specific number of layers and then switching to another material. In this way, up to five variations of the same base resin could be tested, as shown in Fig. 3.116. The inherent property of photoluminescence is preserved in the 3D samples that were printed from the quantum dot nanocomposite, as can be seen in the printed example. Manual intervention is needed to clean the surface of the printed sample on the print platform to avoid the possibility of uncured material mixing at the boundaries between two materials at the first layer printed after the material exchange.

Multiple materials can also be printed simultaneously within one slice, as shown by the schematic in Fig. 3.117. For this simultaneous material curing, it is intuitive to fill all the materials in the sections created by a metal barrier or rim in such a way that the projections can print each material simultaneously. An example of the combined usage of both these methods is demonstrated by the 3D-printed photoluminescent sample shown in Fig. 3.117. In this case, the quantum dot composites were all printed simultaneously, and then the common base cover was printed sequentially onto these QD rods using the base resin only.

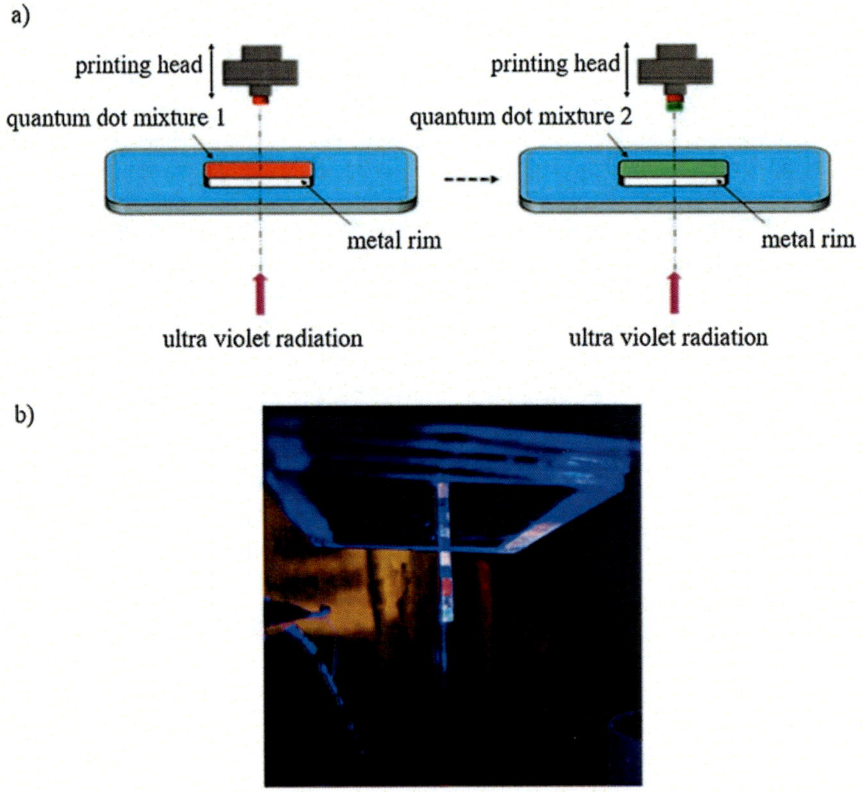

Fig. 3.116 Schematic diagram showing the sequential steps of a multiple material printing process where different materials are introduced into a single print job in the direction of the print axis, i.e., a set of layers of one material is cured on top of a set of layers of another material, sequentially. In the printed sample shown here, quantum dot nanocomposites of different emission wavelengths were printed in a single print job

Using the available commercial 3D printing setup and introducing minor tweaks into the printing process allowed multiple material printing to be implemented for quantum dot nanocomposites. The challenge of barrier diffusion (see Fig. 3.86) arises because the metal barriers used to section the material compartments are not sturdy, since they are only temporarily placed in the resin vat and are removed after the prints are complete. This temporary placement was an alternative to reusing the resin vats for more than one configuration of multiple material prints (for instance, changing the metal barriers depending on the number of materials and hence the number of sections). To achieve stable prints, continued efforts are needed to fabricate customized resin vats that can accommodate multiple materials within the same resin vat with minimal risk of material mixing.

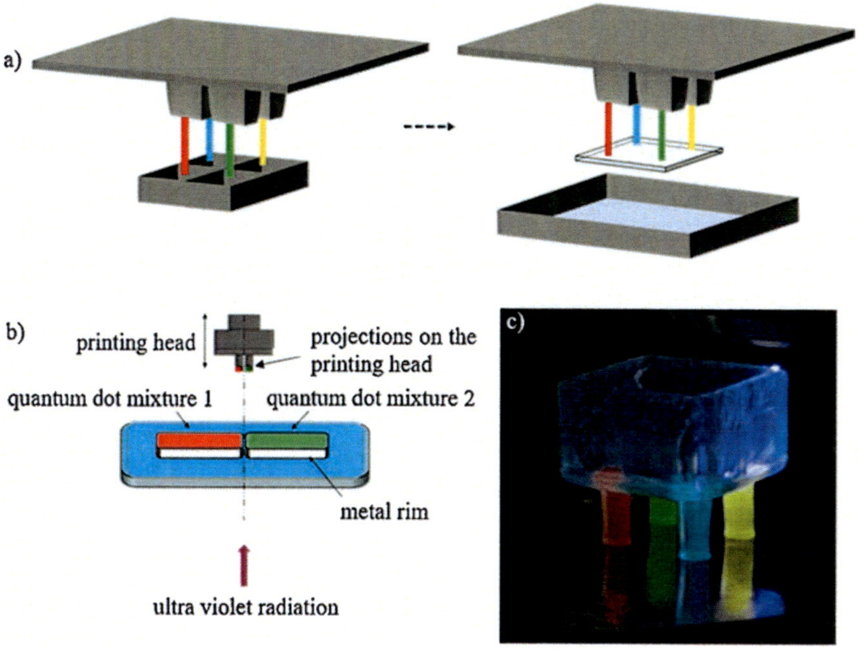

Fig. 3.117 A schematic diagram showing the possibility of printing multiple materials within the same layer, spatially separated from one another. This technique was used to print quantum dot composites of different emission wavelengths and then sequentially print a pure resin base at the bottom to obtain a sample printed with five different variations (four nanocomposites and the pure resin) of the same photopolymer resin

3.8 Summary

This chapter examined the additive manufacturing of active and passive optical components using conventional 3D printing technologies. In general, it established that surface roughness, layered printing, and the associated inhomogeneity during the curing process are at the root of reduced optical properties in macroscopic elements. Nevertheless, there are many possible applications for additively manufactured optical components, especially in the field of lighting. The complex component geometries allowed by additive manufacturing are particularly advantageous.

The 3D printing of microlenses also opens up the possibility to use additive manufacturing for imaging systems, since microlenses have a significantly lower surface roughness and also do not exhibit a layered structure. Furthermore, to achieve greater levels of shape complexity, microlenses can be deformed in their liquid state within an electric field before curing. This provides an additional degree of design freedom.

In addition to passive optical components, active optical components also appear very promising. New functionality can, for example, be integrated into printed optics by incorporating quantum dots, and liquid lenses with an adaptively modifiable freeform surface can be developed.

To further improve the quality of 3D-printed optics in the future, or to integrate additional functionality into optical elements, the manufacturing process needs to be adapted to the requirements of optical elements. A key step is the transition to additive manufacturing with six degrees of freedom, which is made possible by robots. Another key goal in this context is to reduce the lateral structure resolution, which requires new approaches, such as double patterning.

Acknowledgments The authors would like to thank the members of the Center for Optical Technologies at Aalen University, especially Katharina Blapp, Annika Dehm, Sebastian Hägele, Tobias Novak, Andre Horsak, Martina Rosner, Katja Schlichting, Regina Schuster, Anne Suckow, and Thomas Tischler for their contributions. Moreover, the authors would like to thank Rainer Börret, Harald Riegel, Markus Merkel, Verena Richter, and Herbert Schneckenburger from Aalen University for sharing equipment and knowledge. Additionally, the authors would like to thank Dr. Markus Haidl (Carl Zeiss Vision), Lutz Autschbach (Carl Zeiss Jena GmbH) and Dr. Michael Totzeck (Carl Zeiss AG) for supporting our group with fruitful discussions and joint projects. Last but not least, the authors would like to thank the Baden-Württemberg Stiftung (project SOMS3D and NeMFago), the German Federal Ministry of Education and Research (funding program Forschung an Fachhochschulen: project IMiSens (grant no. 03FH048PA4), VAGEB (grant no. 03FH002PX5), RaFoK (grant no. 13FH215PX6)), and the German Research Foundation (DFG 6D Additiv (grant no. HE 3533/7) for funding.

References

1. Ligon, S. C., Liska, R., Stampfl, J., Gurr, M., & Mülhaupt, R. (2017). Polymers for 3D printing and customized additive manufacturing. *Chemical Reviews, 117*(15), 10212–10290.
2. Abu-Abdoun, I.I. (2019). Photopolymerization of epoxide and vinyl monomers by metal carbonyl compounds. In *ACTA Scientific Pharmaceutical Sciences* (issn: 2581-5423), vol. 3, 3–7.
3. Steyrer, B., Neubauer, P., Liska, R., & Stampfl, J. (2017). Visible light photoinitiator for 3D-printing of tough methacrylate resins. *Materials (Basel), 10*(12), 1445.
4. Ikemura, K., Ichizawa, K., Yoshida, M., Ito, S., & Endo, T. (2008). UV-VIS spectra and photoinitiation behaviours of acylphosphine oxide and bisacylphosphine oxide derivatives in unfilled, light-cured dental resins. *Dental Materials Journal, 27*(6), 765–774.
5. Moszner, N., Lamparth, I., Angermann, J., Fischer, U. K., Zeuner, F., Bock, T., Liska, R., & Rheinberger, V. (2010). Synthesis of bis(3-{[2-(allyloxy)ethoxy]methyl}-2,4,6-trimethylbenzoyl)(phenyl)phosphine oxide—a tailor-made photoinitiator for dental adhesives. *Beilstein Journal of Organic Chemistry, 6*, 26.
6. Green, W. A. (2010). *Industrial photo initiators, a technical guide* (pp. 170–171). Boca Raton: CRC Press.
7. Zhanga, J., Zivicb, N., Dumurb, F., Guoc, C., Lic, Y., Xiaoa, P., Graffa, B., Gigmesb, D., Fouassier, J. P., & Lalevéea, J. (2015). Panchromatic photo initiators for radical, cationic and thiol-ene polymerization reactions: A search in the diketopyrrole or indigo dye series. *Materials Today Communications, 4*, 101–108.
8. de Lombard-Watts, T. & Weissman P. (2004) Comparison of performance of acrylate and methacrylate aliphatic urethanes. RadTech e5 Technical Proceedings.
9. Li-Ming, T., Yu, F., Liang, Y., & Zhi-wei, F. (2004). Influence of reactive diluent on UV-curing of acrylate terminated hyperbranched polymers. *Chem. Res. Chinese U., 20*(2), 248–252.

10. Chai, R., Chen, S., & Zhang, J.. (2012). Combined effect of hindered amine light stabilizer and ultraviolet absorbers on photodegradation of poly (vinyl chloride). *Wiley Online Library, Society of Plastics Engineers, Journal of Vinyl & Additive Technology, 18*, 17.
11. Van Landuyt, K. L., Snauwaert, J., De Munck, J., et al. (2007). Systematic review of the chemical composition of contemporary dental adhesives. *Biomaterials, 28*(26), 3757–3785.
12. Schwalm, R. (2001). Photoinitiators and photopolymerization. In Encyclopedia of materials (pp. 6946–6951) (isbn: 978-0-08-043152-9).
13. Jacobs, P. F. & Reid, D. T. (1992). Rapid prototyping & manufacturing: Fundamentals of stereolithography. In Society of Manufacturing Engineers. isbn: 978-0-87-263425-1.
14. Lee, J. H., Prud'homme, R. K., & Aksay, I. A. (2001). Cure depth in photopolymerization: Experiments and theory. *Journal of Materials Research, 16*(12), 3536–3544.
15. Heinrich, A., Börret, R., Merkel, M., & Riegel, H. (2018). Additive manufacturing of reflective and transmissive optics: potential and new solutions for optical systems. In Proc. SPIE 10523, laser 3D manufacturing V, 1052302.
16. Hecht, E. (2016). Optics, global edition. Pearson (Intl), 281. isbn:978-1-29-209696-4.
17. Rank, M. & Heinrich, A.. Measurement and use of the refractive index distribution in optical elements created by additive manufacturing. In Proc. SPIE 10930, Advanced Fabrication Technologies for Micro/Nano Optics and Photonics XII, 109300Z (pp. 58–64) (2019).
18. Odian, G. (2004). *Principles of polymerization* (4th ed.). Hoboken: Wiley. isbn:978-0-47-147875-1.
19. Song, M. S., Choi, H. Y., Seong, J. H., & Kim, E. S. (2015). Matching-index-of-refraction of transparent 3D printing models for flow visualization. *Nuclear Engineering and Design, 284*, 185–191.
20. Gupta, A., Lee, J., & Koshel, R. J. (2001). Design of efficient lightpipes for illumination by an analytical approach. *Applied Optics, 40*, 3640–3648.
21. Greivenkamp, J. E. (2004). Field guide to geometrical optics. SPIE Press.
22. Hui, D., Zhang, M., Geng, Z., et al. (2012). Designs for high performance PAL-based imaging systems. *Applied Optics, 51*(21), 5310–5317.
23. Fang, F. Z., Zhang, X. D., Weckenmann, A., Zhang, G. X., & Evans, C. (2013). Manufacturing and measurement of freeform optics. *CIRP Annals, 62*(2), 823–846.
24. Clayor, N. E., Combs, D. M., Lechuga, O. M., Mader, J. J. (2004). An overview of freeform optics production. In Proc. SPIE 5494.
25. De Chiffre, L., Kunzmann, H., Peggs, G., & Lucca, D. A. (2003). Surfaces in precision engineering, microengineering and nanotechnology. *CIRP Annals, 52*(2), 561–577.
26. Ye, J., Chen, L., Li, X., Yuan, Q., & Gao, Z. (2017). Review of optical freeform surface representation technique and its application. *Optical Engineering, 56*, 11.
27. Hosseinimakarem, Z., Davies, A. D., & Chris, J. E. (2016). Zernike polynomials for mid-spatial frequency representation on optical surfaces. In Proc. SPIE 9961.
28. Chrisp, M. P., Primeau, B., & Echter, M. A. (2016). Imaging freeform optical systems designed with NURBS surfaces. *Optical Engineering, 55*, 7.
29. Sigel, A., Dohmen, M., & Heinrich, A. (2019). Evaluation of a 3D printed freeform reflector and lens. *Optical Design and Fabrication*, 2019.
30. Hui, D., Zhang, M., Geng, Z., Zhang, Y., Duan, J., Shi, A., Hui, L., Fang, Q., & Liu, Y. (2012). Designs for high performance PAL-based imaging systems. *Applied Optics, 51*(21), 5310–5317.
31. Benítez, P., Miñano, J.C., Grabovickic, D., Zamora, P., Buljan, M., Narasimhan, B., & Nikolic, M. (2017). Freeform optics for virtual reality applications. *Optical Design and Fabrication*, 2017.
32. Heinrich, A., Rank, M., Suckow, A., Bauckhage, Y., Rößler, P., Lang, J., Shariff, F., & Pekrul, S. (2016). Additive manufacturing of optical components. *Advanced Optical Technologies, 5*, 293–301.
33. Ries, H., & Muschaweck, J. (2002). Tailored freeform optical surfaces. *Journal of the Optical Society of America, 19*, 590–595.

34. Assefa, B. G., Meuret, Y., Tervo, J., Saastamoinen, T., Kuittinen, M., & Saarinen, J. (2015). Evaluation of freeform lens designs for specific target distributions and fabrication using 3D printing. In The eleventh Finland-Japan joint symposium on optics in engineering.
35. Assefa, B. G, Saastamoinen, T., Pekkarinen, M., Biskop, J., Kuittinen, M., Turunen, J., & Saarinen J. (2018). Design and characterization of 3D-printed freeform lenses for random illuminations. In Proc. SPIE 10554.
36. Williams, K. E., Ulanch, R. N., King, P., Uthoff, R. D., Diaz, L. R., Steidle, J., & Koshel, R. J. (2017). Functional fitting of freeform optics. *Optical Design and Fabrication, 2017.*
37. Diaz, L. R., Uthoff, R. D., Ulanch, R. N., Williams, K. E., Saxton, N., Koshel, R. J. (2017). Solid-state lighting for high-valued artwork. Light, Energy and the Environment.
38. Ochiai, Y., Kumagai, K., Hoshi, T., Rekimoto, J., Hasegawa, S., & Hayasaki, Y. (2015). Fairy lights in femtoseconds: Aerial and volumetric graphics rendered by focused femtosecond laser combined with computational holographic fields. *ACM Transactions on Graphics, 35*, 2.
39. Nayar, S. K., & Anand, V. N. (2007). 3D display using passive optical scatterers. *Computer, 40*(7), 54–63.
40. Willis K., Brockmeyer, E., Hudson S., & Poupyrev I. (2012). Printed optics: 3D printing of embedded optical elements for interactive devices. In Proceedings of the 25th annual ACM symposium on user interface software and technology (pp. 589–598).
41. Zhu, R., Hong, Q., Zhang, H., & Wu, S. T. (2015). Freeform reflectors for architectural lighting. *Optics Express, 23*(25), 31828–31837.
42. Knöchelmann M., Kloppenburg G., & Lachmayer R. (2018). Headlamp innovations: optical concepts for fully adaptive light distributions. In Proc. SPIE 10546.
43. Khan, M. S., Rahlves, M., Lachmayer, R., & Roth, B. (2018). Polymer-based diffractive optical elements for rear end automotive applications: Design and fabrication process. *Applied Optics, 57*, 9106–9113.
44. Khan, S., Rahlves, M., Roth, B., & Lachmayer, R. (2018). Laser based projection system for traffic collision prevention 119. Jahrestagung der DGaO.
45. Serrano-Aguilera, J. J., Valenzuela, L., & Fernández-Reche, J. (2015). Inverse MCRT method for obtaining solar concentrators with quasi-planar flux distribution. *Energy Procedia, 69*, 208–217.
46. Finckh, M., Dammertz, H., & Lensch, H. P. A. (2010). Geometry construction from caustic images. *Computer Vision—ECCV, 2010*, 464–477.
47. Weyrich, T., Peers, P., Matusik, W., & Rusinkiewicz, S. (2009). Fabricating microgeometry for custom surface reflectance. *ACM Transactions on Graphics, 28*, 3.
48. Ries, H. (2012). Optikdesign mit maßgeschneiderten Freiformflächen. *Optik & Photonik, 7*(3), 30–33.
49. Romijn, L. B., Ten Thije Boonkkamp, J. H. M., & IJzerman, W. L. (2019). Freeform lens design for a point source and far-field target. *Optical Design and Fabrication, 2019.*
50. Ten Thije Boonkkamp, J. H. M & IJzerman, W. L. (2019). Illumination freeform design using Monge-Ampère equations. *Optical Design and Fabrication, 2019.*
51. Oliker, V.(2003). Mathematical aspects of design of beam shaping surfaces in geometrical optics. In *Trends in nonlinear analysis* (pp. 193–224). Berlin: Springer. isbn:978-3-662-05281-5.
52. Fournier, F. R., Cassarly, W. J., & Rolland, J. P. (2009). Designing freeform reflectors for extended sources. In Proc. SPIE 7423.
53. Fournier, F. R., Cassarly, W. J., & Rolland, J. P. (2010). Fast freeform reflector generation using source-target maps. *Optics Express, 18*(5), 5295–5304.
54. Canavesi, C., Cassarly, W. J., & Rolland, J. P. (2012). Direct calculation algorithm for two-dimensional reflector design. *Optics Letters, 37*(18), 3852–3854.
55. Canavesi, C., Cassarly, W. J., & Rolland, J. P (2012). Relationship between supporting paraboloids and linear programming for 2D reflector design. *Frontiers in Optics, 2012.*
56. Kochengin, S., & Oliker, V. (2003). Computational algorithms for constructing reflectors. *Computing and Visualization in Science, 6*, 15–21.

57. Kloppenburg, G. & Wolf, A. (2016). Eigenschaften und Validierung optischer Komponenten. In *3D-Druck beleuchtet* (pp. 87–102). Berlin: Springer. isbn:978-3-662-49055-6.
58. Lachmayer, R., Wolf, A., & Kloppenburg, G. (2015). Additive manufacturing of optical components. In SPIE Newsroom.
59. Lachmayer, R., Kloppenburg, G., & Wolf, A. (2015). Rapid prototyping of reflectors for vehicle lighting using laser activated remote phosphor. In Proc. SPIE 9383.
60. Lieberwirth, C., Harder, A., & Seitz, H. (2017). Extrusion based Additive manufacturing of metal parts. *Journal of Mechanics Engineering and Automation, 7.*
61. Schanz, J., Hofele, M., Ruck, S., Schubert, T., Schneider, G., Merkel, M., & Riegel, H.(2017). Metallurgical investigation of laser based post processing of additive manufactured aluminum parts. DGM Metallographie Tagung Aalen.
62. Chen, E., et al. (2014). Design a freeform microlens array module for any arbitrary-shape collimated beam shaping and color mixing. *Optics Communications, 321*, 78–85.
63. Chow, E., et al. (2019). Surface measurement using compressed wavefront sensing. *Photonic Sensors, 9*(2), 115–125.
64. Surdo, S., et al. (2018). Single-shot laser Additive manufacturing of high fill-factor microlens arrays. *Advanced Optical Materials, 6*, 1701190.
65. Youan, W., et al. (2018). Fabrication of microlens array and its application: A review. *Chinese Journal of Mechanical Engineering, 31*, 16. https://doi.org/10.1186/s10033-018-0204-y.
66. Xie, P., et al. (2014). Rapid hot embossing of polymer microstructures using carbide-bonded graphene coating on silicon stampers. *Surface and Coatings Tech., 258*, 174–180.
67. Huang, S., et al. (2018). Fabrication of high quality aspheric microlens array by dose-modulated lithography and surface thermal reflow. *Optics & Laser Tech., 100*, 298–303.
68. Seo, M., et al. (2018). Insect-mimetic imaging system based on a microlens array fabricated by a patterned-layer integrating soft lithography process. *Sensors, 18*(7). https://doi.org/10.3390/s18072011.
69. Peng, Y., et al. (2016). Fabrication of micro-lens arrays based on ink-jet printing and photolithographic hole templates for integral imaging 3-D display. *Journal of Display Technology, 12*(8), 822–872.
70. Biehl, S., et al. (1998). Refractive microlens fabrication by ink-jet process. *Journal of Sol-Gel Science and Technology, 13*(1–3), 177–182.
71. Hoath, D. (2016). Fundamentals of ink-jet printing. Viley Verlag. isbn:978-3-527-33785-9.
72. Eggenhuisen, T. M., Galagan, Y., Biezemans, A. F. K. V., Slaats, T. M. W. L., Voorthuijzen, W. P., Kommeren, S., Shanmugam, S., Teunissen, J. P., Hadipour, A., Verhees, W. J. H., Veenstra, S. C., Coenen, M. J. J., Gilot, J., Andriessen, R., Groen, W. A., & Mater, J. (2015). High efficiency, fully inkjet printed organic solar cells with freedom of design. *Journal of Materials Chemistry A, 3*(14), 7255–7262.
73. Kim, W. H., Mäkinen, A. J., Nikolov, N., Shashidhar, R., Kim, H., & Kafafi, Z. H. (2002). Molecular organic light-emitting diodes using highly conducting polymers as anodes. *Applied Physics Letters, 80*(20), 3844–3846.
74. Köhler, A. & Bässler, H. (2015). Electronic processes in organic semiconductors. Wiley-VCH Verlag GmbH. isbn:978-3-52-768514-1.
75. Oostra, A. J., Blom, P. W. M., & Michels, J. J. (2014). Prevention of short circuits in solution-processed OLED devices. *Organic Electronics, 15*(6), 1166–1172.
76. Roberts, M., Asada, K., Cass, M., Coward, C., King, S., Lee, A., Pintani, M., Ramon, M., & Foden, C. (2010). Fundamental processes governing operation and degradation in state of the art P-OLEDs. Proc. SPIE 7722, Organic Photonics IV, 77220C.
77. Schaer, M., Nüesch, F., Berner, D., Leo, W., & Zuppiroli, L. (2001). Water vapor and oxygen degradation mechanisms in organic light emitting diodes. *Advanced Functional Materials, 11*(2), 116–121.
78. Eder, C., Rank, M., & Heinrich, A. (2020). Additive manufactured organic light-emitting diodes. In Proc. SPIE 11277, Organic Photonic Materials and Devices XXII, 1127705.

79. Wiersma, D. (2008). The physics and applications of random lasers. *Nature Physics, 4*, 359–367.
80. Pisanic, T. R., Zhang, Y., & Wang, T. H. (2014). Quantum dots in diagnostics and detection: Principles and paradigms. *Analyst, 139*(12), 2968–2981, NIH Public Access.
81. Nozik, A. J., Beard, M. C., Luther, J. M., Law, M., Ellingson, R. J., & Johnson, J. C. (2010). Semiconductor quantum dots and quantum dot arrays and applications of multiple exciton generation to third-generation photovoltaic solar cells. *Chemical Reviews, 110*, 6873–6890.
82. Asanol, H., Arai, K., Kita, M., & Omata, T. (2017). Synthesis of colloidal Zn(Te,Se) alloy quantum dots. *Materials Research Express, 4*, 106501, IOP Publishing Ltd.
83. Ng Lee, J., Park, C., & Whitesides, G. W. (2003). Solvent compatibility of poly (dimethyl siloxane)-based microfluidic devices. *Analytical Chemistry, 75*, 23.
84. Suresh Nair, S., Nuding, J., & Heinrich, A. (2018). 3D-printed optical active components. In Proc. SPIE 10529, Organic Photonic Materials and Devices XX, 105290W.
85. Kim, D. Y, Cho, H. N, & Kim, C. Y. (2017). Blue light emitting polymers. Prog. Polym. Sci. 25, Elsevier Science Ltd.
86. Lakowicz, R. (2006). Chapter 6. Principles of fluorescence spectroscopy. In *Solvents and environmental effects* (3rd ed., pp. 219–221). Baltimore: Springer.
87. Schneckenburger, H., Richter, V., & Wagner, M. (2018). Live-cell optical microscopy with limited light doses. In SPIE spotlight series (vol. SL 42), isbn:9781510622593.
88. Zapka, W. (2018). Handbook of industrial inkjet printing. Wiley-VHC. isbn:978-3-527-68720-6.
89. Le, H. P. (1998). Progress and trends in ink-jet printing technology. *Journal of Imaging Science and Technology, 42*, 49–62.
90. Shield, T. W., Bogy, D. B., & Talke, F. E. (1987). Drop formation by DOD ink-jet nozzles: A comparison of experiment and numerical simulation. *IBM Journal of Research and Development, 31*, 96–110.
91. Liu, Y. F., Tsai, M. H., Pai, Y. F., & Hwang, W. S. (2013). Control of droplet formation by operating waveform for inks with various viscosities in piezoelectric inkjet printing. *Applied Physics A, 111*, 509–516.
92. Caglar, U. (2010). Studies of inkjet printing technology with focus on electronic materials. Dissertation, Tampere University of Technology.
93. Shimoda, T., Morii, K., Seki, S., & Kiguchi, H. (2003). Inkjet printing of light-emitting polymer displays. *MRS Bulletin, 28*, 821–827.
94. Cho, D.-W., Lee, J.-S., Jang, J., Jung, J. W., Park, J. H., Pati, F. (2015), Organ printing—Chapter 4: Dispensing-based 3D printing. Morgan & Claypool Publishers. isbn:978-1-6817-4079-9.
95. Zang, Z., Tang, X., Liu, X., Lei, X., & Chen, W. (2014). Fabrication of high quality and low cost microlenses on a glass substrate by direct printing technique. *Applied Optics, 53*(33), 7868–7871.
96. VERMES Microdispensing GmbH, Mikrodosierung, Jetventil, Mikrodosierventil VERMES Microdispensing GmbH. (2019). https://www.vermes.com/.
97. VERMES Microdispensing GmbH, MDV 3280, 2019., https://www.vermes.com/fileadmin/user_upload/MDS_3280_Page_1.png, (2019).
98. VERMES Microdispensing GmbH, Bedienungsanleitung MDS 3280-Series (2018).
99. Dohmen, M. (2019). Microdispensing of UV-curable resin for functionalized optics. Internal research report, Aalen University.
100. van der Bos, A., van der Meulen, M. J., Driessen, T., van den Berg, M., Reinten, H., Wijshoff, H., Versluis, M., Lohse, D. (2014). Physical review applied 1.
101. Campa, R. & Camarillo-Gomez, A. (2018) Unit quaternions: A mathematical tool for modeling, path planning and control of robot manipulators, https://doi.org/10.5772/6197.
102. Universal Robots (2018). *User manual UR3/CB3*.
103. Craig, J. J. (2017) *Introduction to robotics: Mechanics and control*. Pearson. isbn-13:978-0133489798.
104. Papula, L. (2017). *Mathematische Formelsammlung: Für Ingenieure und Naturwissenschaftler*. Springer. isbn-10:3658161949.

105. Zeidler, E., Grosche, G., Ziegler, V., Ziegler, D., Bronštejn, I. N., & Semendjaev, K. A. (2013). *Taschenbuch der Mathematik*. isbn-13:978-3658002848.
106. Bourell, D., Kruth, J. P., Leu, M., Levy, G., Rosen, D., Beese, A. M., & Clare, A. (2017). Materials for Additive manufacturing. *CIRP Annals, 66*(2), 659–681.
107. Berge, B., et al. (2005). Liquid lens technology: principle of electrowetting based lenses and applications to imaging in MEMS 2005 Miami: 18th IEEE International conference on micro electro mechanical systems, 227–230.
108. Suresh Nair, S. & Heinrich, A. (2019). Additive manufacturing of photoluminescent optics. In Proceedings Volume 10915, Organic photonic materials and devices XXI; 1091505. https://doi.org/12.2508574.

Chapter 4
3D Printing of Transparent Glasses

Frederik Kotz, Dorothea Helmer, and Bastian E. Rapp

Abstract Glasses have shaped the field of optics and photonics like no other material—enabling numerous sensing and imaging systems, optical data transfer, and laser systems. Transparent silicate glasses are the material of choice for high-performance optical components as they combine high optical transparency with high thermal, chemical, and mechanical stability. However, precision shaping of glasses is notoriously difficult and mainly limited to grinding and polishing processes for macroscopic objects and hazardous etching processes for the fabrication of microstructures. In recent years, considerable efforts have been made to making glasses accessible to the 3D printing revolution of the twenty-first century. When silicate glasses entered the field of 3D printing, two major directions came up—direct 3D printing of low melting glasses at high temperatures and indirect glass printing of glass precursors using technologies borrowed from the techniques of polymer 3D printing. These precursors can be printed at room temperature and turned into transparent glass in a subsequent heat treatment. In this chapter, we outline the latest developments of 3D printing of transparent silicate glasses.

Keywords Additive manufacturing · 3D printing · Fused silica glass Multicomponent glass · Nanocomposites · Sol-gel · Glassomer · Stereolithography · Direct ink writing · Advanced manufacturing

4.1 Introduction

Glasses are one of mankind's oldest materials and nowadays still one of the most important materials in optical applications, e.g., for lenses, prisms, mirrors, and windows. Glass is the key material for a number of applications, ranging from imaging systems for photography, television, high-resolution microscopy, to telescopes [1]. For all applications which require high transparency and stability, glasses are the first choice. Modern data technology would not be possible without glass fiber

F. Kotz (✉) · D. Helmer · B. E. Rapp
Department of Microsystems Engineering, Laboratory of Process Technology|NeptunLab, University of Freiburg, Freiburg im Breisgau, Germany
e-mail: frederik.kotz@imtek.de; bastian.rapp@imtek.de

A. Heinrich (ed.), *3D Printing of Optical Components*, Springer Series in Optical Sciences 233, https://doi.org/10.1007/978-3-030-58960-8_4

technology, and only 11 years ago, Charles K. Kao received the Nobel Prize in Physics for his work on optical data transfer. Even though glasses have been known for centuries, their technological significance has increased over the centuries.

Glasses are a special class of materials, a fact that is reflected by the ongoing debate about the correct definition of the glassy state—several thousand years after its discovery [2]. Glasses are characterized by a missing long-range order, i.e., they show an amorphous structure and a glass transition temperature. This definition extends beyond the classical pure oxide glasses known from everyday life to, e.g., metallic and polymeric glasses. Of all types of glasses, however, silicate glasses are still the most widely used type in the field of optics. Silicate glasses consist of silicon dioxide (SiO_2) as the network former. A silicate glass consisting of pure SiO_2 is called fused silica glass. These glasses are characterized by a high thermal and chemical stability combined with a high optical transparency in the ultraviolet, visible, and infrared wavelength regime [3]. Therefore fused silica glass is widely used for fibers for optical telecommunication, lenses and photomasks for microlithography [4]. Most silicate glass types further contain network modifiers introduced in the form of, e.g., soda (sodium carbonates) or lime (mostly calcium carbonates). The network modifiers are bound in the ionic state, compensating for the negative charges of the oxygen atoms of the broken Si-O bonds of the silica tetrahedron. This modification can significantly reduce the melting temperature of glass. Whereas fused silica is usually processed above 2000 °C, soda-lime glasses containing additionally sodium and calcium ions can be melted and processed at temperatures of around 1200 °C depending on the mixture. For complex optical imaging systems, optical components with varying refractive index and dispersion are highly sought-after. While some additives are known to influence the optical parameters of glasses, e.g., titanium oxide which can be used to modify the refractive index of the glasses and the thermal expansion coefficient [5], modifications to the glass composition are difficult, with the additives often leading to crystallization and phase separation. Thus, novel ways to achieve glass doping in combination with novel 3D shaping methods could significantly advance the field of glass science.

4.2 Conventional Glass Structuring

On the macroscopic scale, glasses are nowadays mainly structured using high-temperature molding or grinding and polishing processes. For the fabrication of microstructures, wet chemical or dry etching processes are used [6]. However, some alternative technologies have evolved for structuring and microstructuring of glass. Precision glass molding (PGM) has been used for structuring low melting glasses like borosilicate glasses on the macro- and microscale [7, 8]. Here, the glass is heated above the glass transition temperature but below the softening point of the glass before shaping the material by using a high-temperature mold. Structuring of high-purity glasses like fused silica glass using PGM however still remains challenging as it requires temperature-resistant molds made from glassy carbon, which have low wear resistances and are difficult to produce [9]. Laser structuring has become a powerful technology for microstructuring of glasses. Direct ablation and laser-assisted etching technologies have been used for this purpose. In direct laser ablation, the absorption of the laser radiation leads to evaporation of the glass. CO_2 lasers, vacuum UV, and ultrashort pulse lasers have been used to directly fabricate 2D microstructures in glass [6, 10, 11]. However, after evaporation, the material can condensate on adjacent structures. This often results in structures too rough for optical applications [6]. Laser-assisted etching technologies have also been used to structure 2D as well as 3D microstructures [12]. Here, an ultrashort pulse laser is scanned through the bulk of a glass and induces nanocracks along the focal point of the laser [13]. The damaged parts show a higher solubility in etchants and can be selectively removed in hydrofluoric acid. Even though the solubility is increased, this process is still controlled by diffusion and thus slows, especially in confined spaces such as suspended structures. In long channels this can lead to tapered channel geometries [14, 15]. Femtosecond laser direct writing has been used to fabricate three-dimensional objects such as whispering gallery mode optical microcavities or microfluidic channels [12, 16]. However, the relatively high surface roughness (R_q around 100–200 nm) prevents optical applications and necessitates subsequent post-processing [17].

Apart from generating glass structures from molten glass, glasses can also be produced and structured using the sol-gel process [18, 19]. Here, the glass network is formed by hydrolyzing inorganic silicate alkoxide precursors. In a typical sol-gel process, a silicon containing precursor (mostly tetraethyl orthosilicate (TEOS)) is hydrolyzed and condensates to form a gel which is then turned into a transparent glass via drying and sintering. Sol-gel technology has been used to fabricate microstructures like waveguides and thin porous membranes using soft replication processes [18]. However, the drying of the sol-gel components remains a major challenge in these processes, because large amounts of condensation products (e.g., ethanol) produced by the hydrolyzation of the alkoxides must be evaporated from the components resulting in cracks and often destruction of the components [19]. The process is therefore mainly used for thin coatings or replication of microoptical components.

4.3 Evolving Applications in Optics and Photonics

In the past decades, countless novel applications have evolved in the field of optics and photonics as well as in sensing and bioanalysis, most of which require high-precision three-dimensional structuring of optical materials [20, 21]. Three-dimensional photonic crystals, metamaterials, microstructured coupling elements, complex microlens systems, and photonic chip-scale interconnects are just a few examples [22–25]. An exemplary multi-lens objective and a chip-to-chip interconnect fabricated using two-photon direct laser writing (DLW) can be seen in Fig. 4.1.

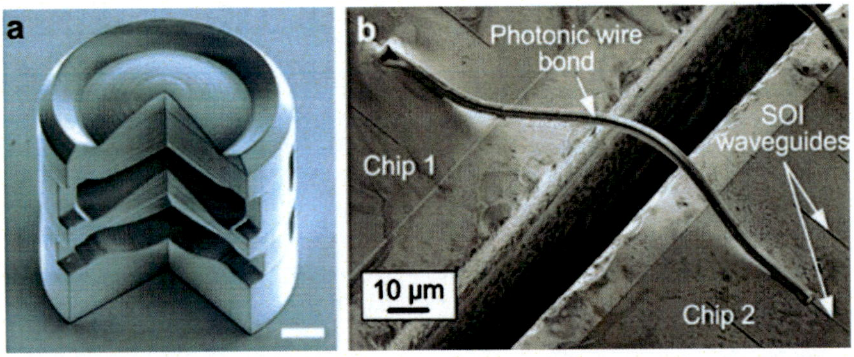

Fig. 4.1 Microoptical components 3D printed in polymeric photoresins: (**a**) multi-lens objective 3D printed using DLW (scale bar, 20 µm). (Reproduced from Gissibl et al. [23]). (**b**) Photonic chip-to-chip interconnect fabricated using DLW. (Reproduced from Lindenmann et al. [22])

Until now, similar structures are inaccessible by conventional glass structuring and are usually fabricated using polymers although, in many cases, glasses would be preferable to their superior material properties. With the advent of high-resolution Additive manufacturing and 3D printing, research and industry has employed polymers to quickly and conveniently design, prototype, optimize, and manufacture intricate structures with resolutions in the single-digit micrometer regime. Technologies like high-resolution stereolithography (SL), inkjet printing, or DLW have emerged [21, 26]. All of these processes have been developed for polymer structuring. In addition to the ease of shaping polymers using these processes, polymers offer great potential for optical applications as they allow convenient tuning of optical properties such as index of refraction or transmission giving optical design significant degrees of freedom. However, polymers suffer from a wide range of disadvantages such as material aging, limited biocompatibility, as well as limited physical, chemical, and thermal robustness. Glasses and especially high-purity glasses like fused silica are also superior to polymers in terms of their optical transparency especially in the UV and the infrared. Three-dimensional shaping of glasses with high-resolution (few micrometers) has therefore become a very active field of research.

In 2014 the first successful attempts to 3D printing of glasses have been demonstrated in the literature [27]. The attempts made since this first breakthrough then can be categorized into two major approaches: direct 3D printing of glass at elevated temperatures and indirect 3D printing using glass precursors such as sol-gel mixtures or nanocomposites. These precursors can usually be printed using 3D printing techniques, e.g., stereolithography, which were originally developed for polymer processing. The precursors are then turned into glass in a subsequent heat treatment process. As such, these processes use shaping methods originally used for structuring polymers and only yield glass after a high-temperature treatment. These are thus often referred to as *indirect glass printing*. In this chapter, we will highlight recent developments in the field of direct and indirect 3D printing of transparent glasses starting from the early approaches to glass printing.

4.4 First Trials for 3D Printing of Glass

Although transparent glasses are one of the most important materials for optics, this class of materials has until recently been an idle bystander in the 3D printing revolution. However many different attempts have been proposed in the past to 3D printing transparent glasses. Different 3D printing techniques such as selective laser sintering (SLS) or selective laser melting (SLM) as well as inkjet printing of glass particle suspensions have been proposed. Klocke et al. were among the first to sinter borosilicate glass powders using SLS in 2004 [28], while Luo et al. studied SLM of soda-lime glass powders in 2014 [29]. However, both processes used micron-sized glass particles which did not allow full densification of the glass powders and thus only yielded porous white and non-transparent glass components (see Fig. 4.2a). The same problem occurred with sintered parts printed from suspensions of micron-sized glass particles using inkjet printing (see Fig. 4.2b) [30]. The reason for the loss in transparency is the incomplete densification of the glass powders, which results in a porous component with increased light scattering at the pore/material interfaces.

4.5 Direct 3D Printing of Transparent Glass

The first successful direct approach to 3D printing of transparent glass was presented by Klein et al. [27] using a modified fused deposition modeling (FDM) printer. A low-temperature melting soda-lime glass powder was heated to a temperature of around 1040 °C, and the melt was extruded through a nozzle (see

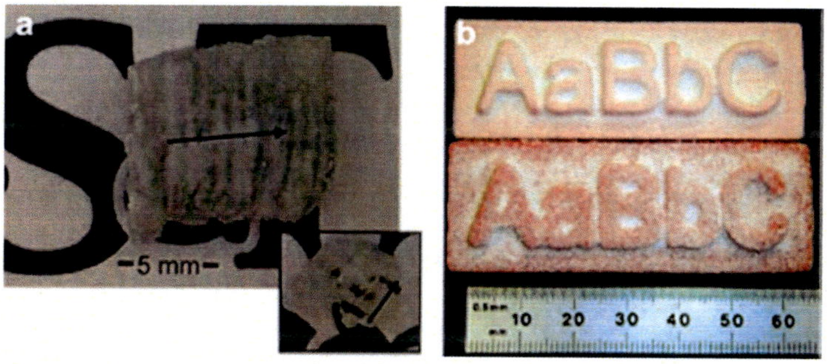

Fig. 4.2 First attempts to 3D printing of glasses: (**a**) selective laser melting of borosilicate glass powders. (Reproduced from Luo et al. [29]). (**b**) Inkjet printing of micron-sized glass particles in a binder matrix and subsequent thermal debinding and sintering. The process results in a porous nontransparent glass part. (Reproduced from Bourell et al. [30])

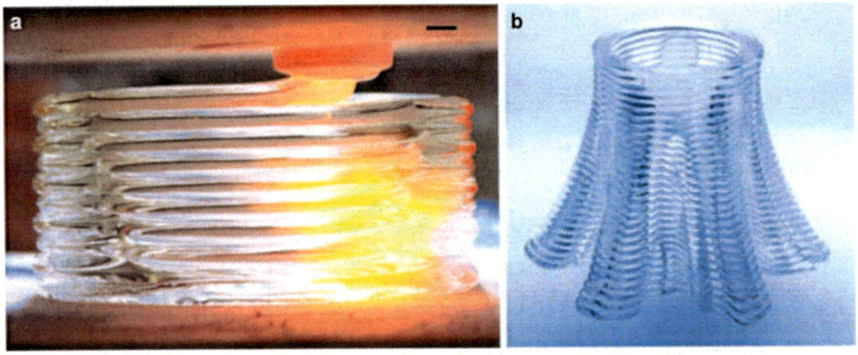

Fig. 4.3 Direct fused deposition modeling (FDM) printing of transparent glass: (**a**) a low melting soda-lime glass powder is melted at a temperature of around 1040 °C and the melt extruded through a nozzle (scale bar, 5 mm). (**b**) Object printed using fused deposition modeling at high-temperatures. (Reproduced from Klein et al. [27])

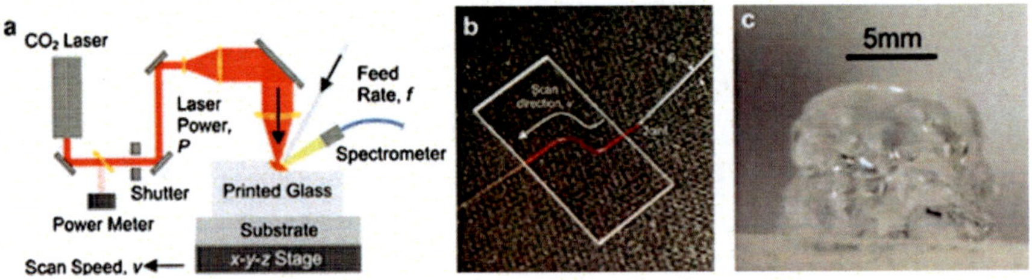

Fig. 4.4 Direct printing of transparent glass using fibers: (**a**) Principle of glass printing via glass fiber melting. (Reproduced from Luo et al. [32]). (**b**) Simple photonic glass fiber waveguide deposited on a glass slide by locally melting the fiber with a CO_2 laser. (Reproduced from Johnson et al. [33]). (**c**) A cube printed using glass fiber melting. The strong geometric deviations during the printing process limit the dimensional accuracy. (Reproduced from Luo et al. [32])

Fig. 4.3a) [27]. This process has been commercialized by the company micron3Dp which modified the process to achieve layer thicknesses of a few 100 μm [31]. However, the process still fabricates parts with resolutions insufficient for optical applications. Due to their high melting points, this direct 3D printing process has only been used for the fabrication of soda-lime glasses. High-purity glasses like fused silica glass cannot be processed using this method due to their high melting temperature of around 2000 °C.

Another direct 3D printing approach has been described by Luo et al. where a glass fiber with a minimum diameter of 100 μm was manually fed into the printing area and melted using a CO_2 laser beam [29, 32, 33]. The process can be seen in Fig. 4.4a. Using this process simple parts have been fabricated in borosilicate or fused silica glass such as a single optical fiber attached on top of a glass substrate (see Fig. 4.4b) [33]. However due to strong thermal variations as a consequence of the uncontrolled heat distribution, the dimensional accuracy suffers. Figure 4.4c shows an exemplary object designed to be a cube, which resulted in an object with only little geometric fidelity.

Until 2017, all direct 3D printing processes resulted in components with resolutions below the requirements for most technical and optical applications. In addition, all processes demonstrated are performed at elevated temperatures requiring special printing equipment.

4.6 Indirect 3D Printing of Transparent Glass

We have recently developed the first indirect high-resolution 3D printing approach to fabricating and structuring high-quality transparent fused silica glass. The process is shown in Fig. 4.5. We have developed liquid nanocomposites (called *Glassomers*) which contain silica nanoparticles (up to 60 vol%) in a photocurable binder matrix [35–37]. These nanocomposites can be printed in benchtop stereolithography printers in a layer-by-layer fashion. The printing process results in a polymeric nanocomposite, the so-called green part. The green part is thermally debound at a temperature of 600 °C, whereby the polymeric binder is completely removed, leaving the so-called brown part. During the final sintering process, the powder is densified at a temperature of 1300 °C to a fully dense and highly transparent fused silica glass. During the sintering process, the part shrinks isotropic in dependence of the solid loading. The sintered glass parts show the same chemical and physical properties like commercial fused silica glass. They show the same high optical transparency in the UV, visible and infrared wavelength region combined with the same high Vickers hardness, bending strength, wetting properties, as well as thermal and chemical stability [34, 36, 37]. These nanocomposites have recently been commercialized by the company *Glassomer GmbH.*

Using a commercially available benchtop 3D printer, structures like the pretzel shown in Fig. 4.6a can be printed. To achieve even higher resolution, microstereolithography (μSL) printers can be used, which allow structuring the Glassomer

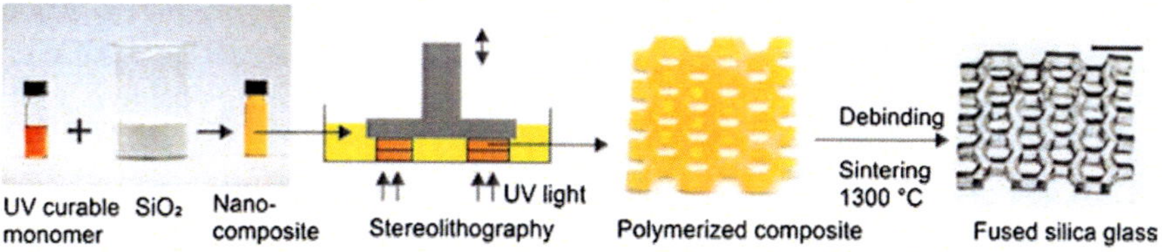

Fig. 4.5 3D printing of transparent fused silica glass using the Glassomer nanocomposites: the nanocomposites consist of fused silica glass nanoparticles in a photocurable binder matrix. These nanocomposites can be printed using high-resolution (micro-)stereolithography printers resulting in polymeric nanocomposites (so-called green part). The binder is removed during a subsequent thermal debinding process at 600 °C and sintered to a transparent fused silica glass at 1300 °C. (Reproduced from Kotz et al. [34])

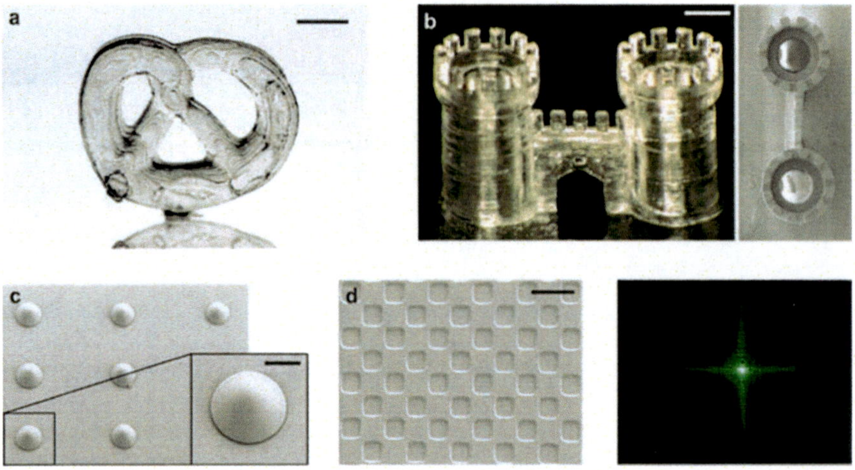

Fig. 4.6 Indirect 3D printing of transparent fused silica glass using Glassomer: (**a**) Glassomer can be printed using benchtop stereolithography printers. A printed and sintered fused silica glass pretzel (scale bar, 5 mm). (**b**) Glassomer nanocomposites can be printed with tens of micron resolution using microstereolithography (μSL). A sintered castle gate printed using μSL (scale bar, 270 μm). (**c**) Optical lenses can be fabricated using grayscale lithography (scale bar, 100 μm). (**d**) Sintered binary diffractive optical element fabricated using direct lithography (scale bar, 100 μm) shown with the optical diffraction pattern (illuminated with a green laser). (Reproduced from Kotz et al. [34])

nanocomposites with lateral resolutions of tens of micrometers. Figure 4.6b shows an exemplary castle gate with pinnacles of 80 μm in width.

Due to its layer-by-layer nature, (μ)SL always results in parts with the so-called staircase defects, showing the individual layers of the print. Although these layers only have a height of a few 10 μm, this is a major problem for many optical applications since the layers reduce the optical transparency along the printing direction. One advantage of the Glassomer nanocomposites is that they can also be used as so-called grayscale resins [34]. Grayscale lithography is a fabrication method whereby a photoactive material is exposed to a light pattern with a gradient in light intensity. Since the cure depth depends on the exposure dose (the product of light intensity and exposure duration), a variation in the exposure dose can be used to generate surface reliefs such as microoptical lenses or microfluidic channel structures [38, 39]. Figure 4.6c shows a fused silica glass microoptical lens array which has been fabricated using grayscale lithography of Glassomer nanocomposites and subsequent thermal debinding and sintering. As can be seen, the lenses show a very smooth surface and do not possess the staircase defects known from stereolithography printing. Besides high-resolution printing, Glassomer nanocomposites can be used as a binary negative photoresist. In this method one layer of the resist is exposed to a binary light pattern resulting in a localized polymerization. An exemplary microoptical diffractive structure with a lateral feature size of 50 μm and a height of 1.5 μm fabricated using direct lithography can be seen in Fig. 4.6d. It further shows the diffractive projection pattern that appears when the structure is illuminated with a green laser. The parts fabricated via direct lithography as well as the top surfaces of the parts printed using μSL showed a surface roughness R_q of around 2 nm, suitable for applications in optics and photonics [34].

Stereolithography printing of Glassomer nanocomposites has been recently used to fabricate optical fiber step-index core preforms [40]. The printed preform was thermally debinded and drawn as the brown part on a commercial drawing tower at a temperature of 1855 °C resulting in an optical fiber with a length of 2.3 km (see Fig. 4.7). UV casting of Glassomer nanocomposites has been recently further used for the reduction of losses in solar modules [41]. Micro-cone textures were used to reduce the front-side reflection losses by ~80% compared to a planar reference. Furthermore embedded freeform surface cloaks which redirect the light from the metallic contact to the solar cell could achieve a cloaking efficiency of up to ~88% [41].

Besides arbitrary three-dimensional freestanding structures, the fabrication of complex hollow microstructures inside fused silica glass is of high interest for the miniaturization of chemical synthesis reactors as well as for the fabrication of, e.g., hollow waveguides in optics and photonics [17, 42]. However the fabrication of hollow microstructures via conventional SL printing is challenging since uncured material remains inside the small channel structures and is partially cured during the printing process [43]. The risk for rough channel walls and even blocked channels is significantly increased when channel dimensions are reduced to a few tens of micrometers. A combination of the Glassomer process and high-resolution 3D printing using DLW of polymeric templates allowed, for the first time, generating truly arbitrary hollow microstructures of centimeter length in bulk fused silica glass [44]. In this process, termed sacrificial template replication (STR), the polymeric template is embedded in the liquid Glassomer nanocomposite, which is consecutively polymerized. The embedded template results in the inverse glass microcavity after thermal debinding and sintering. Figure 4.7a shows a polymeric template which has been fabricated using DLW and the resulting glass microchannels after the thermal heat treatment process. The roughness of the embedded microchannel depends on the roughness of the template. Using templates fabricated by DLW,

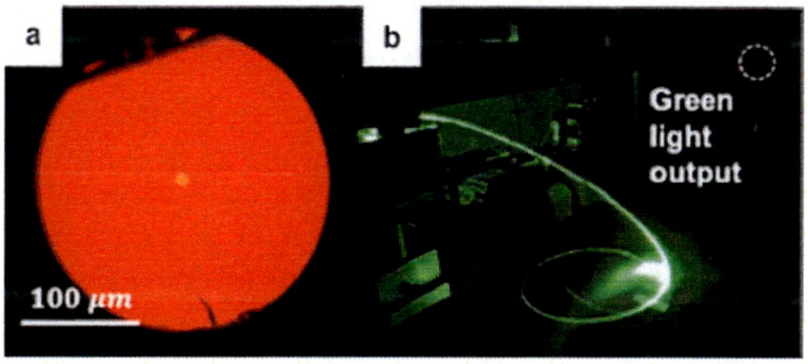

Fig. 4.7 3D printing of optical preforms using the Glassomer process: (**a**) cross-section of a 3D-printed and drawn multimodal fiber. (**b**) Image of the transmission setup of the printed and drawn optical fiber. (Reproduced with permission from Chu et al. [40])

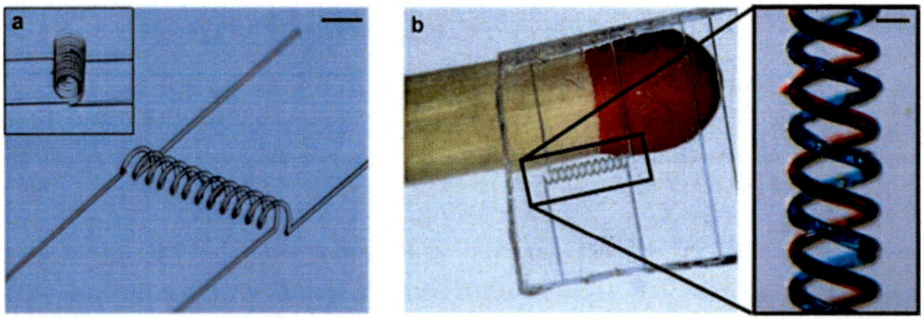

Fig. 4.8 Sacrificial template replication: (**a**) 3D-printed polymeric template structure fabricated using DLW (scale bar, 900 μm) is embedded in Glassomer. (**b**) Resulting inverse intertwined microfluidic glass structure after thermal debinding and sintering. The channels were filled with a red and a blue dye (scale bar, 280 μm). (Reproduced from Kotz et al. [44])

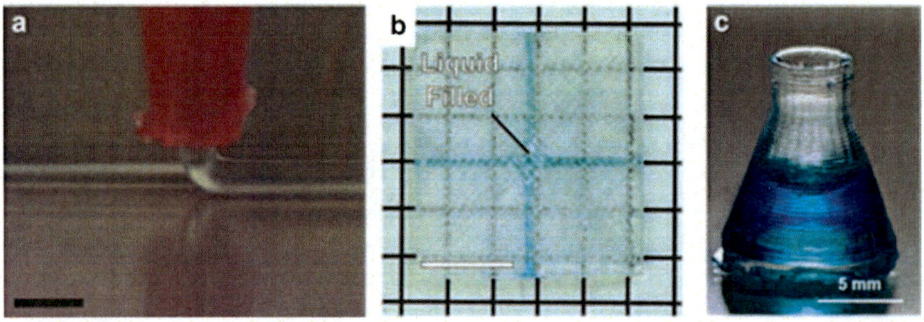

Fig. 4.9 Alternative indirect glass 3D printing approaches: (**a**) a silica-containing ink is extruded through a nozzle (scale bar, 500 μm). (**b**) A microfluidic cross-section with a channel width of 400 μm (scale bar, 4 mm). The channel was filled with dyed water. (a/b reproduced from Nguyen et al. [45]). (**c**) Erlenmeyer flask printed via SL in using photocurable sol-gel precursors. The flask was filled with dyed water. (Reproduced from Cooperstein et al. [46])

microchannels with a surface roughness of R_a ~ 20 nm have been fabricated (Fig. 4.8).

As an alternative technology for indirect 3D printing of transparent silica glasses, direct ink writing of colloidal silica sols and sol-gel based approaches have been described [45, 46]. In direct ink writing, objects are prepared by a filament-by-filament assembly of shear thinning inks which are extruded through a nozzle (see Fig. 4.9a). For the fabrication of fused silica glass, silica nanoparticles have been dispersed in a liquid consisting of tetraethylene glycol dimethyl ether [45]. An alternative approach to 3D print fused silica glass via stereolithography printing was described by Cooperstein et al. [46]. The authors developed a photocurable ink based on sol-gel solutions containing TEOS and a photocurable acryloxypropyl trimethoxysilane which was printed using SL. Both direct ink writing of colloidal sols and the SL printing of sol-gel precursors are converted into fused silica glass by a subsequent heat treatment.

4.7 3D Printing of Multicomponent Silicate Glasses

The addition of metal ions to the glass structure can alter the optical and thermal properties of glasses. For indirect printing of glass based on nanocomposites, the optical properties of printed Glassomer glass can be adjusted by doping the brown part with alcoholic solutions of metal salts. Chromium(III) nitrate and vanadium(III) chloride resulted in a green or blue ionic colored glass. Doping with gold(III) chloride was used to generate gold nanoparticles inside the glass resulting in a colloidal red coloring (see Fig. 4.10a) [34]. Using this approach the absorption properties of the glasses can be tuned as is confirmed by the UV/Vis spectra of the doped glasses shown in Fig. 4.10a. Such doped glasses can be used, e.g., for the fabrication of tailored optical filter components. This doping concept has been further extended to photoluminescent glasses, by doping the brown part with alcoholic solutions containing rare-earth elements Eu^{3+}, Tb^{3+}, and Ce^{3+} (see Fig. 4.10b) [47]. These glass dopants are of interest for optical temperature sensors or laser applications [48, 49]. Examples of such SL printed, doped, and sintered photoluminescent glasses are shown in Fig. 4.10b.

For indirect printing of glass using sol-gel precursors, direct ink writing has been extended to printing of SiO_2-TiO_2 glasses by using adjusted sol-gel feedstocks [50]. SiO_2-TiO_2 glasses have been used in the field of optics and photonics, e.g., for the fabrication of waveguides or gradient index optics [51–53]. These glasses are especially interesting since TiO_2 increases the refractive index and can be further used to reduce the thermal expansion. The printed glasses showed an increase in refractive index similar to commercial TiO_2-SiO_2 glasses. The reduction of the thermal expansion has so far not been demonstrated. Direct ink writing has been furthermore used to fabricate SiO_2-GeO_2 glasses resulting in an increase of the refractive index compared to pure fused silica glass [54].

Moore et al. have presented a modified sol-gel route for the fabrication of porous as well as dense multicomponent glasses in the SiO_2-B_2O_3-P_2O_5 system. Phase separating resins were developed by dispersing sol-gel precursors in an acrylic binder

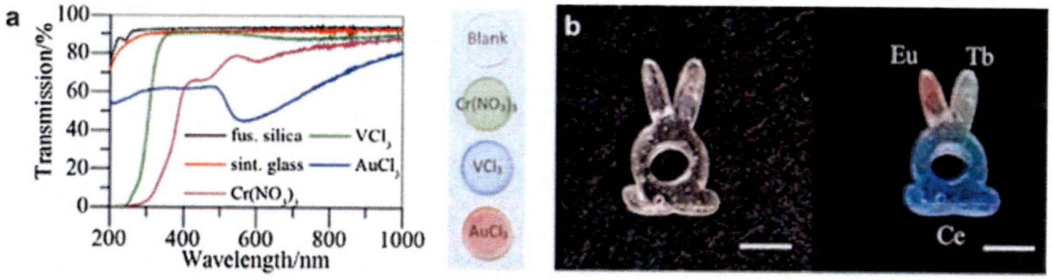

Fig. 4.10 3D printing of colored glasses: (**a**) doping of Glassomer parts with alcoholic solutions of metal salts can be used to fabricate colored glasses. Gold(III) chloride results in a colloidal red color. Vanadium(III) chloride and Chromium(III) nitride results in an ionic blue and green color, respectively. (Reproduced from Kotz et al. [34]). (**b**) Doping of Glassomer with Eu^{3+}, Tb^{3+}, and Ce^{3+} ions to fabricate photoluminescent glasses. The structure under natural light (left) and under illumination by a 254 nm UV lamp (scale bars, 5 mm). (Reproduced from Liu et al. [47])

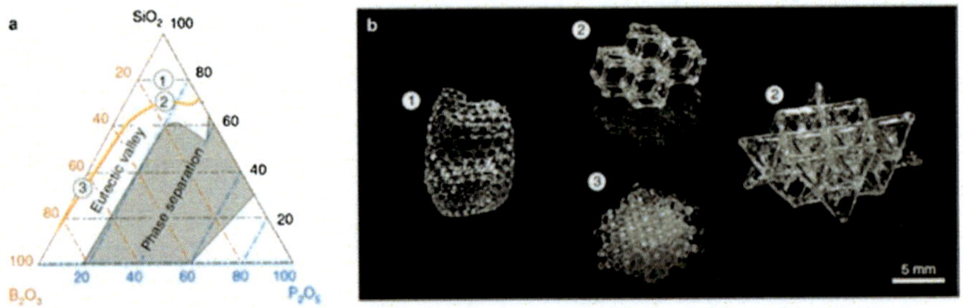

Fig. 4.11 3D printing of multicomponent SiO_2-B_2O_3-P_2O_5 glasses using phase separating resins: (**a**) phase diagram of the SiO_2-B_2O_3-P_2O_5 system. (**b**) 3D printing of phase separating sol-gel precursors after sintering to a multicomponent glass. The compositions are shown in the phase diagram in **a**. (Reproduced with permission from Moore et al. [55])

matrix. Using this process porous glass with an adjustable porosity between a few 100 nm and 1 μm could be printed. Furthermore, the nanoporous objects could be sintered to nearly full density of around 98% resulting in a transparent multicomponent glass with variable composition (see Fig. 4.11).

4.8 Comparison of Indirect Glass 3D Printing Methods

In the following, a comparison of the existing indirect 3D printing methods for the fabrication of glasses in terms of their suitability for optics will be given. One major concern for optical parts lies in the achievable in-plane and out-of-plane resolution of the underlying printing process as well as the resulting roughness. Direct ink writing is capable of printing filaments with a size of a few hundred micrometers. As can be seen from the fluidic chip in Fig. 4.9b, the single filaments can be clearly seen in the final component and significantly reduce the optical transparency. Photocurable sol-gel precursors have so far been printed using SL with a resolution of around 600 μm. The highest lateral resolutions of around 50 μm and a layer thickness of 10 μm have so far been demonstrated using the Glassomer nanocomposites in combination with μSL. Minimum feature sizes of around 7 μm were achieved by the STR approach in combination with DLW. For optical applications, another major issue of the 3D printing process is the visibility of the layers along the z-axis. In order to prevent this, photocurable nanocomposites as well as photocurable sol-gel precursors are promising materials since they have the potential to be used in high-resolution printers such as continuous liquid interface production (CLIP) or DLW. Herein the layers can be reduced into the submicron regime resulting in smooth surfaces [56–58].

A further important aspect when comparing indirect 3D printing processes lies in the subsequent thermal heat treatment. All indirect 3D printing methods involve thermal debinding or drying steps as well as a subsequent sintering process. One significant aspect when comparing these processes is the linear shrinkage which the parts exhibit during this heat treatment process. The shrinkage during the heat treatment depends on the amount of silica, the so-called solid loading in the resins. If the shrinkage is isotropic, the linear shrinkage Y_s can be calculated in dependence of the solid loading Φ, the theoretical density ρ_t, and the final density ρ using the following equation:

$$Y_s = 1 - \left(\frac{\phi}{\frac{\rho}{\rho_t}} \right)^{1/3} \tag{4.7.1}$$

A high solid loading of the precursor materials is especially important for the fabrication of monolithic bulk glass components. A reduction of the amount of binder will reduce the risk of part damage during the thermal debinding or drying process. A high solid loading and low shrinkage is further advantageous to reduce the risk of part bending during the sintering process [59].

A comparison of the achievable solid loadings and the times required for the heat treatment of the different indirect 3D printing techniques can be found in Table 4.1. The solid loading of the colloidal sols used in direct ink writing has been reported to be between 10 and 20 vol% resulting in a high linear shrinkage of up to 50% [45]. The shrinkage of the photocurable sol-gel precursor has been reported to be 40–56% in the x-y plane and 33–48% in the z-direction [46]. This nonlinear shrinkage

Table 4.1 Comparison of relevant methods for high-resolution 3D printing of glass. (Reproduced from Kotz et al. [42])

Method	Resolution	Solid loading [vol%]	Linear shrinkage	Processing time post treatment	Literature
Direct glass 3D printing					
Fused deposition modeling	>1 mm	–	–	–	[60, 61]
Laser melting of glass fibers	>1 mm	–	–	–	[29]
Indirect glass 3D printing					
Stereolithography with nanocomposites	60 µm	40–60 vol%	Isotropic 16–26%	~46 h	[34, 37]
Stereolithography with sol-gel precursor	~600 µm	n.a.	Non-isotropic XY ~ 40%–56% Z ~ 33%–48%	>7 days	[46]
Stereolithography with phase-separating resins	~1 mm	n.a.	~60%	~75 h	[55]
Direct ink writing with colloidal sol	~200 µm	10–20 vol%	Isotropic 41–53%	112 h	[45]

further complicates the designing of the components. The phase-separating sol-gel resins showed a linear shrinkage of even 60% [55]. The Glassomer nanocomposites on the other hand have been demonstrated to achieve solid loadings of up to 60 vol% resulting in a linear shrinkage of only 16% [37].

4.9 Outlook

The convergence of glass material formulation and novel printing methodologies will enable next-generation applications in optics and photonics. Tuning optical properties, e.g., index of refraction, color, or dispersion, will allow developing novel glass compositions in nearly arbitrary shapes. The manufacturing of tailored glass formulations with adjustable optical properties will lead to a paradigm shift in the field of optics and photonics, which until now is mainly limited to a handful of commercially available glass types.

On the other hand, manufacturing techniques evolve further, and the possibility to combine techniques developed predominantly for polymer processing will become available for the generation of high-resolution glass components. This amalgamation of techniques will thus boost the application scope of precision glass components as the manufacturing techniques of glasses and polymers will converge towards the same technology platform. Enhancements which, until recently, were exclusive to polymer processing, e.g., CLIP or DLW, could thus become available for glass processing [56]. Applications, which will directly profit from this next-generation process scope, are, among others, integrated optics, solar cells, optical data science, light conversion, as well as medical gear and biotechnology.

References

1. Bach, H., & Neuroth, N. (2012). *The properties of optical glass*. Berlin: Springer Science & Business Media.
2. Zanotto, E. D., & Mauro, J. C. (2017). *Journal of Non-Crystalline Solids, 471*, 490.
3. Bansal, N. P. & Doremus, R. H. (2013). *Handbook of glass properties* (p. 680). Orlando: Elsevier.
4. Ikushima, A., Fujiwara, T., & Saito, K. (2000). *Journal of Applied Physics, 88*, 1201.
5. Deng, Z., Breval, E., & Pantano, C. G. (1988). *Journal of Non-Crystalline Solids, 100*, 364.
6. Hülsenberg, D., Harnisch, A., & Bismarck, A. (2005). *Microstructuring of glasses* (Materials science) (Vol. 87, 1st ed., p. 323). Berlin: Springer.
7. Takahashi, M., Murakoshi, Y., Maeda, R., & Hasegawa, K. (2007). *Microsystem Technologies, 13*, 379.
8. Takahashi, M., Sugimoto, K., & Maeda, R. (2005). *Japanese Journal of Applied Physics, 44*, 5600.
9. Klocke, F., Dambon, O., Liu, G., & Dukwen, J. (2016). *Production Engineering, 10*, 367.

10. Herman, P. R., Beckley, K. R., Jackson, B. C., Kurosawa, K., Moore, D., Yamanishi, T., & Yang, J. (1997). In Excimer lasers, optics, and applications (International Society for Optics and Photonics (pp. 86).
11. Gattass, R. R., & Mazur, E. (2008). *Nature Photonics, 2*, 219.
12. Lin, J., Yu, S., Ma, Y., Fang, W., He, F., Qiao, L., Tong, L., Cheng, Y., & Xu, Z. (2012). *Optics Express, 20*, 10212.
13. Hnatovsky, C., Taylor, R., Simova, E., Rajeev, P., Rayner, D., Bhardwaj, V., & Corkum, P. (2006). *Applied Physics A, 84*, 47.
14. Kiyama, S., Matsuo, S., Hashimoto, S., & Morihira, Y. (2009). *The Journal of Physical Chemistry C, 113*, 11560.
15. Sugioka, K., & Cheng, Y. (2013). *Femtosecond laser 3D micromachining for microfluidic and optofluidic applications*. Berlin: Springer Science & Business Media.
16. Marcinkevičius, A., Juodkazis, S., Watanabe, M., Miwa, M., Matsuo, S., Misawa, H., & Nishii, J. (2001). *Optics Letters, 26*, 277.
17. He, F., Lin, J., & Cheng, Y. (2011). *Applied Physics B: Lasers and Optics, 105*, 379.
18. Marzolin, C., Smith, S. P., Prentiss, M., & Whitesides, G. M. (1998). *Advanced Materials, 10*, 571.
19. Kajihara, K. (2013). *Journal of Asian Ceramic Societies, 1*, 121.
20. Lambert, A., Valiulis, S., & Cheng, Q. (2018). *ACS Sensors, 3*, 2475.
21. Camposeo, A., Persano, L., Farsari, M., & Pisignano, D. (2019). *Advanced Optical Materials, 7*, 1800419.
22. Lindenmann, N., Balthasar, G., Hillerkuss, D., Schmogrow, R., Jordan, M., Leuthold, J., Freude, W., & Koos, C. (2012). *Optics Express, 20*, 17667.
23. Gissibl, T., Thiele, S., Herkommer, A., & Giessen, H. (2016). *Nature Photonics, 10*, 554.
24. Dietrich, P.-I., et al. (2018). *Nature Photonics, 12*, 241.
25. Deubel, M., Von Freymann, G., Wegener, M., Pereira, S., Busch, K., & Soukoulis, C. M. (2004). *Nature Materials, 3*, 444.
26. Waldbaur, A., Rapp, H., Lange, K., & Rapp, B. E. (2011). *Analytical Methods, 3*, 2681.
27. Klein, J., et al. (2015). 3D printing and Additive manufacturing 2, 92.
28. Klocke, F., McClung, A., & Ader, C. (2004). In *Solid freeform fabrication symposium proceedings, Austin, TX* (pp. 3).
29. Luo, J., Pan, H., & Kinzel, E. C. (2014). *Journal of Manufacturing Science and Engineering, 136*, 061024.
30. Bourell, D., Stucker, B., Marchelli, G., Prabhakar, R., Storti, D., & Ganter, M. (2011). *Rapid Prototyping Journal, 17*, 187.
31. Gal-Or, E., et al. (2019). *Analytical Methods, 11*, 1802.
32. Luo, J., Gilbert, L. J., Bristow, D. A., Landers, R. G., Goldstein, J. T., Urbas, A. M., & Kinzel, E. C. (2016). In SPIE LASE (International Society for Optics and Photonics), p. 97380Y.
33. Johnson, J. E., et al. (2019). In Laser 3D manufacturing VI (International Society for Optics and Photonics), p. 109090Q.
34. Kotz, F., et al. (2017). *Nature, 544*, 337.
35. Kotz, F., Plewa, K., Bauer, W., Hanemann, T., Waldbaur, A., Wilhelm, E., Neumann, C., & Rapp, B. E. (2015). In SPIE BiOS (International Society for Optics and Photonics), p. 932003.
36. Kotz, F., et al. (2016). *Advanced Materials, 28*, 4646.
37. Kotz, F., et al. (2018). *Advanced Materials, 30*, 1707100.
38. Totsu, K., Fujishiro, K., Tanaka, S., & Esashi, M. (2006). *Sensors and Actuators A: Physical, 130*, 387.
39. Atencia, J., Barnes, S., Douglas, J., Meacham, M., & Locascio, L. E. (2007). *Lab on a Chip, 7*, 1567.
40. Chu, Y., Fu, X., Luo, Y., Canning, J., Tian, Y., Cook, K., Zhang, J., & Peng, G.-D. (2019). *Optics Letters, 44*, 5358.
41. Langenhorst, M., et al. (2019). *ACS Applied Materials & Interfaces, 11*, 35015.
42. Kotz, F., Risch, P., Helmer, D., & Rapp, B. E. (2019). *Advanced Materials*, 1805982.

43. Kotz, F., Helmer, D., & Rapp, B. (2018). In Microfluidics, BioMEMS, and medical microsystems XVI (International Society for Optics and Photonics), p. 104910A.
44. Kotz, F., et al. (2019). *Nature Communications, 10*, 1439.
45. Nguyen, D. T., et al. (2017). *Advanced Materials, 29*, 1701181.
46. Cooperstein, I., Shukrun, E., Press, O., Kamyshny, A., & Magdassi, S. (2018). *ACS Applied Materials & Interfaces, 10*, 18879.
47. Liu, C., Qian, B., Ni, R., Liu, X., & Qiu, J. (2018). *RSC Advances, 8*, 31564.
48. Li, C., Dong, B., Li, S., & Song, C. (2007). *Chemical Physics Letters, 443*, 426.
49. Malyarevich, A. M., Denisov, I. A., Yumashev, K. V., Dymshits, O. S., Zhilin, A. A., & Kang, U. (2001). *Applied Optics, 40*, 4322.
50. Destino, J. F., et al. (2018). *Advanced Materials Technologies*, 1700323.
51. Brusatin, G., Guglielmi, M., Innocenzi, P., Martucci, A., Battaglin, G., Pelli, S., & Righini, G. (1997). *Journal of Non-Crystalline Solids, 220*, 202.
52. Kawachi, M., Yasu, M., & Edahiro, T. (1983). *Electronics Letters, 19*, 583.
53. Shingyouchi, K., & Konishi, S. (1990). *Applied Optics, 29*, 4061.
54. Sasan, K., et al. (2020). *ACS Applied Materials & Interfaces*.
55. Moore, D. G., Barbera, L., Masania, K., & Studart, A. R. (2020). *Nature Materials, 19*, 212.
56. Tumbleston, J. R., et al. (2015). *Science, 347*, 1349.
57. Janusziewicz, R., Tumbleston, J. R., Quintanilla, A. L., Mecham, S. J., & DeSimone, J. M. (2016). *Proceedings of the National Academy of Sciences, 113*, 11703.
58. Kawata, S., Sun, H.-B., Tanaka, T., & Takada, K. (2001). *Nature, 412*, 697.
59. Clasen, R., & Oetzel, C. (2009). *Key Engineering Materials, 412*, 45.
60. Zhu, H., Holl, M., Ray, T., Bhushan, S., & Meldrum, D. R. (2009). *Journal of Micromechanics and Microengineering, 19*, 065013.
61. Nagarah, J. M., & Wagenaar, D. A. (2012). *Journal of Micromechanics and Microengineering, 22*, 035011.

Chapter 5
Industrial-Scale Fabrication of Optical Components Using High-Precision 3D Printing: Aspects-Applications-Perspectives

B. Stender, W. Mantei, J. Wiedenmann, Y. Dupuis, F. Hilbert, R. Houbertz, M. von Edlinger, C. Kistner, and J. Koeth

Abstract 3D lithography by means of two- or multiphoton absorption (TPA/MPA) as a special case for direct laser writing has gained considerable attention, particularly in the last two decades. It offers additive, subtractive, and special glass processes. Especially in academics, the possibility to create 3D objects has resulted in many different studies for applications from photonics and microoptics to life science. A special structuring mode in 3D lithography is high-precision 3D printing which is intensely researched all over the world. The combination of high-performance materials with sophisticated fabrication strategies results in a significant reduction of the process time. This leads to a pronounced acceptance of high-precision 3D printing beyond the field of pure research and nowadays into production.

This chapter describes the principles of high-precision 3D printing, with the wide range of materials that can be processed being introduced. It is shown how the subwavelength resolution of the fabrication process enables the manufacturing of not only refractive and diffractive optics but also metaoptics. Individual elements from the sub-μm range to the millimeter range are feasible, and the versatile manufacturing strategies enabled by high-precision 3D printing are discussed. Possibilities to significantly scale up the production are highlighted, among them contouring, beam splitting, and others. Examples are given for the direct use of printed optics in applications, but also for using the printed parts as master for replication.

Keywords High-precision 3D printing · 3D lithography · Microtechnology Hybrid polymers · Microoptics · Hybrid processing · Gas sensors · NIR lasers Nanometer precision · Low cost · Green photonics · Resource-saving

B. Stender (✉) · W. Mantei · J. Wiedenmann · Y. Dupuis · F. Hilbert · R. Houbertz
Multiphoton Optics GmbH, Würzburg, Germany
e-mail: benedikt.stender@multiphoton.de

M. von Edlinger · C. Kistner · J. Koeth
Nanoplus Nanosystems and Technologies GmbH, Gerbrunn, Germany

A. Heinrich (ed.), *3D Printing of Optical Components*, Springer Series in Optical Sciences 233, https://doi.org/10.1007/978-3-030-58960-8_5

5.1 Introduction

3D lithography by means of two- or multiphoton absorption (TPA/MPA) has gained considerable attention, particularly in the last two decades. Since its invention and first experimental proof by Maruo et al. [1] and Sun et al. [2], respectively, 3D lithography as a special case of direct laser writing (DLW) or high-precision 3D printing (HP3DP) has been used by many groups all over the world [3, 4]. Especially in academics, the possibility to create 3D objects has resulted in many different studies for applications in photonics [5–8], microoptics [9–16], microfluidics [13, 17], and life science [13, 18–21]. Pioneering works were reported in literature on the development of specially designed materials which were implemented in photonic crystal structures [22], microoptical elements [23], fabrication of optical waveguides for academic [24] purposes and on an industrial scale [6, 25–27], scaffolds for tissue engineering [18], and drug delivery structures [18, 21, 28, 29]. These works also have comprised parallel fabrication of individual structures [30–35] to increase the fabrication speed and to allow even more applications in various industrial sectors.

Aside from individual structure generation, 3D lithography with its integral part of high-precision 3D printing has been proven as a tool for the fabrication of master tools for replication purposes. These structures can be replicated either individually or in a step-and-repeat process, i.e., an array of the desired structures is replicated step-by-step by moving the master tool across the substrate's surface. Its inherent 3D capability enables the fabrication of components with unique designs and integration of different functions, particularly if multifunctional materials are involved. This enables mass manufacturing of sophisticated components which, for example, allow to create ultra-flat optics to a miniaturization level which has not been demonstrated before.

This distinguishes HP3DP from conventional polymer 3D printing or additive manufacturing technologies such as stereolithography (SLA) [36–38], Digital Light Processing [39–41] (DLP) which is an advanced version of SLA, inkjet printing [42, 43], or fused deposition modeling (FDM) [44, 45]. What all these technologies have in common is that they are working on a layer-by-layer basis, meaning that a two-dimensional workflow is subsequently repeated over and over again to create a three-dimensional structure. Accordingly, this results in certain limitations with respect to freedom in design as free-standing elements cannot be realized or need implementation of special supporting structures to prevent a sinking due to low stability with follow-up processes to create smooth surface finish. However, most of the demonstrations reported so far in high-precision 3D printing rely on the same procedure: structures are formed layer by layer, not using the tremendous advantage of the intrinsic 3D capability of this special fabrication process. In this chapter, we will particularly validate the use of real 3D fabrication, i.e., no layer stacking, but orchestrating the writing laser beam in any direction to form the 3D structures, thus creating an extraordinary user experience for designers and engineers, driving the developments much farther than ever thought of.

Aside from this intrinsic advantage of HP3DP, it has unique advantages over any other 3D printing technology. High-precision 3D printing is particularly suited for directly printing optics on any object or shape with an excellent surface quality. The precision can be simply tuned, enabling optical elements or components which can work from the IR down to the blue wavelength regime. Additionally, the process of HP3DP as it will be demonstrated in this book chapter is independent on the substrate, i.e., HP3DP can be applied to any surface, assembly, or conventionally printed structures, even for very thick or large-scale substrates. It thus can be integrated into any already existing workflow and allows 3D hybrid manufacturing in many different ways with tunable precision. One feature of 3D hybrid manufacturing is the integration of other technologies and processes into the workflow of structure or part formation. Conventional processes on wafer scale such as e-beam writing, UV and grayscale lithography, replication, or processes such as injection molding and inkjet printing can be combined with high-precision 3D printing and conventional 3D printing technologies. Since HP3DP provides scalability from the sub-micrometer to the centimeter range, it thus closes the gap between nano- and microfabrication and large-scale conventional lithography and printing technologies.

This chapter is organized as follows. The possibilities of hybrid manufacturing are briefly explained in Sect. 5.2 and demonstrated by selected examples for different material classes in Sect. 5.3. In Sect. 5.3, selected materials are introduced which are used in HP3DP. Most of these materials have multiple functions integrated on a molecular level and are referred to as "n in 1" materials, allowing the orchestration of the light-induced fabrication process. Orchestration of the laser beam means a real 3D writing process, where 3D structures are fabricated by movements of the laser beam in any direction to form the 3D structures and not just by layer-to-layer stacking.

Section 5.4 introduces general aspects of high-precision 3D printing and different manufacturing strategies, and examples of the high resolution of high-precision 3D printing will be given for selected materials. The production environment will be introduced for 3D hybrid manufacturing which provided significant advantages in processing. Finally, scaling to industrial-scale throughput will be described, supplying a deeper insight into the manufacturing strategies. The transition from micro- to macro optics will be shown with selected examples. HP3DP provides scalability from the sub-micrometer to the centimeter range and thus closes the gap between nano- and microfabrication, standard manufacturing methods for optics, and large-scale conventional 3D printing technologies.

In Sect. 5.5, beam shaping for sensor elements using HP3DP for laser die packaging is presented, where the laser dies are introduced. HP3DP allows a significant miniaturization and cost reduction via direct fabrication of microoptics on the laser dies' facets. Optical and life cycle characterization demonstrate extremely stable performance. All data presented in the following were performed in the LithoProf3D®-GS equipment of Multiphoton Optics GmbH.

5.2 Hybrid Manufacturing

Hybrid manufacturing comprises materials which allow that several processes can be combined in one machine with simple adaptation of the process parameters for the different materials and processes. Basically, any material can be employed in 3D hybrid manufacturing, as individual material or in combination with other materials. In hybrid manufacturing, structures can be additively and subtractively formed, ranging from 1D via 2D to 3D structures. In the following, selected examples will be given for additive and subtractive fabrication using completely different material classes which is a unique feature of the used equipment.

In two-photon absorption (TPA) as a special processing technology for additive manufacturing, structures are created via a light-induced cross-linking of organic moieties of polymer or polymer-like negative-tone materials. This results in solidified structures with a lower solubility compared to its not cross-linked surroundings. The nonexposed material is washed away (also known as development process step) using a suitable solvent, leaving the structures behind (Fig. 5.1). This is exactly the same process as for any conventional negative-tone resist material in laser direct writing or lithography.

3D structures can be additively formed via high-precision 3D printing using polymers or multifunctional inorganic-organic hybrid polymers (see Sect. 5.3.2), allowing also the formation of very complex structures like a complete clockwork of Rolf Lang Dresden [46] or lens stacks without the need of assembling the individual parts in only one writing process [47]. An insight into the clockwork is given in Fig. 5.2. Two gear wheels were selected to demonstrate the high fabrication accuracy (Fig. 5.2a). The size of an individual cog is about 5.1 μm in length and about

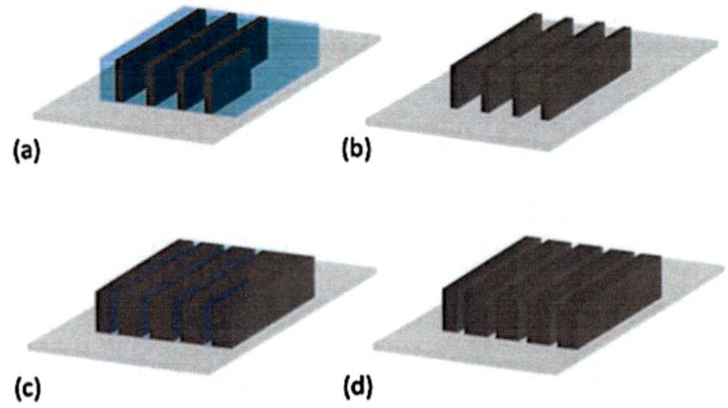

Fig. 5.1 Schematics of structure formation in a negative-tone resist material. (**a**) After exposure and (**b**) after development. The nonexposed areas have a higher solubility than the exposed areas where the resist is cross-linked upon light absorption, and they are washed out in the development step. Structure formation in a positive-tone resist material (**c**) after exposure and (**d**) after development. The exposed areas of the resist have a higher solubility than the nonexposed parts and are washed out in the development step

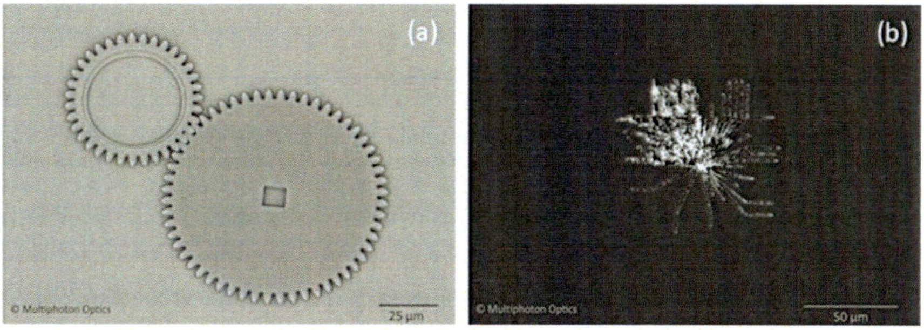

Fig. 5.2 (**a**) Scanning electron microscopy image of gear wheels, as taken from a clockwork (The design file of the clockwork was supplied by Rolf Lang Dresden (2016)), fabricated by HP3DP in an inorganic-organic hybrid polymer. (**b**) Optical microscopy of directly printed Au strip lines

2.55 μm in width. The distance between the cogs of the two gear wheels which are 51 μm and 102 μm in diameter is ≤1 μm.

Aside from using polymer materials in additive manufacturing, high-precision 3D printing also can be used for direct metal structure printing. An example of recently printed gold strip lines is shown in Fig. 5.2b. The typical line width is around 1.35 μm, demonstrating the high potential of the presented processes for many applications such as printed electronics among others.

Subtractive fabrication can be done in positive-tone resist materials using two-photon absorption (TPA) or via conventional metal ablation. For the resist materials, light exposure results in bond breaking, i.e., the solubility is higher for the exposed areas with the nonexposed surrounding material being less soluble (Fig. 5.1). The material in exposed areas is removed in the development step. In most cases, positive-tone resist materials are sacrificial materials, i.e., the structures represent an intermediate state for further processing such as lift-off processing to create metallic 2D and 3D structures on substrate surfaces.

Figure 5.3a shows a parameter search array of subtractively formed contact pads with strip lines in an AZ resist material which was applied on a silicon substrate by spin coating. The parameter search was carried out by means of TPA, varying the average laser power and the laser beam's scan speed. This directly provides suitable parameters for the structure formation in the respective material. The structures were developed with a suited solvent and were subsequently metallized by sputter deposition of Au, followed by a lift-off of the residual resist material. The strip lines have a width of about 2 μm (Fig. 5.3c).

Aside from subtractive processing of positive-tone resist materials via TPA, the tightly focused ultrashort pulsed laser light can be used to structure thin metal films via conventional ablation in the equipment. The metal surface is irradiated by the pulsed laser light which is absorbed by the metal. Dependent on the laser flux, evaporation, sublimation, or plasma formation can occur [48]. Intensity and fluence can be simply adjusted.

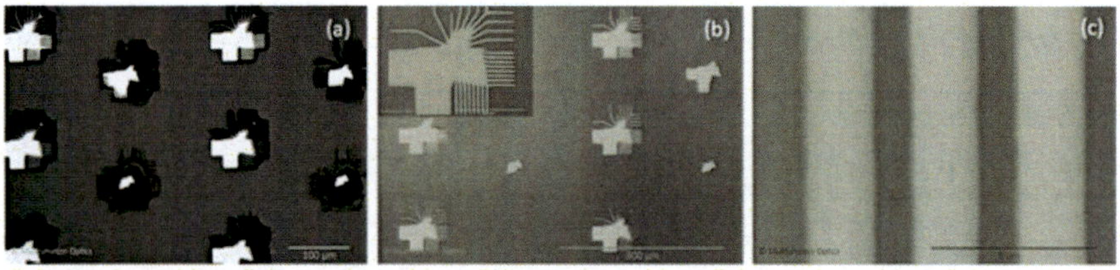

Fig. 5.3 Lift-off processing of contact pads and strip lines on Si(100). (**a**) Parameter search array in AZ resist material; (**b**) metal structures after lift-off. (**c**) Zoom into a structured area (dark areas: AZ lines, light areas: substrate)

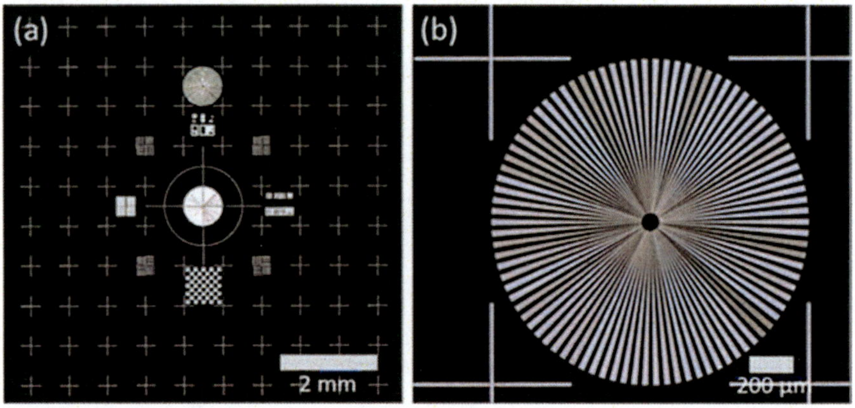

Fig. 5.4 Optical microscopy image of a fiducial fabricated in Ag on a glass substrate via direct laser write (DLW) ablation. (**a**) Overview with different sets of alignment marks arranged on an area of 1 cm^2. (**b**) Zoom into the center alignment mark (Siemens star, 90 spokes)

As an example for an ablated metal structure, Fig. 5.4a shows part of a fiducial of 1 cm^2 in size which was fabricated in an Ag film on a glass substrate. This fiducial is used in the equipment to determine the absolute positions of the XY stages in the implemented alignment system which is used in photonics packaging to identify I/Os (inputs/outputs) of passive and active components. The structure was fabricated by direct laser writing ablation (DLWA) with ablated feature sizes between 1.3 and 1000 µm. Smaller features than 1 µm were not designed; however, it is very likely that smaller ablated features can be achieved once the design is adapted. In Fig. 5.4b, the zoom into the center area of the fiducial shows the well-known Siemens star which was fabricated with 90 spokes. Siemens stars are usually used to determine the resolution of optical instruments, printers, and displays [49]. In conventional lithography, it is used to determine the achievable resolution of resist materials.

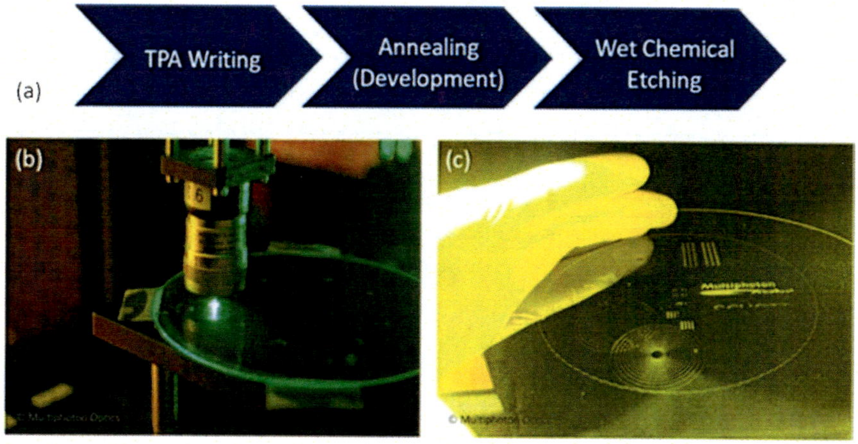

Fig. 5.5 (**a**) Schematics of the structuring process of FOTURAN® glass. (**b**) Direct laser writing on a 6″ FOTURAN® wafer (exposed area, 4″) and (**c**) developed glass wafer after HF etching. The works were carried out in cooperation with Schott AG

A very specialized case of subtractive manufacturing is the structuring of photosensitive glass ceramics such as FOTURAN®.[1] The structuring process is depicted in Fig. 5.5. For demonstration, a 6″ FOTURAN® wafer was exposed to ultrashort pulsed laser light. The structuring process cannot be in situ observed, since no modification of the glass can be detected during the fs light exposure. Light exposure and subsequent annealing process at elevated temperatures initiate a crystallization process in the glass. Upon thermal treatment, the crystallized areas become visible by a color change from transparent to a brownish color. These crystallized areas have a higher solubility in fluoride containing solutions which are used to develop the glass, i.e., to remove the exposed areas, resulting in 2D and 3D structures in glass.

5.3 Materials

5.3.1 Multifunctional Materials

Multifunctional materials are particularly useful for packaging, since they allow to combine multiple functions in one material and thus to reduce the number of materials in a package. This finally has positive impact on the yield, i.e., the less materials are used in multilayer structures, the lower is the probability of failure, and the easier sources of failure can be detected. Inorganic-organic hybrid materials have

[1] FOTURAN® is a brand name for a photosensitive glass ceramics of Schott AG.

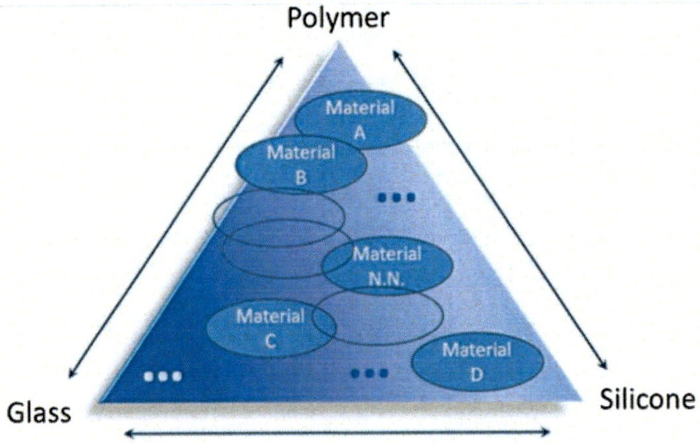

Fig. 5.6 Schematics of the material class of inorganic-organic hybrid polymers, visualizing the infinite amount of possible materials with different properties. Dependent on the synthesis conditions, the materials can be more glass-like, polymer-like, or silicone-like

gained considerable attention during the last two decades, since they synergistically combine properties of glasses, silicones, and polymers (Fig. 5.6).

These hybrids are formed either by physically blending inorganic particles into polymer host matrices [50–52] or by one- or multistep syntheses to form inorganic-organic hybrid polymers [53, 54]. The material class of inorganic-organic hybrid polymers has gained considerable attention during the last two decades due to their wide range of adaptable chemical and physical properties, synergistically resulting from their hybrid structure. There are many different names for inorganic-organic hybrid polymers, for example, ORMOSIL [55, 56] and ORMOCER®.[2] Particularly the latter is excellently suited for industrial purposes, since it combines the material properties of organic polymers with low temperature processing and functionalization with the material properties of glass-like materials such as hardness, chemical and thermal stability, and transparency [57].

Compared to other approaches to synthesize inorganic-organic hybrid polymers, ORMOCER®s are class II hybrids, i.e., the inorganic and organic parts are covalently bonded to each other [58, 59]. This makes these materials superior to other materials such purely organic polymers or particle-polymer composites. Optical and dielectric properties can be simply modified by variation of alkoxysilane precursors and the synthesis conditions in their catalytically controlled syntheses. This allows a precise tailoring of their material properties as required by the application: ORMOCER®s are used for many different applications such as optics [23, 25, 58, 60], microelectronics [61, 62], or biomedical applications [13, 18, 29]. Especially for optics and photonics, the requirements on the materials are more challenging:

[2] ORMOCER®: ORganically MOdified CERamics, registered brand of the Fraunhofer Gesellschaft für Angewandte Forschung e.V.

nm size precision and surface finishes are a prerequisite for functional optical elements. Functionalization via selected moieties is used to create suitable binding sites, particularly for biomolecules and cells for enabling micro- and biomedical applications.

As mentioned above, the synthesis of ORMOCER®s is performed using catalytically controlled hydrolysis and (poly)condensation reactions. Alkoxysilanes are used as precursors. The synthesis reactions can be controlled via adjustment of environmental conditions such as pH value, temperature, or adding of water or other solvents. This results in an organically modified inorganic-oxidic network with the individual parts covalently bonded on a molecular level [63]. Heteroelements, for example zirconium, aluminum, or titanium, can be introduced into the network to further tune the material properties [53, 64], particularly the material's refractive index can be extremely increased [65]. The resulting materials exhibit negative-tone resist behavior, i.e., they are organically polymerized (solidified) initiated by UV light or thermal exposure.

Aside from the precursors used for syntheses, the synthesis conditions have significant influence on the resulting material properties. By varying the kind and/or the amount of alkoxysilane precursors and, in addition, the synthesis conditions such as the type and the amount of the catalyst, the reaction temperature, and the kind and amount of solvent, the materials can be custom-designed with a well-defined parameter profile. This enables their implementation in many different applications. Inorganic-organic hybrid polymers such as ORMOCER®s are well-established on the market and are used, for example, as optical interconnects or waveguides [23, 25, 60, 66, 67], in microoptics [23, 68], in electro-optical applications [25, 69], as dielectric layers [61], and as passivation materials for the encapsulation of microelectronic devices and components [58, 70, 71]. Aside from their use as is, i.e., without fillers, they can be filled with nano- or microparticles for specific applications in electronics or optics [50, 51], even providing more possibilities to create novel designs and products. It has to be mentioned, however, that a change in the synthesis conditions towards a specific material parameter profile in one physical or chemical parameter might influence also other material properties, and a comprehensive know-how is necessary to control this in a well-defined manner.

The organic groups fulfill several functions: they act as network modifiers, as connecting groups, and as polymerizable moieties, resulting in materials with a very broad range of physical and chemical properties. For example, the mechanical and thermomechanical properties of inorganic-organic hybrid materials can be also varied over a broad range, and the Young's modulus and the material's mechanical and thermal stability can be increased by increasing the inorganic content in the hybrid network. Typical Young's moduli are between 10 MPa and 20 GPa [72]. Consequently, this simultaneously leads to a reduction of the coefficient of thermal expansion (CTE). The large variety of employable organic groups allows one to adjust each material exactly to the requirements of an application.

For a controlled fabrication and reliable processing in UV or fs laser lithography such as high-precision 3D printing, a precise control of the organic cross-linking reaction is necessary which is achieved by comprehensive understanding of the light-matter interaction processes. Typically, (oligo-)methacrylate, acrylate, or styryl moieties are used for optical applications. Epoxides usually exhibit parasitic polymerization which has to be avoided in optical applications. They are, however, preferred in electronics as underfill or encapsulation, where parasitic reactions can be tolerated [73]. Alkyl or aryl groups can be connected to the $–[Si–O–]_n$ – network as nonreactive groups. These also might significantly influence the resulting material's properties. For example, increasing their amount in the hybrid polymer can reduce the degree of polymerization due to steric reasons. This then consequently results in a reduction of the material's density which directly impacts the optical or dielectric properties such as the refractive index or the dielectric permittivity.

Particularly with respect to photonics, inorganic-organic hybrid materials such as ORMOCER®s offer tremendous advantages which are often beyond the possibilities of purely organic materials or beyond other hybrids mimicking the ORMOCER®s' material properties. This is related to the precise control in synthesis convoluted with the deep understanding of underlying chemical principles and routes which are employed in syntheses. The processing of ORMOCER®s consists of two steps: prepolymer (resin) synthesis to form an organically modified $–[Si–O–]_n$ – network and formation of the inorganic-organic hybrid network via photochemically and/or thermally initiated organic cross-linking. The second step comprises the presence of photochemically and/or thermally activatable initiators in the prepolymer resin.

In the following, some selected aspects of the materials will be discussed particularly with respect to their application in photonics and their employment in high-precision 3D printing or 3D lithography.

5.3.2 Selected Materials for High-Precision 3D Printing

High-precision 3D printing using LithoProf3D® can be carried out using a vast variety of different materials. For photonics, the use of inorganic-organic hybrid polymers is advantageous, since their optical properties can be precisely controlled via the synthesis conditions and the synthesis paths, and a further fine-tuning is achieved by controlling the cross-linking conditions [5, 58, 74].

For optical applications in the data and telecom regime, i.e., if single- or multimode (SM, MM) waveguides for the near-infrared (NIR) regime have to be processed, the amount of silanol groups with second harmonics around 1438 nm has to be minimized down to diminishing amounts for the telecom regime [23, 58, 60, 75]. This reduces the materials' optical absorption losses. To reduce the amount of silanol groups can be done by using silylating agents or by avoiding access water upon synthesis [76]. This is usually characterized by multinuclear magnetic resonance (NMR) spectra. Aside from this, the amount of aliphatic -CH which is known to absorb around 1300 and 1500 nm can be reduced by replacing them with aromatic substituents. This also increases the refractive index of the resulting material. The material's refractive index can also be increased by introducing heteroelements such as Ti and

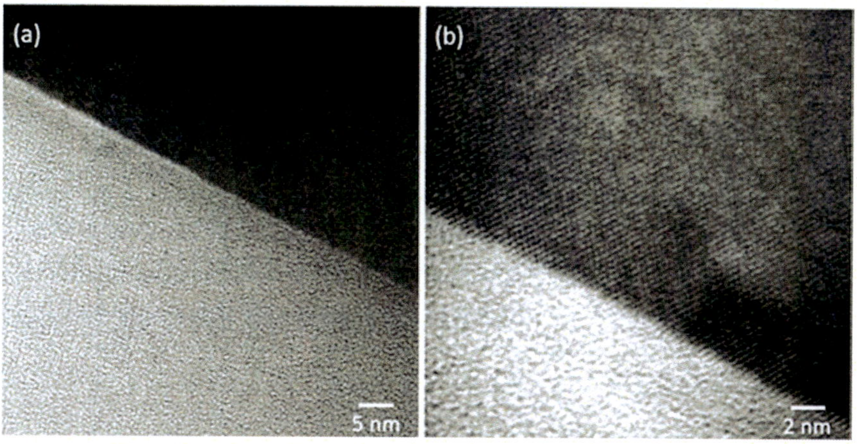

Fig. 5.7 Interface p-Si(100)/ORMOCER® for data- and telecom applications, as prepared by FIB and imaged by HRTEM. (**a**) Large-scale image of the two materials with the ORMOCER® layer as brighter area and the Si as darker area. (**b**) Zoom into the structure exhibiting an atomically flat ORMOCER®/p-Si(100) interface as well as the lattice planes of the silicon substrate [77]

Zr into the $-[Si–O–]_n$ network [22] or by functionalized and highly dispersed nanoparticles [51]. Aside from reducing groups which impact absorption losses, the structure of the hybrid resins and resulting material layers yields organically modified inorganic-oxidic units with structural features in the range of 1–3 nm which was proven by small-angle X-ray scattering (SAXS) measurements [58] and high-resolution transmission electron microscopy measurements [77] of an ORMOCER® which is applied for data- and telecom applications (850, 1310, and 1550 nm). Figure 5.7 shows high-resolution transmission electron microscopy (HRTEM) of an ORMOCER® layer which was applied by spin coating on a p-Si(100) substrate following a process sequence used for waveguide fabrication [58, 60]. Subsequently after processing, focused ion beam (FIB) was used to prepare the sample for HRTEM [77]. Figure 5.7a shows a HRTEM image on a larger scale where the hybrid polymer layer can be clearly distinguished from the silicon substrate, with a well-defined interface between the two materials in the Å range. The lattice planes of the p-Si(100) substrate are clearly visible in the even higher resolved data shown in Fig. 5.7b, also displaying that the ORMOCER®/silicon interface is atomically flat.

The high-precision 3D-printed structures presented in the following chapters were mainly acrylate or methacrylate-based resins. It has to be mentioned that also styryl- or epoxide-based hybrid materials are used for high-precision 3D printing with some constraints related to chemical reactivity and response to the fs laser light, i.e., these materials' reactivities are lower than for acrylate or methacrylate materials and, for epoxides, parasitic polymerization is also very likely [78]. In order to enable fast and reliable processing, the ORMOCER®s which were chosen for HP3DP were formulated with photoinitators typically around 2 wt.-% with varying two-photon absorption cross-sections [5, 79]. The selected materials are transparent for fs laser light which operates in the visible range. Transparency is beyond 90%, and absorption losses are in the range between 0.02 and 0.04 dB/cm. The materials exhibit an extraordinary large process window, which is useful for process development [80]. Aside from hybrid polymer materials, also purely organic materials can be used (see Sect. 5.4.6).

5.4 High-Precision 3D Printing

5.4.1 General Aspects on High-Precision 3D Printing

High-precision 3D printing (HP3DP) uses femtosecond pulsed laser light as manufacturing tool, and it is one of the fabrication modes used in 3D lithography or direct laser writing via TPA. The term high-precision 3D printing is used synonymously to additive manufacturing or 3D lithography with negative-tone resist materials, i.e., materials whose organic moieties are cross-linked upon light exposure. The appeal of this technology is that it allows one to expand 3D lithography to basically any structuring from 1D via 2D to 3D fabrication with tunable precision without losing throughput if properly equipped.

This section briefly describes the physical principle which is already explained in detail in the literature [1, 81–83]. As displayed in Fig. 5.8a, HP3DP excites a

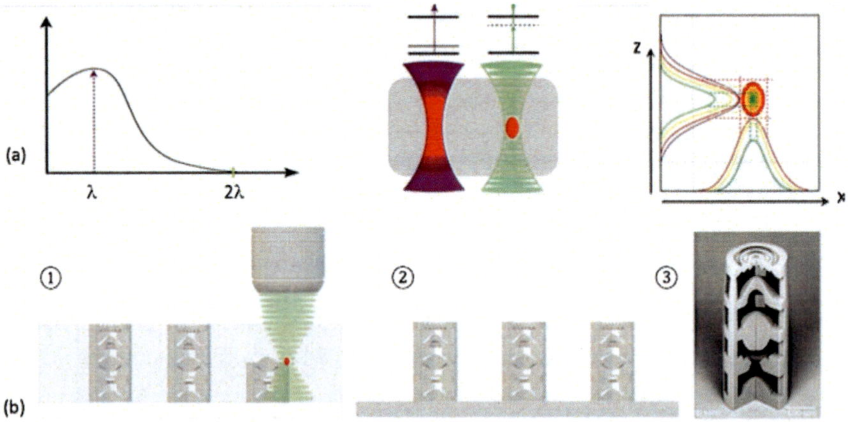

Fig. 5.8 Sketches of (**a**) an absorption curve illustrating the center wavelength of a one-photon absorption process (1PA) and for a two-photon absorption (TPA) process which takes place at 2λ, resulting in a completely polymerized volume for 1PA, while for a TPA process, the process only is initiated by the confined focal volume due to the quadratic dependence of the intensity distribution along the Z and X/Y direction, provided the material is transparent enough. (**b**) Principal sketch of 1-2-3 processing: After coating of a substrate, the structures are fabricated by HP3DP (①), 3D lithography in the additive mode), followed by a development step which removes the nonexposed material analogously to conventional lithography (②), resulting in the 3D structure (here: microlens stack with intrinsically fabricated mounts) which is displayed in the SEM image (③)

photoactive material within a wavelength regime where absorption in general does not occur. In comparison to the exposure with UV light which induces a linear absorption by a single photon (SPA) at λ, a nonlinear absorption of two (or multiple) photons (TPA or MPA) at about twice or multiples of the wavelength ($n \cdot \lambda$, $n = 2, \ldots$) via a virtual state can be triggered [1, 82]. As the probability for TPA is proportional to the square of the laser intensity, the photon density is only high enough to initiate a reaction in a material in the focal volume which thus it is strongly confined. The polymerized volume which is caused by the focal volume is called voxel (volume pixel) which is typically of elliptical shape and is a direct image of the point spread function [84]. If beam shaping is implemented, the elliptical shape can be transformed to a spherically shape [85]. The typical workflow for HP3DP is sketched in Fig. 5.8b and consists of scanning the focal volume through the material to create the 3D structure by organically cross-linking the organic moieties, followed by washing out the unexposed material by a developer solution. An SEM image of the resulting three-dimensional structure is displayed as well, showing a stack of intrinsically mounted microlenses, i.e., no assembling step is necessary as it is provided directly from the fabrication. To allow an insight into the lens stack, part of it was left out.

In principle, the reaction volume which triggers the size of the resulting voxel depends on the technical equipment (positioning system, implemented optical elements, and laser wavelength) and the focusing optics (numerical aperture NA). However, the reaction kinetics of the underlying materials and other processes such as threshold processes related to the material composition also play a crucial role for the minimum size of a resulting voxel. The different interaction volumes which have to be considered are shown in Fig. 5.9. The technical interaction volume (red

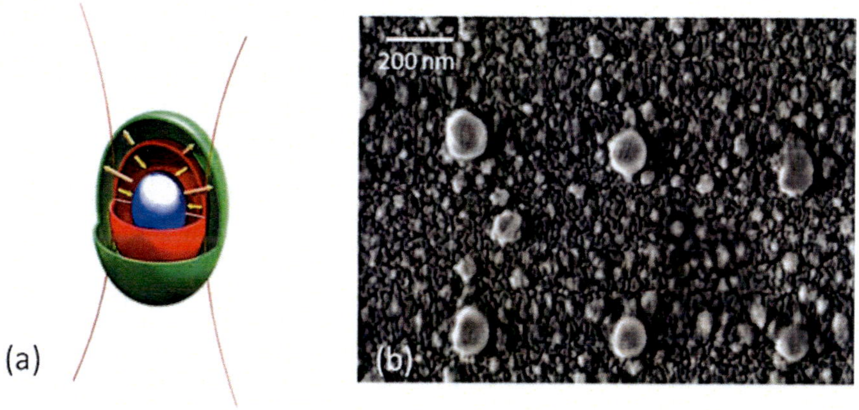

Fig. 5.9 (**a**) Schematics of the different interaction volumes. Technical (red) and chemical (green) interaction volume. The blue interaction volume is smaller as the technical interaction volume and is attributed to the threshold behavior which results in feature sizes below 100 nm [4]. (**b**) Sub-100 nm voxel structures [86], showing the irregular shape deviating from an ideally spherical shape as expected from simulations

voxel) is determined by the employed optics, the stability of the laser, and the stability and accuracy of the positioning system. The chemical interaction volume (green voxel) is much more challenging to minimize, because it depends on many different factors, like the reaction kinetics of the material formulation and, consequently, on absorption, radical formation, diffusion of monomers, propagation and termination reactions in the material. The light-matter interaction is dependent on the technical and the chemical interaction volume. This means that the fabrication parameters, the underlying material's chemistry, and the technical premises determine the material's ability to result in lowest feature sizes and highest optical resolution.

Dependent on the chemical structure of the material, the interaction volume might be similar to the extension of the technical interaction volume (focal ellipsoid) or larger (advancing polymerization) due to a propagating reaction. The latter is strongly dependent on the material's structure. The blue ellipsoid in Fig. 5.9 is attributed to the threshold behavior, where the structural features can be significantly smaller than structures generated with the technical interaction volume. It has to be mentioned, however, that the structures then result in a more irregularly shape once they are in the sub-100 nm range, i.e., circular structures show many kinks. These result from an enhanced diffusion of material components such as initiator molecules, monomers, radicals, oxygen dissolved in the material, etc. in the high-intensity laser beam along the edges of the nanometer size structure. This can be explained analogously to atomistic processes in the early stage of thin-film growth, where in the hit-and-stick diffusion-limited aggregation (DLA) regime, an adatom is arriving at an edge of an island, diffuses along the edge of this island, and then is pinned at a time lower than the time being necessary to relax to a more favorable state. This is accompanied by additional arriving atoms [87]. Thus, it is not expected that ideally spherical shape structures can be created via TPA in the sub-100 nm regime. This is a direct consequence of the diffusion of radicals (of either radicalized initiator molecules, activated monomers, or activated monomer fragments) and of the simultaneously occurring diffusion of solvent molecules, inhibitors, and scavengers which becomes even more pronounced in the threshold regime. In the latter, the amount of moieties which are available for cross-linking is significantly reduced, resulting in fragile structures for most of the experimental materials. Thus, reliable structures which might be industrially relevant using commercially accessible materials are presently limited to about 75–90 nm in diameter [4].The formation of voxels and the diffusion of initiator radicals was simulated using Monte Carlo simulation, supporting these findings [88].

Figure 5.9a shows a schematic of the different interaction volumes as occurring in TPA-initiated direct laser writing with the technical interaction volume (red ellipsoid) defined by the beam waste, the chemical interaction volume (green ellipsoid), and the threshold regime (blue ellipsoid) which can be seen as part of the chemical interaction volume [4]. While the technical interaction volume and the threshold regime limit the voxel size to a certain extent, the chemical interaction volume is a convolution of many different reactions among which are the diffusion of the material's building blocks, initiators, solvent molecules, radicals, scavengers, etc. [89]. Aside from the chemical composition, the rate of cross-linking is strongly influenced by the viscosity of a material facilitating a high degree of organic cross-linking for a low viscosity than for high viscosity due to an enhancement of the diffusion rate. The degree of cross-linking can be precisely controlled upon comprehensive understanding of the underlying material's light-matter interaction. In Fig. 5.9b, sub-100 nm voxel structures of about 75–90 nm diameter are displayed, exhibiting the expected irregular shape deviating from an ideal spherical shape due to DLA.

The reaction can be differently triggered for negative- or positive-tone resist materials, resulting either in the 3D formation of a structure (additive fabrication mode) or a 3D removal of material (subtractive fabrication mode). It has to be mentioned that this process cannot only be used for 3D fabrication, but for any structure formation, i.e., 1D, 2D, and 3D structure fabrication.

Let us now consider the additive fabrication mode of 3D lithography (HP3DP) using polymers or inorganic-organic hybrid polymers. The laser pulses initiate a radical cross-linking of the C=C bonds or a cationic cross-linking of epoxide bonds, dependent on the material's composition. If the focal volume of the laser is scanned in three dimensions through the material, the material is directly cross-linked in 3D along the path of the focal volume. This enables users to fabricate 3D structures in the volume or on a surface of a material with highest precision directly, fast, and reliably from a CAD file in real 3D, i.e., orchestrating the laser beam not only in X-Y direction and stacking 2D layers to form the 3D structure, but scanning the beam also arbitrarily such as in X-Z direction or even along any vector in space which is possible with the equipment used in this study. The degree of cross-linking in a polymer or hybrid polymer and thus the chemical and physical properties of the created material structures can be precisely controlled by different accessible process parameters such as laser power and scan speed, both together supplying the energy dose which is deposited in the material and initiates the cross-linking reaction, but also by the exposure strategies including line and layer density in free space. The cross-linking behavior of various inorganic-organic hybrid polymer materials was extensively investigated using μ-Raman spectroscopy, revealing the precise controllable reactions [90, 91].

The physical and chemical properties of the fabricated structures are directly related to the chosen parameters in the fabrication process. By varying these parameters, physical or chemical properties can be altered to create gradient structural properties which are interesting for many different applications in photonics or biomedicine which is known as 4D printing [92]. For example, the mechanical stability of additively fabricated structures can be modified such that their Young's modulus or mechanical strength differs within the structures or mechanically differently stable structures can be created. This is of particular interest in tissue engineering, since the cells prefer to adhere, differentiate, and proliferate on nontoxic structures which resemble the natural tissue in its mechanical properties [13].

5.4.2 Manufacturing Strategies

Depending on the structure's specifications, i.e., desired dimension and required printing resolution, a range of different manufacturing strategies are available which consist of the following modes. These will be discussed subsequently:

- Exposure mode: exposure setup for the fabrication
- Fabrication mode: applied axes for the scanning of the focal volume
- Structuring mode: full volume vs. contouring
- Trajectory mode: layer by layer or three-dimensional

Please note that all manufacturing strategies work for negative- and positive-tone materials; in the following, for the ease of explanation, only negative-tone resist materials are considered in the additive mode.

The **exposure mode** comprises the properties of the focusing optics (numerical aperture, working distance) as well as its configuration with respect to the workpiece. In most cases, exposure resembles a bottom-up fabrication.

The higher the numerical aperture (NA) of a microscope objective and the shorter the wavelength, the higher is its printing resolution as the laser light can be focused more tightly (smaller voxel size) compared to a low NA objective, and the intensity distribution is more confined in X-Y and Z direction. In a configuration where the objective is moved in vertical direction (parallel to the optical axis of the incoming laser light) while the object itself is moved only in lateral direction, the working distance of an objective sets an upper limit of the maximum possible printing height for standard 3D lithography (see Fig. 5.10a). Low numerical aperture air objectives provide longer working distances, but typically result in lower printing resolution. High NA immersion objectives, however, inherently possess low working distances and thus limit the achievable printing height to some hundred micrometers only.

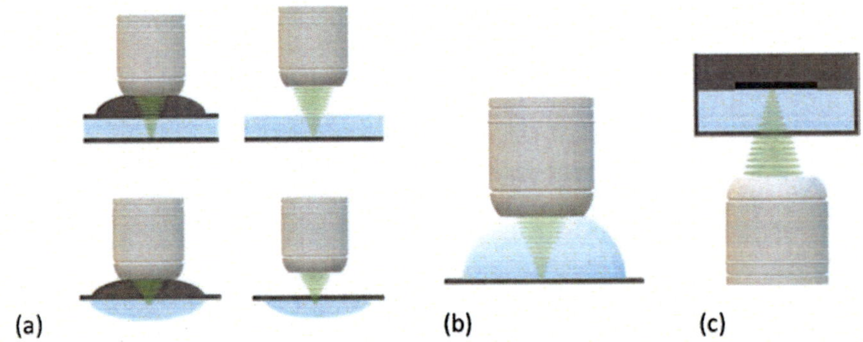

Fig. 5.10 Exposure modes generally applied for 3D lithography. (**a**) Standard 3D lithography with and without cover glass; the structures can be either fabricated on the lower substrate or headfirst on the top substrate. 1D, 2D, 2.5D, and 3D structures can be generated. (**b**) Immersion lithography (dip-in) 3D lithography as special case for a bath setup and (**c**) bath 3D lithography with the workpiece being mounted on the Z axis

This limitation can partially be overcome to some millimeters by using the photoresist as immersion media by dipping the microscope objective into the liquid material (dip-in 3D lithography; see Fig. 5.10b, as demonstrated by Stichel et al. in 2011 [93] and Bückmann et al. in 2012 [94]. However, this method allows only the printing of one single element more pronounced as the corresponding working distance as a subsequent second object will lead to a collision of the objective and the first structure, thus resulting in a lower throughput, as observed for most equipment available in academics. Although same limitation is valid, an additional expansion of the available on-axis printing height to the centimeter range (depending on the traveling range of the mounted axis) is provided by inverted focusing of the laser light into a bath configuration with a fixed position of the focal point along the optical axis and gradually pulling out the substrate out of the bath containing the liquid photoresist [33] (bath 3D lithography; see Fig. 5.10c).

The **fabrication mode** indicates how the focal volume and the material to be organically cross-linked are moved with respect to each other. Conventionally high-precision piezo translation stages are applied to move the sample with respect to the fixed focal point. Due to the piezo's nature, this comes along with the drawback of limited traveling range, creep, and limited scan speed [95]. Alternatively, air bearing stages with lower accuracy than piezo stages allow much higher scan speeds and traveling ranges. However, the difference in accuracy between air bearing and piezo stages does not negatively impact the HP3DP process. This is related to the fact that the typical size of the focal volume used in the fabrication process is in the sub-μm regime. It has to be noted that green light which is used for these works offers a much lower FWHM of the point spread function of about 220 nm (determined for an objective with an NA of 1.4 in the ideal design case) compared to NIR light (780 nm), whose FWHM is about 315 nm for the same design case. Considering this then simply clarifies why piezo stages are significantly overdetermined and not necessary at all for this type of processing, and air bearing stages with a high accuracy can be used. For the threshold range [4], the accuracy of the implemented stages plays a minor role, since only the precise control of the underlying chemical reactions of the laser light with the material to be structured determine the accuracy (see Sect. 5.4.7). In addition, air bearing stages provide stitching-free manufacturing due to long traveling ranges and synchronization with the faster galvo scanner axes in an infinite field-of-view (IFoV) mode. This is particularly important for the fabrication of optical elements larger than the FoV, since no stitching errors occur.

In the following, the fabrication via a movement of the sample with respect to a fixed focal volume is referred to as stage mode (Fig. 5.11a. In HP3DP, scanning of the focal volume is typically performed at constant scanning speed. The axis system has to accelerate to this scan speed. For each structure, an optimum scanning speed vopt can be calculated by the physics of motion according to

$$v_{\text{opt}} = \sqrt{\frac{1}{2}al} \tag{5.4.1}$$

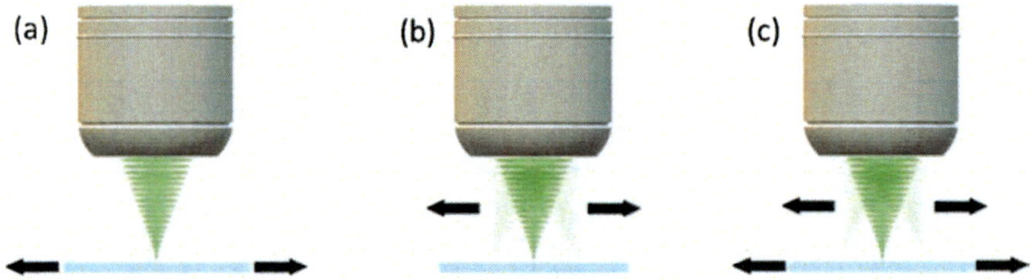

Fig. 5.11 Fabrication modes using high-precision 3D printing in 3D lithography: (**a**) stage-only mode, (**b**) galvo mode, and (**c**) synchronized mode which provides stitching-free fabrication

which results in a minimum printing time t_{min} depending on the acceleration a of the axis system and on a specific line length l which corresponds to a structure's lateral dimension. This formula simply expresses that it is not reasonable to use the highest possible scanning speed for all cases, but to reduce the speed especially for small structures as the effect of acceleration dynamics can be tremendous and may not be neglected.

As translation stages exhibit high inertia, their acceleration is limited. Thus, fabrication using a galvo mode (Fig. 5.11b) with galvo scanners deflecting the laser light by mirrors (low mass and high frequency) with respect to a fixed sample enables higher processing speeds, because the magnitude of the acceleration can be orders of magnitudes higher compared to linear stages. Although the fabrication time can be reduced by applying the galvo mode, the accessible scan area is limited to the field of view (FoV) of the respective microscope objective which is only several hundred micrometers (typically below 0.5 mm) for objectives with an NA > 0.6 [96]. In principle, low numerical aperture objectives with low magnification can be used, but these drastically reduce the possible printing resolution [95]. Structures much more pronounced than the FoV of an objective can in general be fabricated by stitching [97] which, however, leads to joints that are especially unfavorable for optical elements as aberrations (scattering, refraction, etc.) are induced by these artifacts [98]. Synchronizing the movement of translational stages and the galvo scanners with respect to each other (synchronized mode enabled by the controller; see Fig. 5.11c) enlarges the scan field of the galvo scanner to the entire traveling range of the stages as the simultaneous movement of the traveling stage continuously shifts the FoV [99]. By this, the advantages of the different axis systems are combined: high precision and high dynamics of the galvo scanner and long traveling ranges of the translational stages without the necessity of stitching.

The **structuring mode** implies the relation between the volume of a structure and the actually cross-linked volume by HP3DP (Fig. 5.12). Typically, a full-volume scanning by slicing and hatching in a layer-by-layer manner is used by many users of HP3DP. This can be very time-consuming, particularly if larger parts or large arrays of elements will be fabricated. By contouring just the outer shell of an object, for example, a microlens, the production time can be significantly reduced by up to

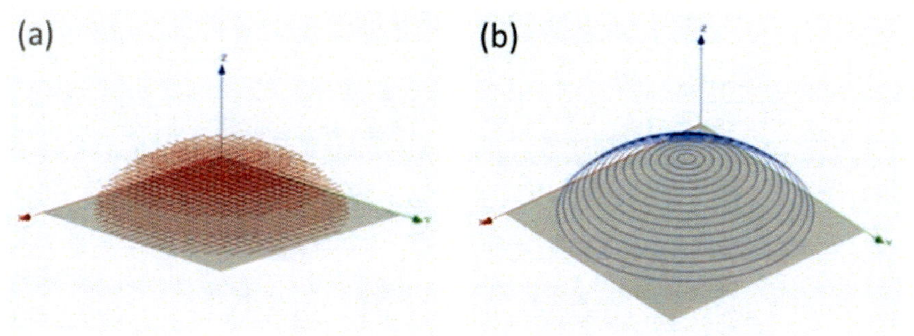

Fig. 5.12 Structuring modes for high-precision 3D printing: (**a**) full-volume scan and (**b**) contouring

95% [81, 82, 100]. Multifunctional materials such as ORMOCER®s (see Sect. 5.3.1) are particularly suited for this approach, since they enable very stable structures even if only one outer layer is fabricated with the inner part of the object still being in the liquid state. If the structure is just a big object where the inner part is not important for its function, notches can be implemented at selected parts of the structure to allow the liquid, i.e., still non-polymerized material to be removed upon the development step [81, 82]. This approach was also used for the lens stack displayed in Fig. 5.8b. For microoptical elements such as microlenses, the residual liquid photoresist enclosed between cross-linked shell and the substrate is polymerized using a short UV flood exposure step subsequently after the development step. This combines conventional 2D processing with 3D processing.

The cross-linking via fs laser light exposure with two-photon polymerization and UV exposure was investigated in more detail. By μ-Raman spectroscopy, it was proven that this procedure results in very homogeneous organically cross-linked hybrid polymer shells and inner parts with no variation in the homogeneity [13, 101], analogously to conventional processing. Using this structuring mode, the fabrication time of a lens with a diameter of 100 μm and a height of 20 μm can be as low as 1 s per lens in a serial production [100, 102]. The achieved surface roughness of Ra between 20 and 30 nm is still suitable for optical applications down to the visible spectral range [103, 104]. In addition, it has to be mentioned that parallel fabrication of structures is also easily achievable, enabling an even higher throughput (see below).

A detailed investigation of the influence of the fabrication mode and the structuring mode with respect to the fabrication time is illustrated in Fig. 5.13, where the serial fabrication time (one focal spot) is displayed in dependence on the used scan speed. The serial fabrication time using one focal spot was exemplarily estimated for a cube of 100 × 100 × 100 μm³ in size. As resulting from equation (Eq. 5.4.1), the printing time exhibits a minimum at a scan speed vopt and increases for further increasing the scan speed. This observation is related to the increasing influence of the acceleration dynamics of the positioning stages. Independently on the fabrication mode, the optimum scan speed is higher for the galvo mode than for the stage mode due to higher acceleration that is intrinsically provided. As demonstrated for

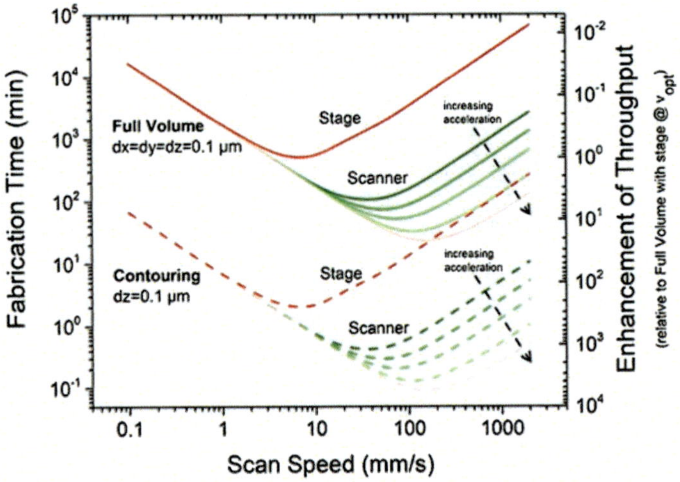

Fig. 5.13 Fabrication times of a 100 µm × 100 µm × 100 µm cube in dependence on the applied scan speed. The fabrication times are estimated for two different structuring modes (full volume and contouring) and fabrication modes (stage and scanner) for serial fabrication with a single laser focus. As obvious, fabrication time does not continuously decrease upon increasing the scan speed but exhibits a certain minimum. This originates from a line length specific optimum scan speed which depends on the underlying acceleration. For the different modes, the enhancement of fabrication throughput is determined with respect to the fabrication time at the optimum scan speed for the stage applying the full-volume structuring mode

lens-like structures, contouring of its shell only can enhance the throughput by more than two orders of magnitude [105].

The **trajectory mode** provides information about the scanning trajectory of the laser focus to fabricate a structure, e.g., a lens as shown in the schematic sketch in Fig. 5.14. Typically, HP3DP makes use of a two-dimensional fabrication scanning layer by layer for both the full-volume and the contouring structuring mode (sequence of subsequent layers in Fig. 5.12). The easiest 2D trajectory is sketched in Fig. 5.14b with equally spaced distances dz between subsequent layers at constant focal volume size. It has to be mentioned that this method can also be modified for equal increments in dr along the radius, in ds along the secant, or in $d\varepsilon$ parameterizing the ellipsoidal radii of the voxel. Recently, a technological development which was named "grayscale two-photon lithography" [106] enabled a 2D trajectory at increased layer distances due to varying focal volume sizes.

This approach is sketched in Fig. 5.14e, and it is suitable for the fabrication of 2D and 2.5D objects as it was already demonstrated in 2014 where tapered ORMOCER® waveguides were fabricated to adiabatically couple on-chip silicon waveguides as shown in Fig. 5.15 [107]. The waveguides were fabricated in a continuous writing mode by dynamically tuning their dimensions varying the exposure parameters along the waveguide's lengths. This was done for waveguides fabricated in a "straight line" writing mode and in a "helical line" writing mode [86]. Data on the

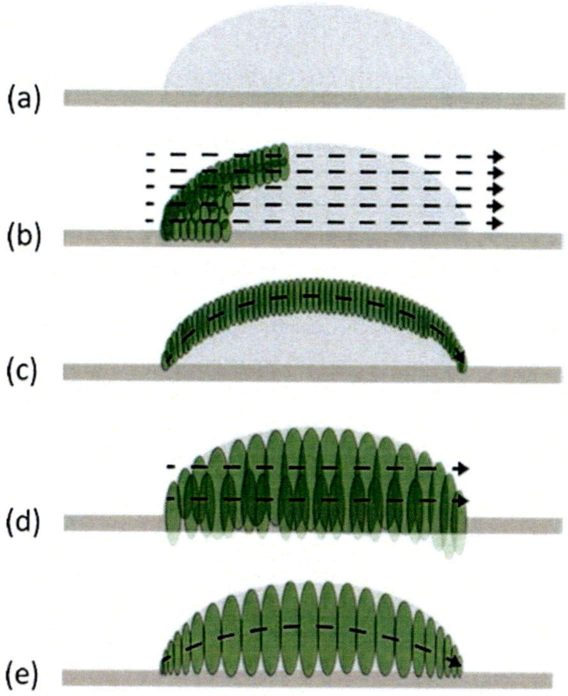

Fig. 5.14 Varying trajectory modes exemplarily illustrated for the fabrication of a lens structure. (**a**) Shape of the lens to be created. Fabrication of the lens structure using (**b**) conventional two-photon lithography via stacking of 2D layers to form the 3D structure; (**c**) laser contouring of the outer shell; (**d**) grayscale two-photon lithography, and (**e**) real 3D Fast Adaptive Laser Contouring Multiphoton Absorption (FALCON 3D)

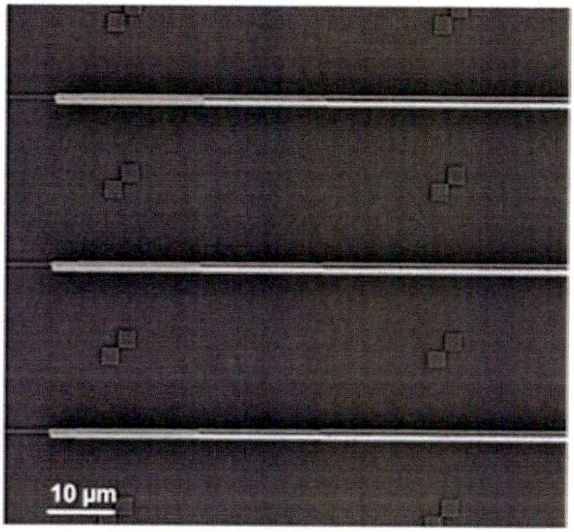

Fig. 5.15 Waveguides fabricated using an ORMOCER® material [86, 107]

losses of these waveguides revealed losses as low as −0.6 to −1 dB for the non-optimized fabrication process [107].

As HP3DP inherently provides real 3D capabilities, also real three-dimensional trajectories are possible in the structuring mode. This type of trajectory scan provides the advantage of a polymerization according to the real 3D design and not just by adding layers on top of each other as approximation of this design. The schematic sketch in Fig. 5.14d illustrates this for the contouring mode at a constant focal volume size. This method is called Real 3D Laser Contouring Multiphoton Absorption (LCon3D MPA). The laser focus is scanned in a 3D trajectory which is predetermined by the topography of a lens. As a result, only one curved layer is necessary instead of several planar layers which results in a reduced printing time. A prerequisite for this type of trajectory is the formation of a stable shell as it is demonstrated by inorganic-organic hybrid polymers.

Additionally, a three-dimensional trajectory at varying focal volume size by simply tuning the laser power can be applied which reduces the traveling height along Z direction, i.e., parallel to the optical axis. This method is called Real 3D Fast Adaptive Laser Contouring Multiphoton Absorption (FALCON 3D MPA). As typically the dynamics of the Z axis is limited due to high loading mass, the reduced height of the trajectory is more suitable for the Z axis to operate. Alternatively, a piggyback system consisting of an air bearing stage and a piezo axis can be used in a synchronized movement, with the piezo performing the three-dimensional trajectory. However, piezo systems with the required dynamical range are currently not on the market, and the ones implemented typically in equipment allow only by far too low dynamics to deploy the full power of a piezo scanner, thus making piezo scanners useless at this stage. On the other hand, piezo stages are used for equipment to enable high resolution in Z. Principally, this is used for methods which show atomic resolution [108, 109], but for 3D lithography in the high-precision 3D printing mode, this is not necessary, since the focal volume in an equipment is usually elongated along the Z axis with values of 200–300 nm at best. Thus, piezo stages which are capable of nm steps are needless from a production point of view. In contrast to the applications of grayscale two-photon lithography, the FALCON 3D MPA trajectory with adaptable voxel size can also be used for the fabrication of 3D structures, as the convolution of 3D trajectory and voxel size can be computed accordingly.

5.4.3 Production Environment

For any process, it is extremely important that the processes can be simply integrated into standard workflows already existing in productive environments. The integration of 3D lithography including high-precision 3D printing and laser ablation into standard workflows is very straightforward, since 3D lithography [110] was developed as direct and complimentary extension of conventional 2D lithography [60], microoptics fabrication via master production and replication [111], and 2.5D lithography such as grayscale lithography [112]. Laser ablation is just an integral part of the equipment[3] used for fabrication of the structures demonstrated in this chapter.

[3] For the fabrication of the structures shown in this book chapter, LithoProf3D-GSII was used as fabrication tool.

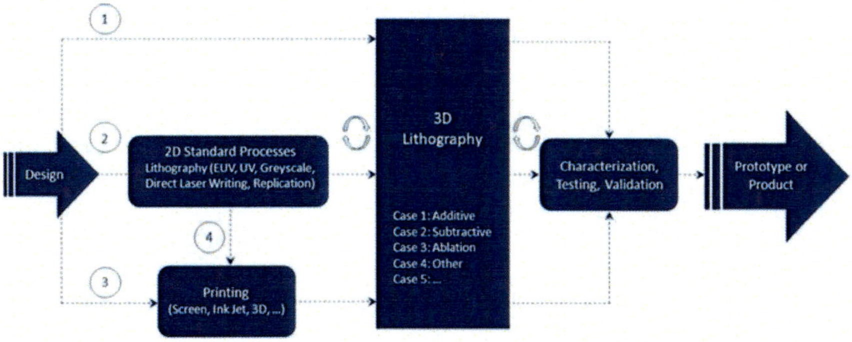

Fig. 5.16 Exemplary implementation of 3D lithography with its special features of high-precision 3D printing (Case 1), subtractive mode (Case 2), ablation mode (Case 3), and other modes (Case 4, e.g., glass structuring) into standard workflows

Figure 5.16 schematically illustrates the implementation of 3D lithography with its special case of high-precision 3D printing for the fabrication of optical elements into standard workflows. Basically any processing route can be taken due to the high sophistication of the used machine (LithoProf3D®-GSII), because an implemented Vision-to-Align System allows to address alignment marks as well as any structure in the range of presently about 1 μm with a very high precision. This is of particular interest if multilayer processing involving 2D and 3D structuring including metallization has to be carried out to create the functional elements. Aside from the fabrication of arbitrary individual optical elements, optical elements can be fabricated on wafer scale [113] and chip level [107, 114], or using any other substrate kind and shape. For the fabrication, any substrate size, shape, and material can be used. Using HP3DP as special case in 3D lithography, the structures are fabricated in three steps: dispensing-additive fabrication-development, in most cases without any further post-processing. After the structures have been processed, they are characterized, tested, and validated as prototypes or products (Sect. 5.5). It has to be mentioned, however, that conventional processing including plasma activation, UV and/or thermal prebake, post-exposure bake, and curing can be also performed as in standard semiconductor processing, if required.

5.4.4 *Scaling to Industrial-Scale Throughput*

As mentioned in Sect. 5.4.2, the structuring mode by contouring [5, 81, 115] significantly reduces the fabrication time. For example, the fabrication time for microlenses with a diameter of 100 μm and a radius of curvature (RoC) of 100 μm is between 1 and 5.5 s compared to several minutes for the full-volume structuring mode [102]. The resulting *rms* roughness of 20–30 nm is still appropriate for optical applications down to the visible. For serial fabrication with one focal spot only, the resulting fabrication time for a (10 × 10) mm^2 lens array consisting of 6400 lenses consequently ranges between 12 and 4 h including all autofoci and translational movements of the axes. Considering multifocal fabrication by (2 × 2) focal spots, the production time can be reduced to less than 1 h.

One way to increase the throughput of HP3DP for industrial-scale production is to parallelize the printing. Since the tool for HP3DP uses a focused laser beam, parallelization can be achieved by using multiple focused laser beams instead of using only a single beam [31–34, 116, 117]. This is usually achieved by generation of multiple parallel beams which all have to enter the microscope objective, thus being focused onto the same plane. Motion of the stages or galvo scanner axes moves all focal volumes of all beams synchronously, and multiple structures are printed in parallel. The most apparent advantage of this kind of parallelization is cutting down the fabrication time for manufacturing a number of structures by the number of parallel beams. For example, the use of four parallel beams would reduce the fabrication time for mass production of similar structures to a fourth, or the use of 100 parallel beams would reduce the fabrication time by a factor of 100. It has to be mentioned that only structures of identical shape can be fabricated this way, since all laser beams are performing the same motion and depositing the same energy dose into the resist material. Since all parallel beams are focused through the same microscope objective, the maximum distribution of the focal spots is limited to the field of view (FOV) of the respective microscope objective. For the fabrication of structures with extension LX and LY in X and Y direction, the amount of parallelization NX in X direction and NY in Y direction is limited to NX = FOV/LX and NY = FOV/LY, respectively. The most common techniques for beam splitting are the use of diffractive optical elements (DOEs) [31, 33, 116] and spatial light modulators (SLM) [31, 32, 34, 117, 118].

A DOE is a transparent element with a microstructured surface. This surface consists of areas with at least two different height levels. Light passing a DOE travels varying optical path lengths, depending on the level height of each area and depending on the refractive index of the material. This leads to a phase modulation of the incoming beam, creating interference of the beam behind the DOE. Constructive interference occurs only at specific angles, and thus multiple beams emanate from the DOE. The specific angle of each beam results in a defined displacement of this beam's focus in the focal area of the microscope objective, as shown in Fig. 5.17. The intensity of each single beam can be varied by the design of the DOE, according to the requirements of the specific application. The benefits of DOEs are relatively low prime costs and simple implementation into existing lithography or HP3DP systems, respectively. However, a standard DOE creates a fixed pattern of foci. Changing the pattern of foci requires the use of a different DOE, which hampers the use of DOEs in systems that produce rapidly changing designs.

SLMs are optical devices that are most commonly based on liquid crystals (LC) which alter phase, amplitude, or polarization of a laser beam via a pixel grid [160, 161]. An SLM that changes the phase is referred to as phase SLM, while an SLM that changes the amplitude is referred to as amplitude SLM. Commercially available SLMs are available in a reflective or transmissive layout. Reflective SLMs

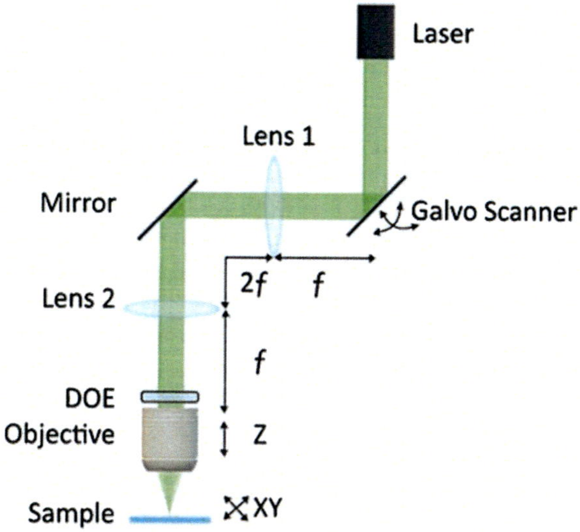

Fig. 5.17 (**a**) Diffractive optical element (DOE) splits an incoming laser beam into multiple beams. Each beam is focused by the microscope objective to a different lateral position, depending on the beam's diffraction angle. This way, the fabrication can be parallelized for the simultaneous fabrication of multiple similar structures

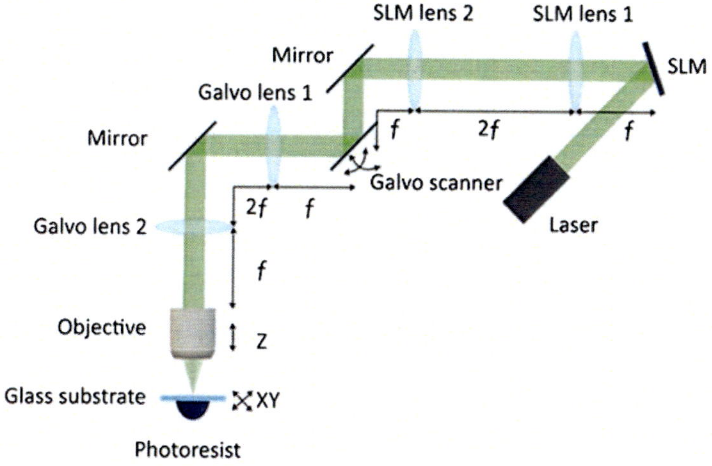

Fig. 5.18 A spatial light modulator (SLM)—shown here in reflective layout—splits an incoming beam into multiple beams. Unlike a DOE (Fig. 5.17), an SLM is able to switch between different diffraction patterns and thus can be used as versatile parallelization technique for varying applications

mostly rely on liquid-crystal-on-silicon (LCoS) technology, while liquid-crystal-on-glass (LCoG) technology is widely used for transmissive SLMs. Similar to a DOE, the SLM splits the incoming beam into multiple beams, which are then focused by the microscope objective for printing. Figure 5.18 depicts parallelization with a reflective SLM. An SLM requires electronic control of the individual pixels. It allows on-the-fly modification of the phase mask. This, in contrast to a DOE, enables the realization of different focal patterns and applications with a single

SLM. The pattern can be changed from application to application as well as also during a single fabrication job. This enables the simultaneous fabrication of nonperiodic structures and to speed up the fabrication of an individual structure [117, 119]. These additional possibilities of an SLM compared to a DOE come along with higher cost for purchase and implementation. Another drawback of SLM is the usually strong zeroth-order intensity. This undesired zeroth order is often blocked with complex setups [117, 118, 120].

In the following, focus will be on the parallelization using DOEs, while all other cases will be described elsewhere. In contrast to an SLM, DOEs can be manufactured so that they produce a phase pattern of much better quality. This results in the possibility to generate much more beams with a DOE than with an SLM. Typically, SLMs are used to generate a few beams up to no more than 36 beams [32, 117, 118]. The zeroth-order beam is often so strong that it needs to be blocked to prevent undesired illumination. This results in low efficiency of the overall setup. DOE can be used to generate 121 beams and more, as shown in the Horizon 2020 project PHENOmenon [121]. Skilled DOE manufacturers are able to precisely control the zeroth-order beam, enabling high efficiency. Figure 5.19 shows 121 pyramids fabricated in parallel with a 11 × 11 beam splitting DOE35. Fabrication time for this array was 70 s. Serial fabrication without the DOE would have taken 121 times longer, resulting in a fabrication time of approx. 2.5 h (Fig. 5.19).

Recent research indicates that multiple beams in close vicinity may cause a proximity effect, which causes a decrease of the polymerization threshold, especially in the center of the array [35, 116]. It was demonstrated that this effect depends on the choice of the objective, the photoresist, and the photoresist layer thickness [116]. Thus, a fabrication with a large number of beams is possible when suitable

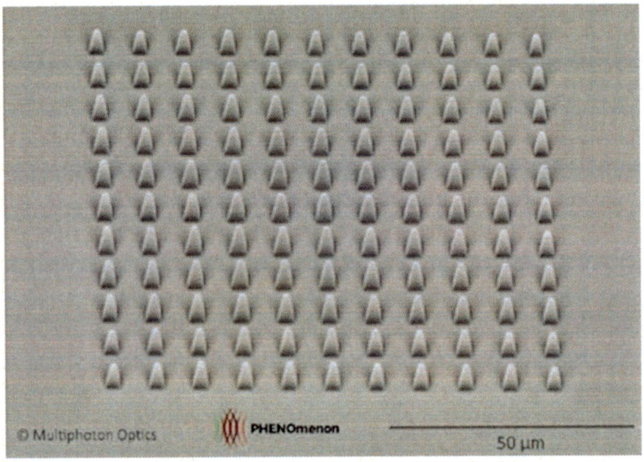

Fig. 5.19 Pyramid-like structures of 4 µm × 4 µm × 12 µm in size, fabricated in parallel with a DOE-generated (11 × 11) beam array. The total fabrication time was 70 s. Compared to serial fabrication, i.e., with a single laser beam, the fabrication would have taken 2.5 h at the same scan speed [35, 121]

measures are taken to account for the proximity effect [35, 116, 121]. Further works are under progress to better understand the underlying effects in multiple beam exposure and to further increase the number of involved beams.

5.4.5 *From Micro- to Macro Optics*

Although the technology of commercially available printing equipment using multiphoton polymerization is continuously improving, the fabrication mainly focuses on structures with dimensions on the micro- and mesoscale. This is due to the limitations imposed by the size of the FoV which is typically around 0.5 mm or much lower for high NA microscope objectives dependent on the magnification, the available traveling range of traveling stages, and individual structures without taking the needs of packaging technologies into account.

Larger structures have been demonstrated by stitching, however, inducing joints which are disadvantageous, especially for optical applications [98]. In some cases, a post-processing can be applied to reduce these stitching artifacts [122], but not for direct fabrication on active and passive devices [123]. For this purpose, a detailed knowledge of useful manufacturing strategies, as described in Sect. 5.4.2, provides the efficient fabrication routines to create structure from the sub-micrometer (see Fig. 5.20a) to the centimeter range (see Fig. 5.20b) without relying on stitching. The minimum feature sizes in the array structures shown in Fig. 5.20a are about 300–400 nm laterally and about 100 nm in thickness. The (2 × 2) cm^2 structure in

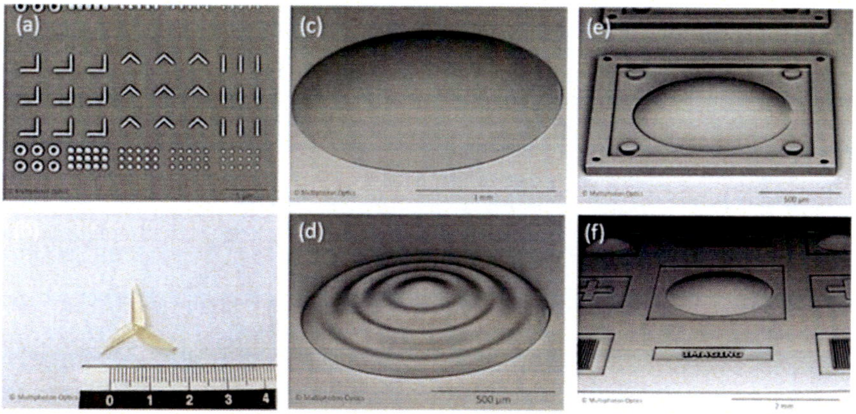

Fig. 5.20 SEM images of different structures fabricated by HP3DP. All structures were fabricated in one fabrication step without any stitching. (**a**) Meta-structure study. The smallest dots are around 300–400 nm in diameter with a height of ≈100 nm. (**b**) Fully sliced and hatched propel of 2.5 cm in diameter, fabricated in a bath setup. (**c**) shows a microlens of 2 mm in diameter and (**d**) a free-form lens structure. (**e**) Microlens with a diameter of 1 mm in situ fabricated with alignment structures and (**f**) larger-scale area with microlenses, grating and alignment structures, and text features as replication master which is part of a 7.5 mm × 7.5 mm element

Fig. 5.20b was fabricated in a full-volume mode with coarser printing resolution using the bath 3D lithography exposure mode (Fig. 5.10c). This represents a scalability in size of up to six orders in magnitude—from sub-micrometer to centimeter—and thus demonstrates that HP3DP as special case of 3D lithography closes the gap between conventional 3D printing and conventional nano-/microfabrication.

As it is obvious to achieve shorter fabrication times, the printing resolution could be reduced, supporting the fabrication of larger structures in reasonable printing times. The fabrication of large-scale optical elements with smooth surface is extremely challenging, and reducing the printing resolution is often not an option in industrial manufacturing. In general, a full-volume structuring mode in combination with a stage-only fabrication mode can be applied. But due to high printing resolutions necessary to achieve the optical surface quality in one process, the resulting fabrication time may be uneconomic. For this reason, special trajectory modes like LCON 3D and FALCON 3D (Fig. 5.14) are perfectly suitable to create mm-sized lenses.

As all 3D printing technologies enable complex designs, HP3DP additionally offers the advantage of fabrication with printing resolution and manufacturing modes on demand depending on the required shape accuracy and surface quality. For microoptics, ultrasmooth surface finishes can be achieved. Aside from this, additionally features can be fabricated within a single process step and no consecutive mounting which is exemplarily demonstrated for a boxed mm lens in Fig. 5.20e. This design inherently provides alignment features for further processes, for example, step-and-repeat fabrication in imprint lithography. Additionally, even more complex feature for branding (text fields) or security reasons (gratings) can be in situ fabricated in one single writing process as shown in Fig. 5.20f which shows part of a 7.5 mm × 7.5 mm structure including optical elements up to a size of 2 mm in diameter, also with much finer grating features (not shown).

For optoelectronic applications, microlens arrays (MLAs), diffractive optical elements (DOEs), diffusers, and prisms are an essential part in sensing purposes (illumination, imaging, and homogenization). Especially the capability of HP3DP to manufacture complex designs in a single process step allows novel lens systems without the need for complex and costly alignment procedures of the individual lenses with respect to each other [47]. Besides overall dimensions of a single element in the macro range, large-area fabrication is also possible. This is demonstrated for microprisms in Fig. 5.21a which were fabricated on an area of 4 mm^2 in total; a zoom into the structure exhibiting extraordinary homogeneity is depicted in Fig. 5.21b. Figure 5.21c and d show homogeneous and customized MLAs with a size of 1 cm^2 each for microlenses of 1 mm in diameter at the same height of 280 μm (Fig. 5.21c. Lenses in Fig. 5.21d are of varying size and shape with diameter ranging from 0.5 to 1.0 mm and height from 22 to 270 μm. Due to the extraordinary high degree of freedom in structural features, sizes, and fabrication routes to create a 3D structure in a timely fashion, the additive mode of 3D lithography is an ideal tool not only for the prototyping phase where a design can be iteratively optimized but also for production where individual 3D elements or complete masters are fabricated to allow small series up to volume production. As completely compatible with

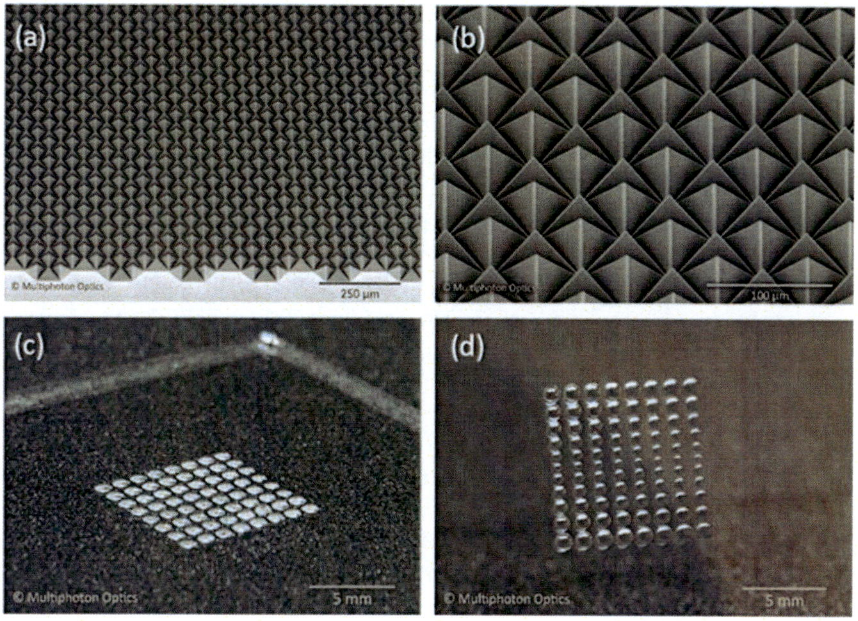

Fig. 5.21 Retroreflector master as fabricated via 3D lithography in its high-precision 3D printing mode. (**a**) Section out of the master structure which is 4 mm^2 in total dimension. (**b**) Zoom into the structure which was used further on to fabricate a Ni-shim via electroplating (not shown here)

replication methods (UV-NIL, hot embossing, galvanization), the masters produced by HP3DP are an integral part of the supply chain and can be used as a tool for mass manufacturing of volume manufacturing tunable from 1D to 3D structures.

5.4.6 From Curved Optics to Flat Optics

Like no other fabrication technology, 3HP3DP enables high-quality fabrication of curved refractive optical elements/lenses, thin diffractive optical elements, and extremely flat optics which are also known as metaoptics, metamaterials, or metasurfaces. Classical manufacturing techniques for each individual kind of optics exist to some extent, but they are either not suited for fabrication of the required designs, or they can fulfill the requirements to some extent, but need extremely high effort, are time-intensive, and are costly. 3D lithography with its integral part of HP3DP provides a complimentary technology approach which can be easily integrated in any processing line (Sect. 5.4.3) and provides the necessary tool set to manufacture all these different kinds of optics within a single equipment, far exceeding the limits of any fabrication technology by covering several orders of magnitude in size range from the nm up to the cm range.

The lensmaker's equation for thin lenses describes the relation between the curvatures R_1 and R_2 of the two sides of a conventional curved refractive lens, its refractive index n, and its focal length f [124]:

$$\frac{1}{f} = (n-1)\left(\frac{1}{R_1} - \frac{1}{R_2}\right) \tag{5.4.2}$$

From this equation, it can be seen that the focal length f of a refractive lens depends on the curvatures R_1 and R_2 of the lens' surfaces. The smaller f, the smaller are the lenses curvature radii R_1 and R_2. Thus, refractive lenses with short focal length f end up relatively thick, depending on the required diameter of the lens. Refractive lenses are widespread in a vast amount of applications, since they are easy to manufacture with classical fabrication techniques like molding, diamond turning, grinding, and polishing [125, 126]. HP3DP can be used for the direct fabrication of classical curved lenses with highest surface quality and without the need for subsequent grinding or polishing. This is particularly useful in cases where lenses are fabricated inside a module or assembly, or when lenses of different size and shape are combined. Thus, HP3DP is complimentary to conventional ultraprecision processing.

However, there are applications where lenses should be as flat as possible and hence classical lenses are unsuitable due to their thickness (Fig. 5.22a). The concept of a Fresnel lens was developed in the nineteenth century and is named after the French physicist Augustin-Jean Fresnel [127]. Nowadays, Fresnel lenses are well-known in optics [128]. A Fresnel lens is a refractive optical element based on ray optics, like a conventional refractive lens. Since the focal length is determined only by the surface curvature of the lens, traveling through bulk material does not add to the optical function. The idea of a Fresnel lens is to reduce the thickness of a lens by greatly removing bulk material while keeping the curvature of the lens (Fig. 5.22b). This way, the lens is reduced in thickness, but has still the same optical properties as a conventional curved lens. However, the fabrication of Fresnel-type lenses, especially for microlenses, is difficult with standard fabrication methods like diamond turning, because of the narrow high aspect ratio grooves which require the

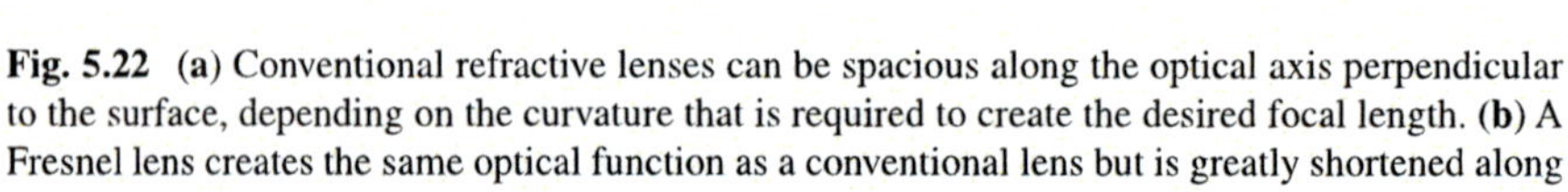

Fig. 5.22 (**a**) Conventional refractive lenses can be spacious along the optical axis perpendicular to the surface, depending on the curvature that is required to create the desired focal length. (**b**) A Fresnel lens creates the same optical function as a conventional lens but is greatly shortened along the optical axis. It has the same radius of/surface curvature as a conventional lens, while bulk material below the surface is reduced. (**c**) Metamaterials can be designed to show the same or even improved optical functions compared to conventional lenses. Since metamaterials consist of subwavelength elements (so-called metaatoms), these are ultra-flat optical elements with a thickness of 2 μm or smaller, for example, for VIS light

corresponding tools. HP3DP overcomes this issue and allows the fabrication of highly precise Fresnel-type microoptics [12, 129].

Diffractive optical elements (DOEs) rely on the diffraction of light and can be designed for various applications, like beam splitting and beam shaping. A comprehensive insight on how a DOE works is given in reference [130]. Standard UV photolithography processes are commonly used for the fabrication of DOEs. The diffraction efficiency of a DOE rises with the number of fabricated levels. However, the fabrication of a multilevel DOE with standard UV photolithography processes requires the subsequent illumination with multiple different, expensive photomasks and a complex alignment procedure for each mask. Since HP3DP is a mask-less technique, the printing of DOE [131] saves the user the expensive production of masks and the complex alignment procedure alongside with a much higher degree of freedom in optical design [15]. This allows much faster and cheaper development of DOEs with regard to the iteration process of design, test, and redesign.

Flat optics is a term that describes optical elements, whose thickness is in the order of magnitude of the operating wavelength and are also referred to as **metaoptics** or **metamaterials** [132–136]. In contrast to classical DOEs, flat optics are composed of subwavelength-sized elements, also known as metaatoms. Metaatoms are typically shaped like lines or cylinders [137], and they are arranged such that a desired phase change is imposed on the incoming light to exhibit a specific optical function, for example, focusing or defocusing. A crucial requirement for the fabrication of metaoptics is the possibility to make metaatoms that cover the whole range of phase changes between 0 and 2π. The fabrication of flat metaoptics is demanding due to the fact that the lateral extension of the required metaatoms is well below 1 μm for visible light. However, recent results clearly demonstrate that HP3DP is a suitable tool for metaoptics fabrication in one single processing step [138]. Since HP3DP is based on two-photon absorption (TPA), a resolution far below the diffraction limit of the laser wavelength can be achieved (see also Sect. 5.3.1).

Figure 5.23a shows the worldwide first metalens designed for visible light ($\lambda = 633$ nm) that was fabricated by 3D lithography in the additive working mode in one single writing step in a polymer material. It has a total diameter of 60 μm, and the metaatoms which compose the lens have diameters ranging from 100 to 450 nm and heights ranging from 300 to 1000 nm. The fabrication time was set to be 10 min which is extraordinary slow, but directly supplied the full metaoptics in the very first step. Compared to e-beam writing, this fabrication time is extremely fast and easy. The focal length was very roughly determined to be around 100 μm, and further characterization is presently taking place which will provide more precise values and which will be described elsewhere. In Fig. 5.23b, a (2 × 2) array of metalenses having the same design as shown in Fig. 5.23a is displayed which clearly demonstrates the reproducibility of the fabrication. Metaoptics which are fabricated with classical semiconductor fabrication processes usually end up with a fixed height of all metaatoms. In contrast, HP3DP allows to individually adjust the diameter and height for each metaatom. This provides a much higher degree of freedom in the design of metaoptics and will enable extremely flat optics for camera optics for cell phones or for AR/VR application, among others.

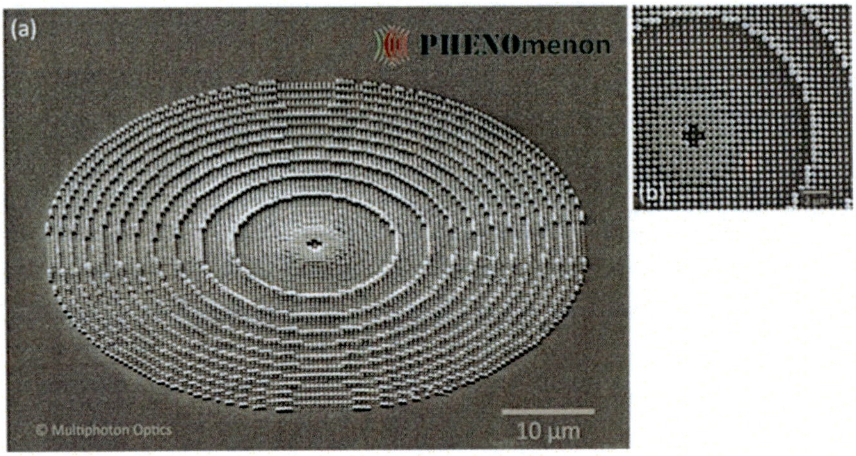

Fig. 5.23 SEM images of the worldwide first metalens for visible light ($\lambda = 633$ nm), fabricated in one step via 3D lithography in the additive working mode HP3DP using a purely organic material [121]. (**a**) Full metalens with metaatoms are arranged as pillars with diameters ranging from 100 to 450 nm and heights ranging from 300 to 1000 nm. (**b**) Zoom into the structure shown in (**a**), revealing the excellent reproducibility of the structures and their arrangement

5.4.7 Resolution

This section gives a brief overview over the spatial and lithographic resolution which can be achieved in 3D lithography using TPA as a fabrication mode. These resolution definitions play an important role in classical industrial manufacturing of devices, for example, in CMOS production. The same definitions apply to 3D lithography but are unfortunately often confused in literature.

Resolution in the literature dealing with TPA and 2PP often is referred to as "structural feature size." The terminology which is used to describe the results will be defined in the following to allow a distinction between features and resolution, describing:

(a) Feature sizes
(b) Optical resolution
(c) Lithographic resolution

A feature is a structural element which has a certain size, for example, a length, a height, a diameter, and an aspect ratio. Optical and lithographical resolutions are strongly related to each other. In a very simple picture, the optical resolution is given by the fact that two features can be separately recognized as two individual features (Rayleigh resolution limit). The lateral resolution is in the ideal case given by $k\lambda/\mathrm{NA}$, where k is a constant ranging between 0.4 and 0.6 depending on the cut-off definition, λ is the wavelength of the system, and NA is the numerical aperture of the optical system. From this it can be clearly deducted that the smaller the

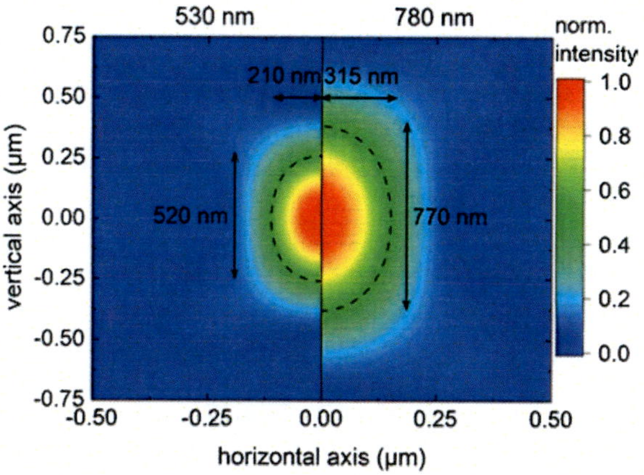

Fig. 5.24 Comparison of the normalized PSF intensity simulated for a 1.4 NA oil immersion objective at the bottom of a 170-µm-thick Zeiss cover glass (design case) for 530 (left) and 780 nm (right) laser light wavelength. The dashed lines indicate the FWHM. Clearly, green light provides a smaller fabrication volume, since the Rayleigh resolution is proportional to the laser wavelength

wavelength, the better is the optical resolution, i.e., green light is better suited for high-resolution fabrication than NIR light. Figure 5.24 shows the simulation of the point spread function (PSF) for an ideal design case using an oil immersion objective with an NA of 1.4 and a Zeiss cover glass of a thickness of 170 µm. For better illustration, only half of the PSF is shown for each wavelength. For a TPA process, the FWHM of the PSF is $0.37\lambda/\mathrm{NA}$.

The same is true for the lithographic resolution, but with one further restriction: the lithographic resolution is strongly influenced by the aspect ratio of the neighboring features; the higher the aspect ratio, the higher are capillary forces which prevent the development of the structures after exposure if they are arranged very close to each other. In the following, examples for minimum feature sizes and optical and lithographic resolution will be given, while for the latter also, the aspect ratio of a structure will be taken into account.

Feature sizes around or below 100 nm are reported in the literature [4, 13, 139–142]. Most of the results so far have been achieved with commercial or experimental materials, and some of the materials may include cumbersome post-processing procedures or may often not suited to be implemented in industrial processing, since they are only accessible if certain equipment will be used. Aside from this, they are often not tested concerning process and environmental reliability in Telcordia (Bellcore) testing. Especially for optical materials which as used in system-in-package (SiP) or on-chip packaging, the materials have to fulfill many different properties in the various application wavelength regimes, among which are no yellowing, high transparency or low absorption, suitable refractive index, and high mechanical, thermal, and chemical stability according to Telcordia (Bellcore) reliability testing. A material class which has gained a tremendous attention in industrial manufacturing is the class of inorganic-organic hybrid materials such as

ORMOCER®s (see Sect. 5.3.2) which provides superior materials with highest reliability in industrial manufacturing, suited for the fabrication of ultra-flat optics and optical lenses on the micro-, meso-, and macroscale (Fig. 5.20).

Let us first consider the formation of voxels with the underlying parameters using an inorganic-organic hybrid polymer. Voxels and continuous lines are the smallest building blocks which are used to create a structure. The formation of voxels has been widely investigated in the past using the ascending scan method [79, 139, 143]. In general, the size of a single voxel depends on various parameters and the way how they are created. Typically, most of the methods create the voxels in a "stop-and-write" process, i.e., the laser beam is focused on the location where the voxel is to be written followed by the exposure, and then the next voxel is written by moving to the next position. This procedure is repeated until the entire array is finished. A much smarter way of creating large arrays is to write the voxels "on the fly" among other procedures which is much faster and requires an excellent understanding of the underlying chemical processes occurring in light-matter interaction and on the engineering of the 3D lithography tool. Fabrication "on the fly" is integrated in the equipment (LithoProf3D®) used for structuring and allows the access to an extraordinary high throughput. As for classical lithography, the precise control of the exposure dose (product of power P times the exposure time ton per exposed area) is an important prerequisite to fabricate the required structures with high throughput and a high degree of control. Thus, it is important to understand the light-matter interaction of the given materials in dependence of the average laser power and the on-time ton to allow a precise control of the formation process of the resulting structures.

Voxel sizes are typically fabricated using the ascending scan method [79, 139, 144] which was used to determine the feature sizes of the inorganic-organic hybrid polymer OrmoComp in its commercially available formulation[4] in dependence of the average laser power and the on-time ton. The Z position of the focal volume (writing layer) is varied by 100 nm from line to line, and the voxel is printed partially on the substrate depending on the Z position and the focal volume length. For a very low Z position, most of the voxel is written in the substrate while only a small portion of the elliptical voxel appears at the surface. The substrate is not affected by this procedure. Then, the Z position is changed for 100 nm, and the fabrication of a voxel line is done again. This is continued until the voxels start to flip. If the Z position is chosen such that the entire focal intensity distribution does not touch the substrate's surface anymore, the voxels are mainly washed away by the developer. Only those which are still covalently linked to the substrate can be used to determine the voxel length. This method was also employed for quantifying spherical aberrations by experimentally mapping the optical point spread function which directly manifests itself in the shape and size of an individual voxel, and this was used to improve the fabrication of macroscopic 3D scaffolds for regenerative medicine [145].

An exemplary SEM image is displayed in Fig. 5.25a. The structures were fabricated using an average laser power P of 6 mW and an on-time ton of 1 ms with the

[4] Licensed by the Fraunhofer Gesellschaft für Angewandte Forschung e.V. to micro resist technology GmbH.

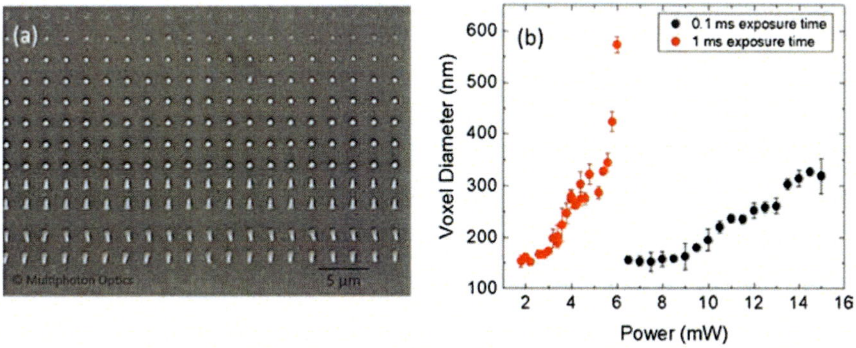

Fig. 5.25 (**a**) Exemplary SEM image of a voxel array fabricated at an average power of 6 mW and an on-time of 1 ms as a result from the ascending scan method for the determination of voxel diameter and length. (**b**) Voxel diameter in dependence of the average laser power for two on-times: 0.1 and 1 mW. The experiments were carried out using OrmoComp in its standard formulation (1 wt.-% Lucirin TPO)

Z height of the focal volume at its lowest level at the top of the array with increasing height to the bottom of the imaged array. At a certain height, the voxels flip over, and once they are lying on the substrate, the voxel length becomes accessible. Finally, Fig. 5.25b shows the dependence of the voxel diameter for two different exposure times ton for voxels fabricated by the ascending scan method. By choosing a certain on-time for each individual voxel (exemplarily set to 0.1 and 1 ms, respectively) and varying the laser power, the size of the voxel can be varied, whereas it has to be mentioned that the voxel length is typically much more altered than the voxel diameter due to the elliptical shape of the focal volume provided that no beam shaping is considered and the influence of the photoinitiator system is neglected. As obvious from Fig. 5.25b, the voxel size for the given material system can be dynamically adjusted between about 150 nm for the smallest diameter up to about 300 to 600 nm for the two different on-times. Note that the on-time of 1 ms results in a dynamic range for the voxel diameter which is about twice as large as for the lower on-time of 0.1 ms at a much broader average power interval, i.e., resulting in a larger fabrication window at the given parameters. For a range of average laser powers, the size of the voxels is saturated on the minimal diameter of approximately 150 nm, indicating that not the threshold volume is the limiting factor in feature size, but the chemical interaction volume [4]. Data on selected material systems are already reported in the literature [4, 79], and more comprehensive investigations will be described elsewhere.

By using highly efficient photoinitiators formulated into base matrix resins or certain post-processing procedures, further reduction in feature size well below 100 nm is possible [142, 146, 147]. Houbertz et al. [4] have demonstrated that sub-100 nm structures can be created in styryl-based materials without post-processing, if not desired [86] (Fig. 5.9). It has to be mentioned, however, that post-processing, for example, thermal treatment, might result in a slight increase in

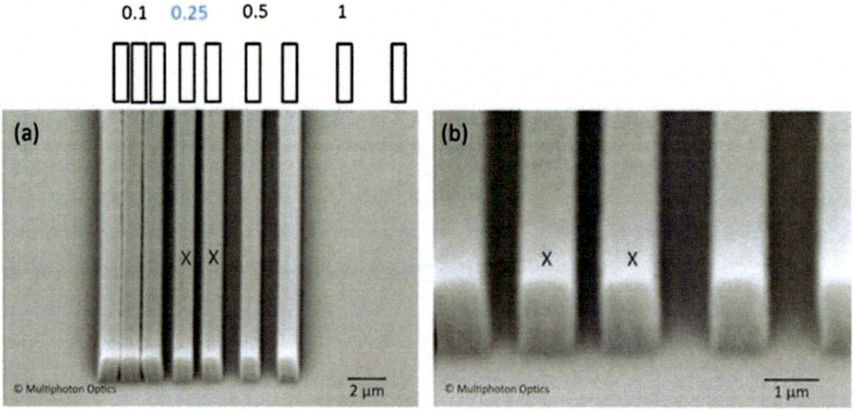

Fig. 5.26 Resolution pattern of ORMOCER® structures with a designed aspect ratio of 2:1, fabricated 0.5 µm below the substrate's surface, i.e., resulting in an aspect ratio of 1.5:1. (**a**) shows an overview over the structure defined by the schematic sketch with gaps between the structures ranging from 0.1, 0.25, 0.5, to 1 µm, showing an optical resolution of about 100 nm and a lithographic resolution of about 250 nm or better. The structures with 1 µm gap were not fabricated. (**b**) Zoom into the structure marked with two crosses in (**a**) and (**b**)

cross-linking dependent on the material system, thus affecting the refractive index which is then also increased to a certain extent [90].

Let us now consider structures which have a certain size and aspect ratio, being formed by continuous laser light exposure while moving the beam and the sample in the 5-axes equipment. Figure 5.26a shows the schematic sketch of a resolution structure with lines of several 10 µm in length and a line width of 1 µm. The lines are arranged such that there are gaps of 100, 250, 500, and 1000 nm between them. Since the resolution was found to be much better than 1000 nm, lines with this spacing have not been fabricated.

The height of the extended structures was set to 2 µm by design which would result in an aspect ratio of 2:1. Since the structure was positioned 500 nm below the substrate, the aspect ratio is a bit altered to 1.5:1 for the given line structures. Please note that these structures are completely compliant with classical 2D lithography such as UV or grayscale lithography, since 3D lithography allows one to create 1D, 2D, 2.5D, and 3D structures by simply changing the process parameters and undercuts and specially formed structures can be fabricated much easier and without the limits of UV lithography and without the constraints of grayscale lithography.

The result of the line pattern is shown with the SEM image in Fig. 5.26b, clearly demonstrating that individual structures can be identified even for a distance between the structures of 100 nm which might be described in the Sparrow limit [148]. At this distance, the interaction forces (capillary forces) between the structures are so high that they touch each other as a result from the development step which is employed to remove the nonexposed material out of the gaps between the structures. At a distance of 250 nm between the structures, the structures do not touch each other anymore, and the gap between them is clearly visible which is also

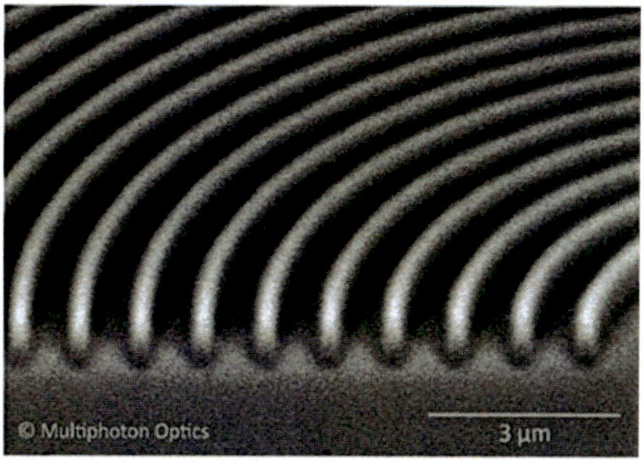

Fig. 5.27 Grating with bended ringlike structures of about 900 nm height and 300 nm µm lateral extension

shown in the zoom into the structure in Fig. 5.26b, where one clearly can distinguish between optical and lithographic resolution: for a distance of 100 nm between the structures, the optical resolution in the Sparrow limit is 100 nm, while there is no lithographic resolution, i.e., the structures cannot be developed such that the gaps between the lines are free of material. For a distance of 250 nm between the structures, the optical and the lithographic resolution is 250 nm, since the gap is free of residual material and the structures can clearly be distinguished. A close inspection of the structures shows that the gap might be further decreased to some extent without losing neither optical nor lithographic resolution. Other material systems might reveal even better performance which will be published elsewhere.

Finally, an array of circularly shaped lines of 300 nm in width and about 900 nm in height, resulting in an aspect ratio of 3:1, was fabricated (Fig. 5.27). These special types of gratings have attracted considerable attention and will be employed in industrial processing and will be further investigated.

5.5 Beam Shaping for Sensor Products

5.5.1 NIR Laser Dies for Gas Sensing

Tunable diode laser absorption spectroscopy (TDLAS) is a more and more important sensor technique that enables to quantitatively measure the presence of extremely low levels of gases even in harsh environments. The method is based on single-mode tunable diode laser sources, e.g., a distributed feedback laser matched to a characteristic absorption line of a specific gas, a gas cell with the absorbing medium, a detector, as well as supporting optics for laser beam guidance. The laser's emission wavelength is tuned over the absorption line of the gas under investigation which leads to a reduction of the transmitted laser intensity that is recorded with a photodiode. Dedicated electronics and wavelength-modulation spectroscopy with

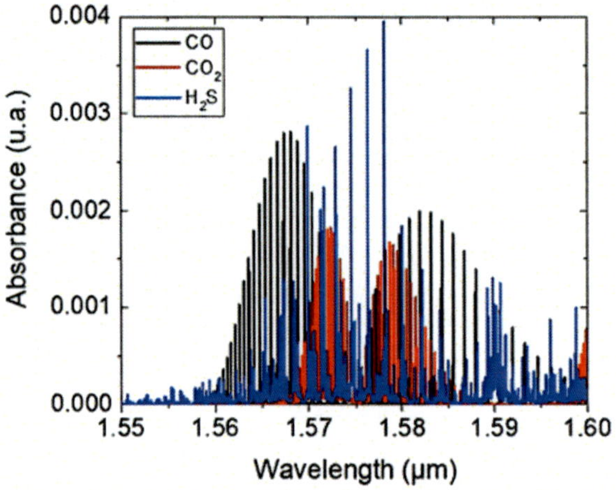

Fig. 5.28 Absorption features of CO, CO_2, and H_2S in the wavelength region between 1.55 μm and 1.60 μm [25]

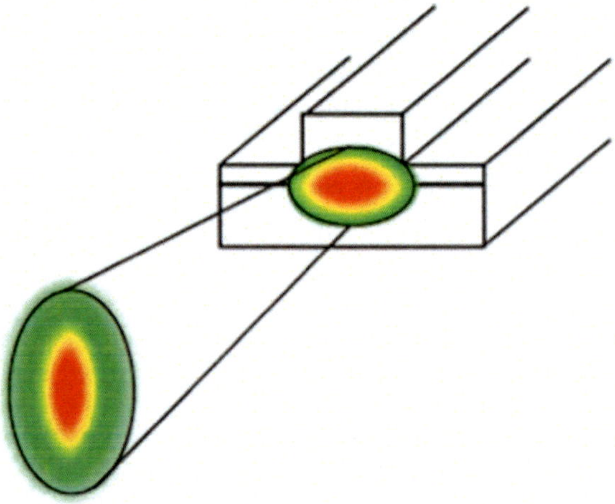

Fig. 5.29 Schematic of near- and far-field emission characteristic of an edge-emitting InP-based DFB laser device

lock-in technique is often employed to enhance the sensitivity. Typical sensitivities go down to *ppm* and even *ppb* fractions [149]. Figure 5.28 shows the characteristic fingerprint region of several industrially relevant gases such as CO, CO_2, and H_2S in the wavelength region from 1.55 to 1.60 μm.

Edge-emitting single-mode distributed feedback lasers are widely used for TDLAS spectroscopy over a wide wavelength range, e.g., 760 nm up to 16 μm [150]. Due to their intrinsic laser structure, the mode profile of the emitted light is highly divergent in the vertical direction, which leads to the need of costly beam shaping optics for spectroscopy. Figure 5.29 schematically depicts an edge-emitting DFB laser and its emission characteristic. In the vertical direction, the light is diffraction limited by the thickness of the epitaxial layers (typically around 1–2 μm). Due to the small aperture size, the divergence in this direction is relatively high and is called fast axis. The lateral light mode distribution is determined by the ridge waveguide geometry of the laser with less confinement in comparison to the vertical direction. Consequently, the beam divergence in the far field is much weaker compared to the vertical direction, and the axis is called slow axis. Typical values for fast- and slow-axis beam divergence for DFB lasers used in this study are 40° to 50° and 15° to 25°, respectively. Thus, the beam profile is highly elliptical. This limitation of the edge-emitting DFB devices can be overcome by direct laser processing of a 3D-printed lens as described in the following chapter.

The lasers used for this study are metal-grating DFBs emitting around 1590 nm based on InP epitaxial material. For device fabrication, the ridge wave guides are processed on the epitaxially grown laser crystal by several consecutive lithography and deposition processes. Single-mode operation is achieved by structuring a metal grating along the ridge. The laser processing is completed by deposition of metallic top and bottom contacts followed by a galvanic top layer for optimized head removal. Finally, the lasers are singularized, and the back facet is coated with a highly reflective layer, whereas a simple passivation layer is applied onto the front facet on which the 3D lens will be structured. For optimized handling and head removal, the lasers are soldered to C-mounts. Details on laser fabrication can be found, e.g., in Ref [150, 151].

5.5.2 *High-Precision 3D Printing for Laser Die Packaging*

Conventional lithography and 3D printing technologies are limited in the type of substrate which can be employed for structure processing. In classical lithography, substrates are typically restricted to flat substrates such as wafers. Structures in 3D printers are usually printed on flat mounts (e.g., a platform) and build up in a layer-by-layer fashion. For example, in SLA printing the structures are printed from a CAD file in a layer-by-layer fashion: a layer is coated by a blade, and the first layer is printed. The platform is then lowered, the second layer is coated by the blade, and the procedure is repeated until the entire 3D structure is formed. Classical 3D printing such as SLA, SLM, or FDM is not precise enough to account for optical surface quality.

As 3D lithography with its intrinsically available HP3DP mode provides real 3D printing capabilities without the need for layering with deposition of new material by wiping sequences at a very high precision, the process is not limited to special types of substrates. For this reason, optical elements can be printed directly on almost any kind of substrate formats, mounts, and assemblies [6]. Particularly, if optical functional devices have to be fabricated, ease in fabrication is generated by the fact that optics can be directly fabricated on pre-configured substrates, for example, printed circuit boards [6, 25], mounted laser dies (see Sect. 5.5), and LED chips [152], among other devices. Typically, in photonic packaging integrated optics are assembled on optoelectronic devices by active and passive alignment which is a complex task and relies on several process steps [153]. As HP3DP is not limited to certain substrates, many applications were impressively demonstrated for integrated optics [6, 107, 114, 129, 152, 154–157], and the number of previously necessary process steps can be reduced as assembly is not necessary anymore.

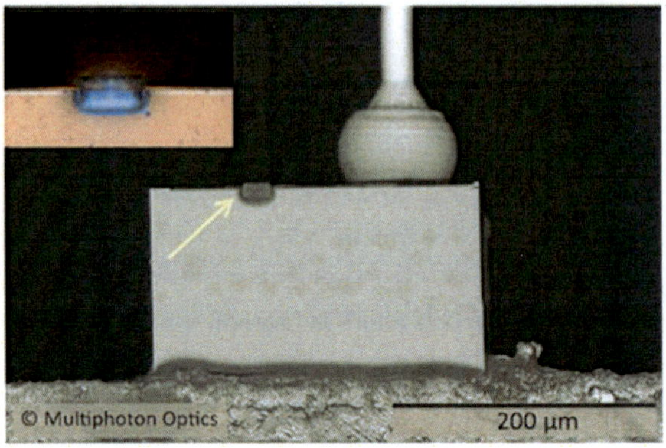

Fig. 5.30 SEM image of a laser die with a cylindrical microlens (marked by an arrow) inherently mounted on the die's facet by direct writing using HP3DP. The inlay of the right picture shows a scanning confocal microscopy image of the lens

By means of 3D lithography, microlenses were directly written on InP laser dies (see Sect. 5.1) whose facets are about 1 µm × 3 µm in size (Fig. 5.30). The process provides an inherent mounting procedure without the necessity of typical assembling steps, for example, optical bonding. Costly and bulky beam shaping optics which are nowadays used for beam shaping can be then avoided. The laser dies were diced and mounted on C-mounts; the dimensions of the total assembly were roughly 0.7 cm × 0.65 cm × 0.2 cm without contact wiring. This assembly is used as substrate for the HP3DP process for the fabrication of cylindrical microlenses which were designed such that the divergence angle of the fast axis is reduced to the value for the divergence angle of the slow axis in order to create a circular beam profile in the far field. The overall size of the optics is reduced by orders of magnitude as a result of the microlens whose thickness is typically below 20 µm. Compared to the macroscopic optical elements which are usually employed for packaging, the size of the package can be also significantly miniaturized.

For faster fabrication, a set of assemblies is placed into the 3D lithography system, and automated routines such as detection of the laser die's facets, translation and rotation of the microlens design, and the printing process are performed by the printer itself. Due to the nature of the fabrication process, only passive alignment is used. This leads to significant reduction of total fabrication time by avoiding active alignment and thus results in much lower costs. At the same time, the functionality is enhanced, and the drastic miniaturization leads to much more flexibility in the packaging process.

5.5.3 *Optical and Life Cycle Characterization*

In order to investigate the influence of the printed lenses on the performance of the DFB lasers, the key characteristics of these devices were studied including output power, far field, as well as long time stability.

Two fundamental characteristics of a laser device can be derived from the dependence of the optical output power *P* on the driving current *I*, i.e., the laser threshold and the efficiency. As described in Sect. 5.1, the facets of the InP-based DFB lasers are typically treated with a highly reflective coating on the back facet and a passivation layer on the front facet. However, the reflectivity of the front facet will be influenced by the printing of a lens onto it, since the contrast between the reflective indices changes. The lens is expected to work as an antireflective coating. The resulting reflectivity is highly dependent on the effective thickness of the lens. Simulations approximating the lens by a homogenous layer indicate resulting reflectivities between 3 and 27%. Figure 5.31 shows the PI curves of a DFB laser device before and after printing of the lens. An increase in the threshold current from 19.2 to 22.7 mA as well as an improvement of the efficiency from 0.18 to 0.23 W/A can be observed. This is in accordance with the expected effect of an antireflective coating caused by the lens. Assuming no changes in the internal losses and a constant reflectivity of the back mirror, a decrease of the front facet's reflectivity from 30 to 28% was calculated which is within the range that is expected from the simulations.

As discussed in Sect. 5.1, a narrow and circular far field is beneficial for many DFB laser applications. An InP-based edge-emitting laser typically emits an elliptic far field with FWHM angles in the fast and slow axis of 40° to 50° and 15° to 25°, respectively. The printed lenses described in Sect. 5.5.2 were designed to reduce the angle of the fast axis without influencing the slow axis. The goal was to create a far field that is as close to circular as possible. Figure 5.32 compares the far fields of InP-based lasers with and without printed lens. Figure 5.32a shows the intensity distribution of the slow axis with and without printed microlens; Fig. 5.32b shows the same for the fast axis. As depicted, the angle of fast axis can be reduced to 19° by a well-designed printed lens, whereas the angle of the slow axis is only affected minimally.

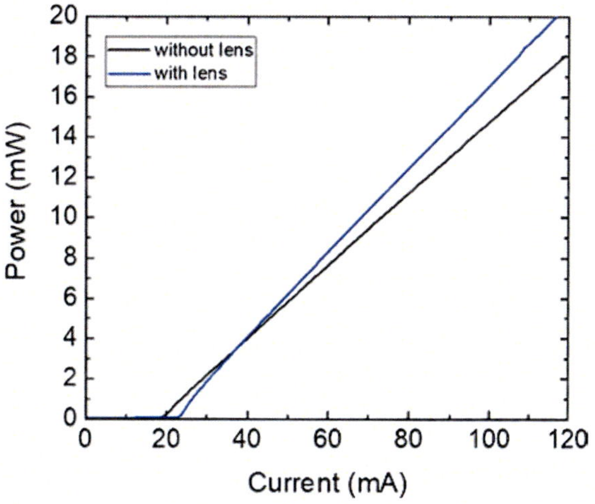

Fig. 5.31 PI curves of an InP-based DFB laser device with and without printed lens on facet

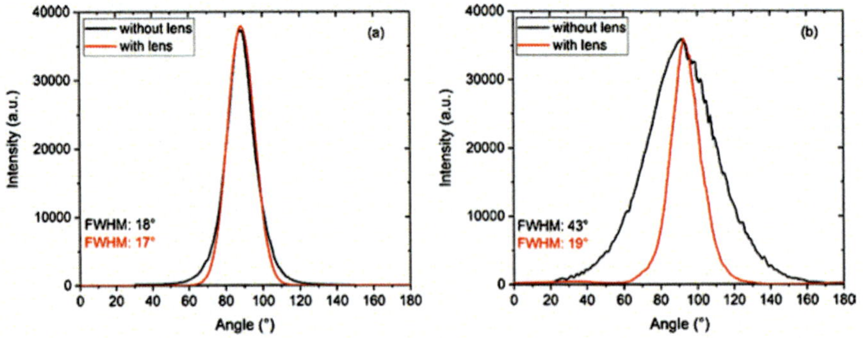

Fig. 5.32 Far-field characteristics of InP-based DFB laser devices with and without printed lens. (**a**) Slow axis and (**b**) fast axis

Fig. 5.33 Effect of printed lens on the polarization of an InP-based laser device

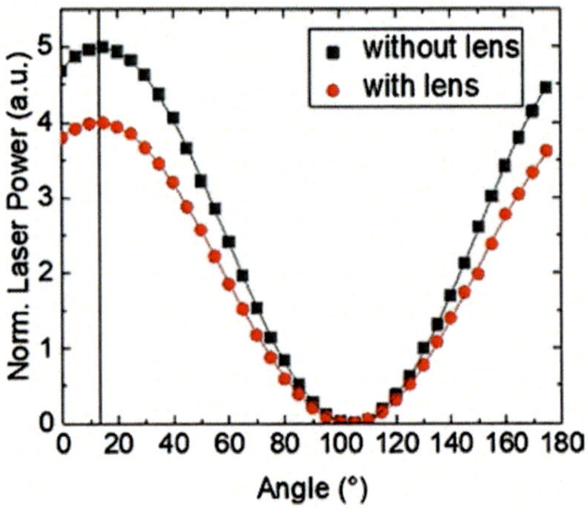

The polarization of the laser light is relevant in certain applications, e.g., if beam splitters are involved. Therefore, the influence of a printed lens on the polarization was investigated. Figure 5.33 depicts the output power of a DFB laser device before and after printing of a lens measured through a polarizing filter. In order to improve the readability, the curves were normalized to different values. As can be seen in Fig. 5.33, the lens has only a minimal effect on polarization.

Another important aspect that needs to be considered when implementing 3D-printed on-chip optics is the possible impacts on the overall system reliability and lifetime. One common cause for laser breakdown is catastrophic optical damage (COD) [158]. Small deteriorations on the facet can result in optically induced thermal destruction of large areas of the facet. This is due to the increased absorption of the propagating light, which heats the facet and can result in melting of the facet, which will lead to a decrease of laser performance up to a total failure of the device. It is therefore of high interest to investigate the influence of the printed lenses on the COD threshold. The InP-based laser devices used in this experiment

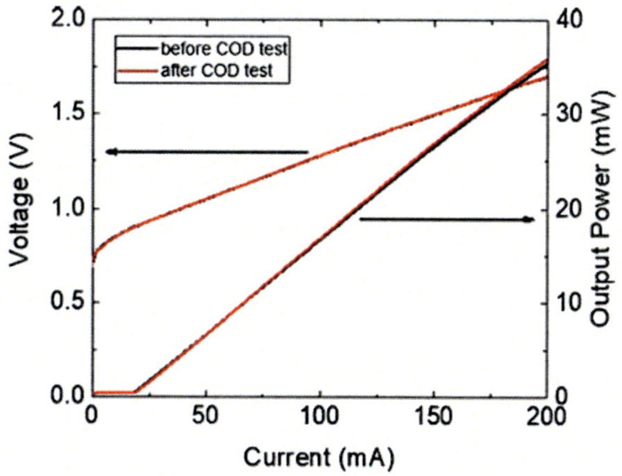

Fig. 5.34 PI curves of a DFB laser with printed lens on facet before and after COD testing

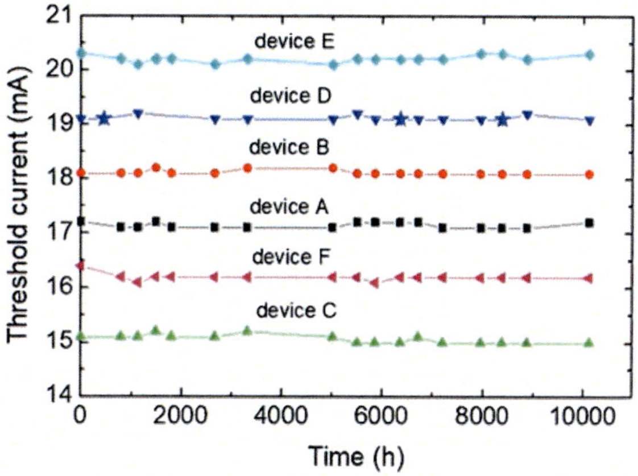

Fig. 5.35 Threshold currents of six DFB lasers with printed lens on facet as a function of operation time. The star indicates a COD measurement

are typically operated at currents below 150 mA. In order to study the likelihood of a COD-based breakdown, the driving current of multiple laser devices was tuned from 0 to 500 mA in several cycles, and PI curves were recorded before and after the test. Figure 5.34 compares the results of an exemplary laser device before and after the COD tests. No negative influence of the COD test on the laser performance could be observed as there are no significant changes in the electro-optical characteristics observable.

The other important reliability aspect to be considered is the lifetime of the lasers with the on-chip lens. Regular DFB laser devices in this wavelength region are specified for >10,000 hours of operation. An indicator for laser degradation and reduced lifetime is rising internal losses, resulting in increasing threshold currents and decreasing efficiency [159]. Figure 5.35 depicts the evolution of the threshold currents of six lasers with on-chip printed lenses over the course of more than 10,000 h of continuous operation at 25 C and 120 mA. The laser device D was also used for the COD measurements mentioned in the previous paragraph. The occurrence of this test is marked by a star in the graph. No degradation could be observed, indicating no negative influence of the lenses on the life cycle of the laser devices.

5.6 Summary

3D lithography with its integral fabrication mode high-precision 3D printing has been demonstrated to provide a universal tool for industrial fabrication of functional optical structures ranging from 1D via 2D and 2.5D to 3D structures. The material class of inorganic-organic hybrid polymers (ORMOCERs) was used to create microoptical elements which can be arbitrarily arranged and shaped, directly influencing the optical performance of a module. The packaging process can be significantly simplified by drastically reducing the amount of necessary steps to create an optical package. This results in lower fabrication cost and consumption of resources. In particular, optical elements can be directly printed on any substrate kind and substrate shape. Aside from the more classical approaches by using glass, wafers, or optical fibers as substrates, complete assemblies can be used in 3D lithography. This was shown with cylindrical lenses which were directly printed on InP laser dies which have revealed extraordinary performance in device and life cycle testing, along with a significant miniaturization of the optical package. Green light is used for the fabrication, and sub-μm as well as sub-100 nm structures were demonstrated. The data show that a well-defined and very controlled fabrication can be achieved for industrial-scale device fabrication and that high-quality surface finishes are routinely delivered upon request. For the first time worldwide, organic polymers were used to create a metalens for the VIS regime by direct laser writing without the necessity of costly processes such as e-beam lithography and subsequent vacuum processes. 3D lithography offers additive, subtractive, and glass processes on any kind of substrate, and a huge variety of materials can be processed. The combination of high-performance materials with sophisticated fabrication strategies results in a significant reduction of the process time. Thus, 3D lithography and particularly its integral part of HP3DP lead to a pronounced acceptance of HP3DP beyond the field of pure research and into nowadays production.

Acknowledgments One of us (R.H.) would like to thank all students and people who supported our works in the last decades, especially S. Steenhusen, Th. Stichel, and M. Timpel. Part of the works resulted from a German Science Foundation Program SPP1327 (grant HO 2475/3-1) project which was carried out in the time of 2009–2014. The cooperation with KIT during the time from 2013 to 2014 is gratefully acknowledged. Finally, we are part of a wonderful consortium in a Horizon 2020 project PHENOmenon (grant 780278) which also supported part of the works.

References

1. Maruo, S., Nakamura, O., & Kawata, S. (1997). Three-dimensional microfabrication with two-photon-absorbed photopolymerization. *Optical Letters, 22*, 132–134.
2. Sun, H. B., Matsuo, S., & Misawa, H. (1999). Three-dimensional photonic crystal structures achieved with two-photon-absorption photopolymerization of resin. *Applied Physics Letters, 74*(6), 786–788.
3. Houbertz, R. (2005). Laser interaction in sol–gel based materials—3-D lithography for photonic applications. *Applied Surface Science, 247*(1–4), 504–512.
4. Houbertz, R., Steenhusen, S., Stichel, T., & Sextl, G. (2010). Two-photon polymerization of inorganic-organic hybrid polymers as scalable technology using ultra-short laser pulses. Coherence and Ultrashort Pulse Laser Emission (pp. 583–608).
5. Serbin, J., Egbert, A., Ostendorf, A., Chichkov, B. N., Houbertz, R., Domann, G., Schulz, J., Cronauer, C., Fröhlich, L., & Popall, M. (2003). Femtosecond laser-induced two-photon polymerization of inorganic–organic hybrid materials for applications in photonics. *Optics Letters, 28*(5), 301.
6. Wüchter, C., Langer, G., Satzinger, V., Schmidt, V., Houbertz, R., Wolter, H., & Kuna, L. (2011). Optical waveguides embedded in PCBs—a real world application of 3D structures written by TPA. *MRS Proceedings, 1054*, 1–12.
7. Gansel, J. K., Thiel, M., Rill, M. S., Decker, M., Bade, K., Saile, V., Von Freymann, G., Linden, S., & Wegener, M. (2009). Gold helix photonic metamaterial as broadband circular polarizer. *Science, 325*(5947), 1513–1515.
8. Aristov, A. I., Manousidaki, M., Danilov, A., Terzaki, K., Fotakis, C., Farsari, M., & Kabashin, A. V. (2016). 3D plasmonic crystal metamaterials for ultra-sensitive biosensing. *Scientific Reports, 6*, 25380.
9. Gissibl, T., Thiele, S., Herkommer, A., & Giessen, H. (2016). Two-photon direct laser writing of ultracompact multi-lens objectives. *Nature Photonics, 10*(8), 554–560.
10. Liberale, C., Cojoc, G., Candeloro, P., Das, G., Gentile, F., De Angelis, F., & Di Fabrizio, E. (2010). Micro-optics fabrication on top of optical fibers using two-photon lithography. *IEEE Photonics Technology Letters, 22*(7), 474–476.
11. Belazaras, K., Žukauskas, A., Piskarskas, A., Malinauskas, M., Tikuišis, K., Gadonas, R., & Purlys, V. (2012). Laser fabrication of various polymer microoptical components. *European Physical Journal Applied Physics, 58*(2), 20501.
12. Zhai, X., Xiao, S., Guo, R., Xia, A., Li, J., & Huang, W. (2006). Micro lens fabrication by means of femtosecond two photon photopolymerization. *Optics Express, 14*(2), 810.
13. Burmeister, F., Steenhusen, S., Houbertz, R., Asche, T. S., Nickel, J., Nolte, S., Tucher, N., Josten, P., Obel, K., Wolter, H., Fessel, S., Schneider, A. M., Gärtner, K. H., Beck, C., Behrens, P., Tünnermann, A., & Walles, H. (2015). Two-photon polymerization of inorganic-organic polymers for biomedical and microoptical applications. In Optically induced nanostructures: Biomedical and technical applications (pp. 239–265).
14. Gruetzner, G., Schleunitz, A., Houbertz, R., Vogler, M., & Klein, J. J. (2015). Towards high precision manufacturing of 3D optical components using UV-curable hybrid polymers. In *Optical Interconnects XV*, vol. 9368, 93680E.
15. Steenhusen, S., Burmeister, F., Eckstein, H.-C., & Houbertz, R. (2015). Two-photon polymerization of hybrid polymers for applications in micro-optics. In Laser 3D Manuf. II, vol. 9353, 93530K.
16. Langer, G., Brodoceanu, D., & Bäuerle, D. (2006). Femtosecond laser fabrication of apertures on two-dimensional microlens arrays. *Applied Physics Letters, 89*(26), 87–90.
17. Eaton, S. M., De Marco, C., Martinez-Vazquez, R., Ramponi, R., Turri, S., Cerullo, G., & Osellame, R. (2012). Femtosecond laser microstructuring for polymeric lab-on-chips. *Journal of Biophotonics, 5*(8–9), 687–702.

18. Stichel, T., Hecht, B., Houbertz, R., & Sextl, G. (2010). Two-photon polymerization as method for the fabrication of large scale biomedical scaffold applications. *Journal of Laser Micro/Nanoengineering, 5*(3), 209–212.
19. Böhrnsen, F., Krier, M., Grohmann, S., Hauptmann, N., Frant, M., Liefeith, K., & Schliephake, H. (2019). MSC differentiation on two-photon polymerized, stiffness and BMP2 modified biological copolymers. *Biomedical Materials, 14*(3).
20. Danilevicius, P. (2012). Micro-structured polymer scaffolds fabricated by direct laser writing for tissue engineering. *Journal of Biomedical Optics, 17*(8), 081405.
21. Doraiswamy, A., Patz, T., Narayan, R., Chichkov, B., Ovsianikov, A., Houbertz, R., Modi, R., Auyeung, R. and Chrisey, D. B. (2004). Biocompatibility of CAD/CAM ORMOCER polymer scaffold structures. *MRS Proceedings 845*, AA2.4.
22. Houbertz, R., Declerck, P., Passinger, S., Ovsianikov, A., Serbin, J., & Chichkov, B. N. (2007). Investigations on the generation of photonic crystals using two-photon polymerization (2PP) of inorganic-organic hybrid polymers with ultra-short laser pulses. *Physica Status Solidi (A) Applications and Materials, 204*(11), 3662–3675.
23. Houbertz, R., Serbin, J., Dannberg, P., Wolter, H., & Uhlig, S. (2006). Advanced packaging materials for optical applications: Bridging the gap between nm-size structures and large-area panel processing. *Photonics Packaging and Integration VI, 6126*, 612605.
24. Hillerkuss, D., Leuthold, J., Lindenmann, N., Freude, W., Koos, C., Balthasar, G., Schmogrow, R., & Jordan, M. (2012). Photonic wire bonding: A novel concept for chip-scale interconnects. *Optics Express, 20*(16), 17667.
25. Leeb, W., Houbertz, R., Schmid, V., Langer, G., & Satzinger, V. (2008). Optoelectronic printed circuit board: 3D structures written by two-photon absorption. In Org 3D Photonics Mater Devices II, vol. 7053, 70530B.
26. Riester, M. B. K., Houbertz-Krauss, R., & Steenhusen, S. (2013). Chip-to-board interconnects for high-performance computing. *Optoelectron Interconnects XIII, 8630*, 863002.
27. Riester, M. & Houbertz, R. (2014). Level 1 optical interconnects—A pathway to volume manufacturing. In Proc. World Telecommun. Congr., VDE, Berlin, Germany.
28. Ovsianikov, A., Ostendorf, A., & Chichkov, B. N. (2007). Three-dimensional photofabrication with femtosecond lasers for applications in photonics and biomedicine. *Applied Surface Science, 253*(15), 6599–6602.
29. Doraiswamy, A., Jin, C., Narayan, R. J., Mageswaran, P., Mente, P., Modi, R., Auyeung, R., Chrisey, D. B., Ovsianikov, A., & Chichkov, B. (2006). Two photon induced polymerization of organic-inorganic hybrid biomaterials for microstructured medical devices. *Acta Biomaterialia, 2*(3), 267–275.
30. Formanek, F., Takeyasu, N., Tanaka, T., Chiyoda, K., Ishikawa, A., & Kawata, S. (2006). Three-dimensional fabrication of metallic nanostructures over large areas by two-photon polymerization. *Optics Express, 14*(2), 800.
31. Kelemen, L., Valkai, S., & Ormos, P. (2007). Parallel photopolymerisation with complex light patterns generated by diffractive optical elements. *Optics Express, 15*(22)), 14488–14497.
32. Gittard, S. D., Nguyen, A., Obata, K., Koroleva, A., Narayan, R. J., & Chichkov, B. N. (2011). Fabrication of microscale medical devices by two-photon polymerization with multiple foci via a spatial light modulator. *Biomedical Optics Express, 2*(11), 3167–3178.
33. Stichel, T. (2014). *Die Herstellung von Scaffolds aus funktionellen Hybridpolymeren für die regenerative Medizin mittels Zwei-Photonen-Polymerisation.*
34. Kelemen, L., Ormos, P., & Vizsnyiczai, G. (2011). Two-photon polymerization with optimized spatial light modulator. *Journal of the European Optical Society Rapid Publications, 6*, 32.
35. Hilbert, F., Wiedenmann, J., Stender, B., Mantei, W., Houbertz, R., Carlier, Q., Perez Covarrubias, L. A., Heggarty, K., Arnoux, C., Monnereau, C., & Baldeck, P. (2020). Impact of massive parallelization on two-photon absorption micro- and nanofabrication. In H. Helvajian, B. Gu, & H. Chen (Eds.), Laser 3D Manuf. VII 8264(1), vol. 4. SPIE.
36. Melchels, F. P. W., Feijen, J., & Grijpma, D. W. (2010). A review on stereolithography and its applications in biomedical engineering. *Biomaterials, 31*(24), 6121–6130.

37. Halloran, J. W. (2016). Ceramic stereolithography: Additive manufacturing for ceramics by photopolymerization. *Annual Review of Materials Research, 46*(1), 19–40.
38. Bertsch, A., Jiguet, S., Bernhard, P., & Renaud, P. (2003). Microstereolithography: a review. Materials Research Society Symposium Proceedings 758, LL1.1.
39. Patel, D. K., Sakhaei, A. H., Layani, M., Zhang, B., Ge, Q., & Magdassi, S. (2017). Highly stretchable and UV curable elastomers for digital light processing based 3D printing. *Advanced Materials, 29*(15), 1–7.
40. Kim, S. H., Yeon, Y. K., Lee, J. M., Chao, J. R., Lee, Y. J., Seo, Y. B., Sultan, M. T., Lee, O. J., Lee, J. S., Yoon, S. I., Hong, I. S., Khang, G., Lee, S. J., Yoo, J. J., & Park, C. H. (2018). Precisely printable and biocompatible silk fibroin bioink for digital light processing 3D printing. *Nature Communications, 9*(1), 1–14.
41. Dean, D., Wallace, J., Siblani, A., Wang, M. O., Kim, K., Mikos, A. G., & Fisher, J. P. (2012). Continuous digital light processing (cDLP): Highly accurate additive manufacturing of tissue engineered bone scaffolds. *Virtual and Physical Prototyping, 7*(1), 13–24.
42. Derby, B. (2015). Additive manufacture of ceramics components by inkjet printing. *Engineering, 1*(1), 113–123.
43. Derby, B. (2010). Inkjet printing of functional and structural materials: Fluid property requirements, feature stability, and resolution. *Annual Review of Materials Research, 40*(1), 395–414.
44. Mohamed, O. A., Masood, S. H., & Bhowmik, J. L. (2015). Optimization of fused deposition modeling process parameters: A review of current research and future prospects. *Advanced Manufacturing, 3*(1), 42–53.
45. Lieneke, T., Denzer, V., Adam, G. A. O., & Zimmer, D. (2016). Dimensional tolerances for additive manufacturing: Experimental investigation for fused deposition modeling. *Procedia CIRP, 43*, 286–291.
46. "Mikroproduktion.", Mikroproduktion 5, title page (2018).
47. Stender, B. & Houbertz, R. (2017). Skalierbarkeit des hochpräzisen 3D-Drucks. *Mikroproduktion*, 10–11.
48. Shirk, M. D. (1999). Ablation of diamond and graphite with experimental verification. Iowa State University Digital Repository, 178.
49. "Siemens Star" (n.d.).
50. Domanti, G., Houbertz, R., Bahr, J. & Spitzlei, C. (2012). Improvements of inorganic-organic hybrid polymers as dielectric and passivation material. In 2012 IEEE 14th Electronics Packaging Technology Conference (pp. 72–76). IEEE.
51. Popall, M., Houbertz-Krauss, R., Cochet, S., Wolter, H., Ohmori, K., & Sato, T. (2016). Polymerizable compositions, cured products obtained therewith, and use of these materials. US 9.434,818 B2.
52. Hasselmann, S. (2012). Untersuchung magnetisch modifizierter Hybridmaterialien.
53. Haas, K.-H., & Wolter, H. (1999). Synthesis, properties and applications of inorganic–organic copolymers (ORMOCER®s). *Current Opinion in Solid State & Materials Science, 4*(6), 571–580.
54. Wojcik, A. B., & Klein, L. C. (1995). Transparent inorganic/organic copolymers by the sol-gel process: Copolymers of tetraethyl orthosilicate (TEOS), vinyl triethoxysilane (VTES) and (meth)acrylate monomers. *Journal of Sol-Gel Science and Technology, 4*(1), 57–66.
55. Li, C.-Y., Tseng, J. Y., Morita, K., Lechner, C. L., Hu, Y., & Mackenzie, J. D. (1992). ORMOSILS as matrices in inorganic-organic nanocomposites for various optical applications. In J. D. Mackenzie (Ed.) *Proc. SPIE* (pp. 410–419).
56. Budunoglu, H., Yildirim, A., Guler, M. O., & Bayindir, M. (2011). Highly transparent, flexible, and thermally stable superhydrophobic ORMOSIL aerogel thin films. *ACS Applied Materials & Interfaces, 3*(2), 539–545.
57. Haas, K.-H. (2000). Hybrid inorganic-organic polymers based on organically modified Si-Alkoxides. *Advanced Engineering Materials, 2*(9), 571–582.

58. Wächter, C., Streppel, U., Dannberg, P., Martin, H., Fröhlich, L., Popall, M., Domann, G., Cronauer, C., Bräuer, A., Schmitt, A., Park, J.-U., Buestrich, R., & Houbertz, R. (2003). Inorganic–organic hybrid materials for application in optical devices. *Thin Solid Films, 442*(1–2), 194–200.
59. Sanchez, C., Julián, B., Belleville, P., & Popall, M. (2005). Applications of hybrid organic–inorganic nanocomposites. *Journal of Materials Chemistry, 15*(35–36), 3559.
60. Streppel, U., Dannberg, P., Waechter, C. A., Braeuer, A. H., Nicole, P., Froehlich, L., Houbertz, R., & Popall, M. (2001). Multilayer optical fan-out device composed of stacked monomode waveguides. *Mater. Devices Photonic Circuits II, 4453*, 61–68.
61. Haas, U., Haase, A., Satzinger, V., Pichler, H., Leising, G., Jakopic, G., Stadlober, B., Houbertz, R., Domann, G., & Schmitt, A. (2006). Hybrid polymers as tunable and directly-patternable gate dielectrics in organic thin-film transistors. *Physical Reviews B, 73*(23), 1–7.
62. Fröhlich, L., Houbertz, R., Jacob, S., Popall, M., Mueller-Fiedler, R., Graf, J., Munk, M., & Von Zychlinski, H. (2002). Inorganic-organic hybrid polymers as photo-patternable dielectrics for multilayer microwave circuits. *Materials Research Society Symposium Proceedings, 726*, 349–354.
63. Haas, K.-H. & Wolter, H. (2001). Properties of polymer-inorganic composites. [Encyclopedia of Materials: Science and Technology], Elsevier, Oxford, 7584–7594.
64. Helmerich, A., Raether, F., Peter, D., & Bertagnolli, H. (1994). Structural studies on an ORMOCER system containing zirconium. *Journal of Materials Science, 29*(5), 1388–1393.
65. Declerck, P. (2010). Synthesis and technological processing of hybrid organic-inorganic materials for photonic applications. Doctoral Thesis.
66. Streppel, U., Dannberg, P., Wächter, C., Bräuer, A., Fröhlich, L., Houbertz, R., & Popall, M. (2003). New wafer-scale fabrication method for stacked optical waveguide interconnects and 3D micro-optic structures using photoresponsive (inorganic-organic hybrid) polymers. *Optical Materials, 21*(1–3), 475–483.
67. Uhlig, S. (2006). [Micro-processing applied to optical interconnects and high-frequency packaging].
68. Braeuer, A., Dannberg, P., Mann, G., & Popall, M. (2001). Precise polymer micro-optical systems. *MRS Bulletin, 26*(7), 519–522.
69. Robertsson, M. E., Hagel, O.-J., Gustafsson, G., Dabek, A., Popall, M., Cergel, L., Wennekers, P., Kiely, P., Lebby, M., & Lindahl, T. (1998). O/e-MCM packaging with new, patternable dielectric and optical materials. In Proceedings of the 48th Electronic Components Technology Conference (pp. 1413–1421). IEEE.
70. Houbertz, R., Fröhlich, L., Schulz, J., & Popall, M. (2001). Inorganic-organic hybrid materials (ORMOCERs) for multilayer technology—Passivation and dielectric behavior. *MRS Proceedings, 665*, C8.16.
71. Houbertz, R., Schulz, J., Fröhlich, L., Domann, G., & Popall, M. (2003). Inorganic-organic hybrid materials for polymer electronic applications. *Materials Research Society Symposium Proceedings, 769*, 239–244.
72. Haas, K. H., & Rose, K. (2003). Hybrid inorganic/organic polymers with nanoscale building blocks: Precursors, processing, properties and applications. *Reviews on Advanced Materials Science, 5*(1), 47–52.
73. Schweigart, H. (2018). [Guidelines for the application and utilisation of protective coating for electronic assemblies]. 4th ed. GfKORR—Gesellschaft für Korrosionsschutz e.V., Frankfurt/Main.
74. Houbertz, R., Domann, G., Schulz, J., Olsowski, B., Fröhlich, L., & Kim, W. S. (2004). Impact of photoinitiators on the photopolymerization and the optical properties of inorganic-organic hybrid polymers. *Applied Physics Letters, 84*(7), 1105–1107.
75. Rousseau, A. & Boutevin, B. (1992). Synthesis of low absorption halogenated polymers for POF. In Proc. Plast. Opt. Fibers Conf. (pp. 33–37). Paris.

76. Buestrich, R., Kahlenberg, F., Popall, M., Dannberg, P., Müller-Fiedler, R., & Rösch, O. (2001). ORMOCER®s for optical interconnection technology. *Journal of Sol-Gel Science and Technology, 20*, 181–186.
77. Timpel, M., Houbertz, R., Nube, M., Reinert, F., & Römer, M. (2008). *Fabrication of micro- and nanostructures in hybrid polymer materials by focused ion beams*. Luzern: DGM FIB Work.
78. van Krevelen, D. W. & te Nijenhuis, K. (2009). [Properties of polymers]. 4th ed. Elsevier.
79. Steenhusen, S., Stichel, T., Houbertz, R., & Sextl, G. (2010). Multi-photon polymerization of inorganic-organic hybrid polymers using visible or IR ultrafast laser pulses for optical or optoelectronic devices. In Advanced fabrication technologies for micro/nano optics and photonics XIV 7591, 759114.
80. Schleunitz, A., Klein, J. J., & Houbertz, R. (2017). Novel hybrid polymers for fabrication of advanced micro-optics, 2–5.
81. Kawata, S., Sun, H. B., Tanaka, T., & Takada, K. (2001). Finer features for functional microdevices. *Nature, 412*(6848), 697–698.
82. Serbin, J., Houbertz, R., Fallnich, C., & Chichkov, B. N. (2003). Three-dimensional microfabrication with femtosecond laser pulses. In Laser micromachining for optoelectronic device fabrication, vol. 4941, p. 73.
83. Serbin, J., Chichkov, B. N., & Houbertz, R. (2003). Three-dimensional nanostructuring of hybrid materials by two-photon polymerization, In *Nanocrystals, and organic and hybrid nanomaterials* (p. 171), vol. 5222.
84. Chon, J. W. M., Maeda, M., Kawata, S., Gu, M., Sun, H.-B., & Takada, K. (2003). Experimental investigation of single voxels for laser nanofabrication via two-photon photopolymerization. *Applied Physics Letters, 83*(5), 819–821.
85. Schmidt, V., Kuna, L., Satzinger, V., Houbertz, R., Jakopic, G., & Leising, G., (2007). Application of two-photon 3D lithography for the fabrication of embedded ORMOCER waveguides. In Optoelectronic Integrated Circuits IX, vol. 6476, 64760P.
86. Houbertz, R. & Steenhusen, S.. Sub-100 structuring using ultra-short laser pulses, German Science Foundation Priority Program SPP1327 (grant: HO 2475/3-1).
87. Zhang, Z., & Lagally, M. G. (1997). Atomistic processes in the early stages of thin-film growth. *Science, 276*(5311), 377–383.
88. Steenhusen, S. (2008). Mikrostrukturierung von Hybridpolymeren mit Zwei-Photonen-Absorption.
89. Ye, C., Kamysiak, K. T., Sullivan, A. C. & McLeod, R. R. (2012). Single mode 3D diffusive photopolymer optics for optical integrated circuits. In Integrated photonics research, silicon and nanophotonics, IPRSN (pp. 3–6).
90. Fodermeyer, V. (2009). Spektroskopische Untersuchung zur lichtinduzierten Vernetzung von Hybrid- polymeren.
91. Bock, A., Pieper, T., Houbertz, R., & Sextl, G. (2009). Impact of technological processing conditions on optical properties of inorganic-organic hybrid polymers. In Photonics packaging, integration, and interconnects IX, vol. 7221, 72210K.
92. Khoo, Z. X., Teoh, J. E. M., Liu, Y., Chua, C. K., Yang, S., An, J., Leong, K. F., & Yeong, W. Y. (2015). 3D printing of smart materials: A review on recent progresses in 4D printing. *Virtual and Physical Prototyping, 10*(3), 103–122.
93. Houbertz, R., Stichel, T. & Steenhusen, S. (2011). Device and method for producing three-dimensional structures. EP 2 905 121 B1.
94. Bückmann, T., Stenger, N., Kadic, M., Kaschke, J., Frölich, A., Kennerknecht, T., Eberl, C., Thiel, M. and Wegener, M. (2012). Tailored 3D mechanical metamaterials made by dip-in direct-laser-writing optical lithography (pp. 2710–2714).
95. Zhou, X., Hou, Y., & Lin, J. (2015). A review on the processing accuracy of two-photon polymerization. *AIP Advances, 5*(3), 030701.

96. Malinauskas, M., Gilbergs, H., Žukauskas, A., Purlys, V., Paipulas, D., & Gadonas, R. (2010). A femtosecond laser-induced two-photon photopolymerization technique for structuring microlenses. *Journal of Optics, 12*(3), 035204.
97. Dehaeck, S., Scheid, B., & Lambert, P. (2018). Adaptive stitching for meso-scale printing with two-photon lithography. *Additive Manufacturing, 21*, 589–597.
98. Li, J., Fejes, P., Lorenser, D., Quirk, B. C., Noble, P. B., Kirk, R. W., Orth, A., Wood, F. M., Gibson, B. C., Sampson, D. D., & McLaughlin, R. A. (2018). Two-photon polymerisation 3D printed freeform micro-optics for optical coherence tomography fibre probes. *Scientific Reports, 8*(1), 1–9.
99. Jonušauskas, L., Gailevičius, D., Mikoliunaite, L., Sakalauskas, D., Šakirzanovas, S., Juodkazis, S., & Malinauskas, M. (2017). Optically clear and resilient free-form μ-optics 3D-printed via ultrafast laser lithography. *Materials (Basel), 10*(1), 1–18.
100. Stender, B., Mantei, W., Houbertz, R., & Hall, B. (2017). From lab to fab—High-precision 3D printing towards high throughputs and industrial scalability. *Laser Technik Journal, 14*(2), 20–23.
101. Steenhusen, S. (2018). Untersuchungen zur sub-100 nm Strukturierung von Hybridpolymeren mittels Zwei-Photonen Absorption und Anwendungen.
102. Houbertz, R., Stender, B., & Krupp, A. (2017). Closing the gap—High-precision 3D printing, 634. WE -Hereaus Seminar: Merging Micro- and Nanooptics: 3D printing for advanced and functional optics.
103. Born, M. & Wolf, E. (1980) [Principles of optics], 6th ed. Pergamon Press, Oxford.
104. Sinzinger, S. & Jahns, J. (2003) [Microoptics], 2nd ed. Weinheim: Wiley-VCH GmbH & Co. KGaA.
105. Stender, B., Mantei, W., & Houbertz, R. (2017). From lab to fab - high-precision 3D printing. *Laser Technik Journal, 14*(2), 20–23.
106. Rodríguez, S. (2020). Redefining microfabrication of high-precision optics. *Photonics Views, 17*(1), 36–39.
107. Lindenmann, N., Hoose, T., Steenhusen, S., Billah, M. R., Köber, S., Houbertz, R., & Koos, C. (2014). Photonic wire bonding as an enabling technology for multi-chip photonic systems, SPIE vol. 8991—Opitcal Interconnects XIV, SPIE—The International Society for Optical Engineering.
108. Houbertz, R., Memmert, U., & Behm, R. J. (1998). On the potential-dependent etching of Si(111) in aqueous NH4F solution. *Surface Science, 396*(1–3), 198–211.
109. Houbertz, R., Weber, U. and Hartmann, U. (1998). Scanning tunneling spectroscopy on Au thin film structures deposited on highly oriented pyrolitic graphite. *Applied Physics A—Materials Science & Processing, 66*(Suppl. 1).
110. Houbertz-Krauss, R., Schulz, J., Fröhlich, L., Popall, M., Chichkov, B., & Serbin, J. (2001). Verfahren zum Erzeugen dreidimensionaler Formkörper oder Oberflächen aus organopolysiloxanhaltigen Ausgangsmaterialien durch Laser-Bestrahlung und deren Verwendung. DE 101 52 878 B4.
111. Houbertz, R., Fröhlich, L., Popall, M., Streppel, U., Dannberg, P., Bräuer, A., Serbin, J., & Chichkov, B. N. (2003). Inorganic-organic hybrid polymers for information technology: From planar technology to 3D nanostructures. *Advanced Engineering Materials, 5*(8), 551–555.
112. Banerji, S., Meem, M., Majumder, A., Vasquez, F. G., Sensale-Rodriguez, B., & Menon, R. (2019). Imaging with flat optics: Metalenses or diffractive lenses? *Optica, 6*(6), 805.
113. Houbertz, R. (2020). Multiphoton optics, 6″ and 4″ fully populated wafers, SPIE Photonics West Exhibition.
114. Houbertz, R., Esslinger, M., Steenhusen, S., Preve, G. B., & Shaw, M. (2016). High-precision 3D printing: fabrication of micro-optics and integrated optical packages. Opt. Interconnects XVI—Nov. Opt. Waveguide Interconnect Technol., SPIE—The International Society for Optical Engineering.

115. Tanaka, T., Sun, H. B., & Kawata, S. (2002). Rapid sub-diffraction-limit laser micro/nanoprocessing in a threshold material system. *Applied Physics Letters, 80*(2), 312–314.
116. Pérez Covarrubias, L. A., Arnoux, C., Carlier, Q., Khaldi, A., Baldeck, P. and Heggarty, K. (2020). Proximity effect in parallelized microfabrication using two-photon polymerization. *26*.
117. Obata, K., Koch, J., Hinze, U., & Chichkov, B. N. (2010). Multi-focus two-photon polymerization technique based on individually controlled phase modulation. *Optics Express, 18*(16), 17193.
118. Hahn, V., Kiefer, P., Frenzel, T., Qu, J., Blasco, E., Barner-Kowollik, C., & Wegener, M. (2020). Rapid assembly of small materials building blocks (voxels) into large functional 3D metamaterials. *Advanced Functional Materials, 30*, 1907795.
119. Vizsnyiczai, G., Kelemen, L., & Ormos, P. (2014). Holographic multi-focus 3D two-photon polymerization with real-time calculated holograms. *Optics Express, 22*(20), 24217.
120. Zandrini, T., Shan, O., Parodi, V., Cerullo, G., Raimondi, M. T., & Osellame, R. (2019). Multi-foci laser microfabrication of 3D polymeric scaffolds for stem cell expansion in regenerative medicine. *Scientific Reports, 9*(1), 1–9.
121. www.phenomenonproject.eu. (2020).
122. Chidambaram, N., Kirchner, R., Fallica, R., Yu, L., Altana, M., Schift, H. (2017). Selective surface smoothening of polymer microlenses by depth confined softening. *Advanced Materials Technologies, 2*(5).
123. Freude, W., Troppenz, U., Blaicher, M., Dangel, R., Hofmann, A., Caer, C., Billah, M., Dietrich, P.-I., Hoose, T., Offrein, B., Koos, C., Reuter, I., & Moehrle, M. (2018). In situ 3D nanoprinting of free-form coupling elements for hybrid photonic integration. *Nature Photonics, 12*(4), 241–247.
124. Hecht, E. (2016). [Optics]. 5th ed. London: Pearson Education Limited.
125. Borelli, N. F. (2005). [Microoptics technology: fabrication and applications of lens arrays and devices]. 2nd ed. Milton Park: Taylor & Francis Ltd.
126. Suratwala, T. I. (2018). [Materials science and technology of optical fabrication]. 1st ed. Hoboken: Wiley.
127. Levitt, T. H. (2013). [A short bright flash: Augustin Fresnel and the birth of the modern lighthouse]. 1st ed. New York: W. W. Norton & Company.
128. Born, M., Wolf, E., Bhatia, A. B., Clemmow, P. C., Gabor, D., Stokes, A. R., Taylor, A. M., Wayman, P. A., & Wilcock, W. L. (1999). [Principles of optics: Electromagnetic theory of propagation, interference and diffraction of light]. 7th ed. Cambridge: Cambridge University Press.
129. Juodkazis, S., Sakellari, I., Gilbergs, H., Gadonas, R., Malinauskas, M., Gaidukevičiūtė, A., Purlys, V., Piskarskas, A., Farsari, M., Momot, A., Paipulas, D., Žukauskas, A., & Belazaras, K. (2010). Femtosecond laser polymerization of hybrid/integrated micro-optical elements and their characterization. *Journal of Optics, 12*(12), 124010.
130. O'Shea, D. C., Suleski, T. J., Kathman, A. D., & Prather, D. W. (2004). [Diffractive optics: Design, fabrication, and test]. 1st ed. SPIE—The International Society for Optical Engineering.
131. Jia, B., Serbin, J., Kim, H., & Lee, B. (2007). Use of two-photon polymerization for continuous gray-level encoding of diffractive optical elements, 1–3.
132. Zentgraf, T., Liu, Y., Mikkelsen, M. H., Valentine, J., & Zhang, X. (2011). Plasmonic Luneburg and Eaton lenses. *Nature Nanotechnology, 6*(3), 151–155.
133. Chen, X., Huang, L., Mu, H., Li, G., Bai, B., Tan, Q., Jin, G., Qiu, C., Zhang, S., & Zentgraf, T. (2012). Dual-polarity plasmonic metalens for visible light, 1–6.
134. Genevet, P., Capasso, F., Aieta, F., Khorasaninejad, M., & Devlin, R. (2017). Recent advances in planar optics: From plasmonic to dielectric metasurfaces. *Optica, 4*(1), 139.
135. Capasso, F. (2018). Review article. The future and promise of flat optics: a personal perspective, 953–957.

136. Metalenses at visible wavelengths: Diffraction-limited focusing and subwavelength resolution imaging. (2016).
137. Genevet, P., & Capasso, F. (2015). Holographic optical metasurfaces: A review of current progress. *Reports on Progress in Physics, 78*(2), 24401.
138. Tellal, A., Ziane, O., Jradi, S., Stephan, O., & Baldeck, P. L. (2019). Quadratic phase modulation and diffraction-limited microfocusing generated by pairs of subwavelength dielectric scatterers (pp. 1051–1061).
139. Sun, H. B., Tanaka, T., & Kawata, S. (2002). Three-dimensional focal spots related to two-photon excitation. *Applied Physics Letters, 80*(20), 3673–3675.
140. Reinhardt, C., Peterhänsel, S., Ovsianikov, A., Frenner, K., Obata, K., Osten, W., Chichkov, B., Paz, V. F., Emons, M., & Morgner, U. (2012). Development of functional sub-100 nm structures with 3D two-photon polymerization technique and optical methods for characterization. *Journal of Laser Applications, 24*(4), 042004.
141. Zheng, L., Kurselis, K., El-tamer, A., Hinze, U., Reinhardt, C., Overmeyer, L., & Chichkov, B. (2019). Nanofabrication of high-resolution periodic structures with a gap size below 100 nm by two-photon polymerization, vol. 14.
142. Haske, W., Chen, V. W., Hales, J. M., Dong, W., Barlow, S., Marder, S. R., & Perry, J. W. (2007). 65 nm feature sizes using visible wavelength 3-D multiphoton lithography (pp. 3426–3436).
143. Malinauskas, M., Farsari, M., Piskarskas, A., & Juodkazis, S. (2013). Ultrafast laser nanostructuring of photopolymers: A decade of advances. *Physics Reports, 533*(1), 1–31.
144. Sun, H. B., & Kawata, S. (2003). Two-photon laser precision microfabrication and its applications to micro—nano devices and systems. *Journal of Lightwave Technology, 21*(3), 624–633.
145. Stichel, T., Hecht, B., Steenhusen, S., Houbertz, R., & Sextl, G. (2016). Two-photon polymerization setup enables experimental mapping and correction of spherical aberrations for improved macroscopic structure fabrication. *Optics Letters, 41*(18), 4269.
146. Xing, J. F., Dong, X. Z., Chen, W. Q., Duan, X. M., Takeyasu, N., Tanaka, T., & Kawata, S. (2007). Improving spatial resolution of two-photon microfabrication by using photoinitiator with high initiating efficiency. *Applied Physics Letters, 90*(13), 31–33.
147. Seniutinas, G., Weber, A., Padeste, C., Sakellari, I., Farsari, M., & David, C. Beyond 100 nm resolution in 3D laser lithography—post processing solutions.
148. Sparrow, C. M. (1916). On spectroscopic resolving power. *The Astrophysical Journal, 44*, 76.
149. Optical gas sensing: a review—IOPscience.
150. Zeller, W., Naehle, L., Fuchs, P., Gerschuetz, F., Hildebrandt, L., & Koeth, J. (2010). DFB lasers between 760 nm and 16 μm for sensing applications. *Sensors, 10*(4), 2492–2510.
151. Kamp, M., Hofmann, J., Schäfer, F., Reinhard, M., Fischer, M., Bleuel, T., Reithmaier, J. P., & Forchel, A. (2001). Lateral coupling—a material independent way to complex coupled DFB lasers. *Optical Materials, 17*(1–2), 19–25.
152. Multiphoton Optics GmbH. Optische chips—preiswert durch 3D-Laserlithographie. Laser Mag. 17–19.
153. Barwicz, T., Lichoulas, T. W., Taira, Y., Martin, Y., Takenobu, S., Janta-Polczynski, A., Numata, H., Kimbrell, E. L., Nah, J. W., Peng, B., Childers, D., Leidy, R., Khater, M., Kamlapurkar, S., Cyr, E., Engelmann, S., Fortier, P., & Boyer, N. (2018). Automated, high-throughput photonic packaging. *Optical Fiber Technology, 44*(February), 24–35.
154. Schumann, M., Bückmann, T., Gruhler, N., Wegener, M., & Pernice, W. (2014). Hybrid 2D-3D optical devices for integrated optics by direct laser writing. *Light: Science & Applications, 3*(December 2013), 1–9.
155. Smowton, P. M., Thomas, R., Ladak, S., Barrow, D., & Li, J. (2018). In situ fabricated 3D micro-lenses for photonic integrated circuits. *Optics Express, 26*(10), 13436.

156. Rodt, S., Kaganskiy, A., Ristok, S., Thoma, A., Srocka, N., Reitzenstein, S., Giessen, H., Herkommer, A., Heindel, T., Strittmatter, A., Schlehahn, A., Thiele, S., Gissibl, T., & Fischbach, S. (2017). Single quantum dot with microlens and 3D-printed micro-objective as integrated bright single-photon source. *ACS Photonics, 4*(6), 1327–1332.
157. Cojoc, G., Liberale, C., Candeloro, P., Gentile, F., Das, G., De Angelis, F., & Di Fabrizio, E. (2010). Optical micro-structures fabricated on top of optical fibers by means of two-photon photopolymerization. *Microelectronic Engineering, 87*(5–8), 876–879.
158. Cobley, R. J., Teng, K. S., Brown, M. R., Wilks, S. P., & Rees, P. (2007). Direct real-time observation of catastrophic optical degradation in operating semiconductor lasers using scanning tunneling microscopy. *Applied Physics Letters, 91*, 081119.
159. Bonati, G. (2003). Exploring failure probability of high-power laser diodes. *Photonics Spectra.*
160. Neff, J. A., Athale, R. A., & Lee, S. H. (1990). Two-dimensional spatial light modulators: A tutorial. *Proceedings of the IEEE, 78*(5), 826–855.
161. Harriman, J., Serati, S. and Stockley, J. 2005. Comparison of transmissive and reflective spatial light modulators for optical manipulation applications. In Optical trapping and optical micromanipulation II, 59302D.

Chapter 6
3D-Printed Microoptics by Femtosecond Direct Laser Writing

Simon Thiele and Alois Herkommer

Abstract Femtosecond direct laser writing offers unique possibilities for the fabrication of microoptical components and systems. Almost unrestricted 3D design freedom, high resolution and accuracy, as well as alignment-free assembly enable direct printing of complex structures with optical functionality. Various imaging and illumination optics are presented and discussed (Most of the contents shown are a directly translated from the PhD thesis "Design, Simulation und Prozessoptimierung für das 3D-Laserdirektschreiben von Mikrooptiken" by Simon Thiele, University of Stuttgart (2019)).

Keywords Laser direct writing · Microoptical components · Freeform design High-resolution printing · Imaging optics · Illumination optics · Two-photon lithography · 3D lithography

6.1 Introduction

Microoptical components have nowadays found their way into almost all areas of our lives and have become indispensable in technology sectors such as communications technology, medical technology, sensor technology, or consumer electronics. Examples include microlenses for beam shaping in fiber optics, small endoscope lenses for minimal invasive surgery, microlens arrays for the homogenization of laser illumination, and the tiny imaging lens systems in the cameras of our smartphones.

The microoptics of today mainly emerged from semiconductor manufacturing in the 1970s and 1980s and would most likely not be as widespread without this important technological driver. Some paradigms of this industry, such as high parallelism, high throughput, and cost reduction through miniaturization, have influenced microoptics and clearly distinguish it from the traditional categories of the optical industry.

S. Thiele · A. Herkommer (✉)
Institute of Applied Optics (ITO) and Research Center SCoPE, Stuttgart, Germany
e-mail: alois.herkommer@ito.uni-stuttgart.de

A. Heinrich (ed.), *3D Printing of Optical Components*, Springer Series in Optical Sciences 233, https://doi.org/10.1007/978-3-030-58960-8_6

An interesting mixture of both worlds can be observed in the lenses of today's smartphones. Here, too, highly automated work is carried out with low-cost materials (polymers) and low-cost processes (injection molding/injection stamping), although the result is complex and consists of many precisely aligned individual components. The optical performance requirements in the case of smartphone cameras demand complex meniscus aspheric shapes, not accessible through wafer-based processes as they allow only limited surface complexity and have so far not been able to assert themselves despite their advantageous cost structure. However, there are clear downward limits to the methods of assembling replicated lens components, as alignment and assembly become more and more difficult with increasing miniaturization. New methods must be developed in order to transfer the complexity of smartphone lenses and other microoptical systems to smaller scales.

One of the most promising candidates for the fabrication of next-generation microoptics is additive manufacturing using multiphoton lithography as it is inherently alignment-free and allows for extreme miniaturization as well as almost unrestricted design freedom. Figure 6.1 compares all three methods based on the assessment of the authors.

The generative fabrication of microoptics in the sense of 3D printing is a comparatively young field of research. Classical methods, such as direct laser writing, have been used for decades to fabricate microoptics, but the layer-by-layer printing

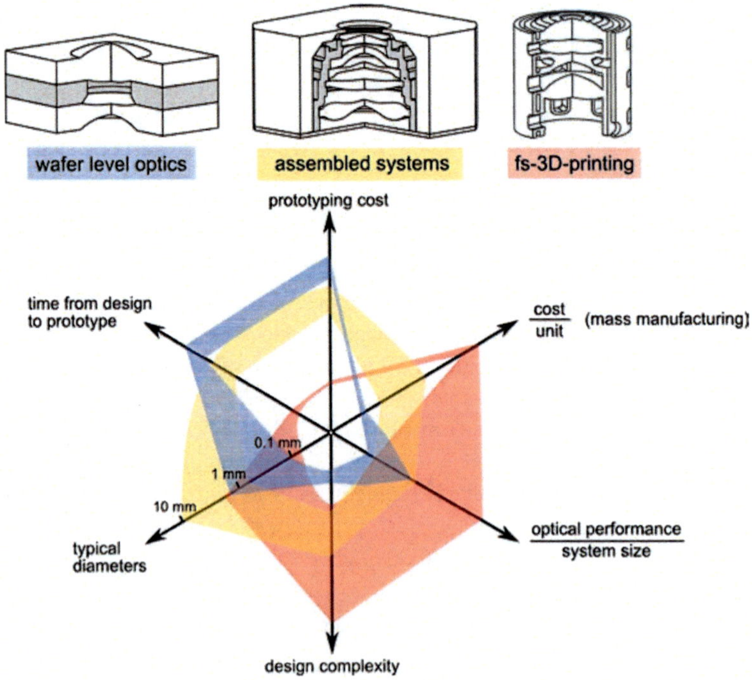

Fig. 6.1 Comparison of different technologies for the fabrication of microoptical systems according to the author's assessment

of three-dimensional structures did not become established until the mid-2000s. An important driver of this technology was the development of multiphoton lithography with first publications in the late 1990s [1–3].

As some of the first structures with optical functionality photonic crystals were fabricated [4–13]. In order to use these crystals in the visible spectrum with high efficiency, finest structure sizes must be realized. Therefore, the method was further improved by using stimulated emission depletion (STED) approaches to reach voxel sizes down to the range of a few 10 nm [14, 15]. To produce smooth surfaces and monolithic components, however, the voxel size is less important, and photoresists with a stronger proximity effect are desired, since rounding attenuates staircase effects. The latter can also be achieved by a specific surface treatment after printing [16].

First microoptical components such as microprisms, waveguides, or microlenses [17–21] were demonstrated in the mid-2000s. In the following years, a number of research papers were published presenting diffractive lenses [22], microlenses (and arrays) [23–27], diffractive-refractive hybrid lenses [28], vortex lenses [29, 30], multifocal lenses [31], pattern generators [32], ring resonators [33, 34] or freeform lenses [35, 36].

Multielement systems in which the components are arranged along the optical axis have also been published several times [37–41]. Due to their built-in microscope, multiphoton lithography setups usually allow a very accurate adaption and alignment of the writing process to the substrate. Therefore, soon the first elements were printed on the tip of optical fibers. In addition to refractive and reflecting surfaces for beam shaping [39, 42–44], photonic crystals [37, 45], phase plates [46, 47], or moth-eye structures for reflex suppression [48] were written directly on fibers. As an alternative to direct printing, a clip-on approach can also be chosen [49]. Other substrates such as optical crystals [50] or image sensors [40] can also be used for printing.

While classical photopolymers are typically used as writing materials, hybrid organic-inorganic polymers [51], proteins [52], or even glass [53] is also used.

Other generative manufacturing processes comparable to multiphoton lithography have so far been limited to inkjet-based approaches [54] and ablative processing of photosensitive glass [55]. In terms of optical functionality, however, the demonstrated components have so far been less mature in comparison.

The fabrication of planar optical elements by single-photon laser direct writing goes back to works from the early 1980s [56]. At that time, semiconductor processes were adapted in order to lithographically generate optical structures such as microlens arrays or diffractive elements directly without any mask. The employed photosensitive (UV) resist typically shows a linear relation between light intensity and chemical structural change. The photon absorption can lead to either a polymerization (negative resist) or to a local de-bonding (positive resist) and thus together with a development step planar structures can be fabricated.

In addition to linear single-photon absorption, nonlinear absorption mechanisms, also known as two- or multiphoton absorption, exist which were first described in the early 1930s [57]. However, due to the extreme photon densities required, this approach required the invention of the laser to be realized experimentally. The development of the titanium-sapphire femtosecond laser [58, 59], which fulfilled the requirements for novel experiments for the first time, is regarded as an important breakthrough for this technology. The dependence of the transition rate on the squared intensity is a great advantage with regard to three-dimensional lithography, since absorption processes can be confined to a very small area (focus).

In the case of two-photon lithography, light from the near infrared is typically used together with a photoresist, which is photosensitive at $\lambda/2$, i.e., mostly in the UV spectral range. The simultaneous absorption of two photons (see Fig. 6.2—right) produces free radicals from so-called photoinitiator molecules, which break double bonds and thus initiate polymerization. This is temporally and spatially limited and results in a so-called voxel of polymerized material at the focal spot.

Usually a voxel has the shape of an ellipsoid and has, depending on the numerical aperture with which it is focused, a varying size and a varying aspect ratio. Even in the case of strong focusing in immersion (e.g., with a numerical aperture of 1.4), an aspect ratio of at least 2.7 remains without further manipulation of the wavefront. The minimum dimensions of the voxels are typically about 100 nm in width and 270 nm in length. Complex three-dimensional objects are created line by line and layer by layer in a mostly liquid photoresist, which remains transparent in the visible spectrum after polymerization. Figure 6.2 schematically shows the functional principle of this method.

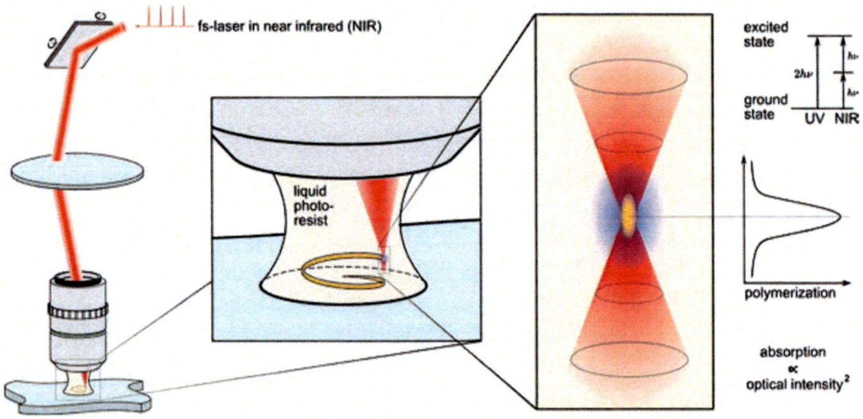

Fig. 6.2 Schematic setup for two-photon lithography. A microscope is dipped into a photoresist fluid. The laser focus is then moved via the galvo scanner within the fluid to locally polymerize the material

6.2 Design Rules for 3D-Printed Microoptics

The above-described printing process and the current limitations of the printing device result in some general and specific design rules, which should be considered during optical design, structural design, and process development.

Limited Printing Volume

The maximum printable volume is limited by the lithography device architecture and by the writing time, which is usually critical due to the small voxel size. Therefore, the size of the solid printed optical elements should be minimized as far as possible. Usually, the total size of the printed system is in the range of below 1 mm in each dimension.

Geometry Restrictions

Due to the small voxel size in vertical direction the printing process allows comparably small lens thicknesses of down to 5 μm. Edge thicknesses close to the structural mount should be thicker. For air gaps between elements, larger distances must be maintained to ensure that remaining photoresist can be reliably washed out.

Elimination of Alignment Tolerances

The complete optical system, including all elements and mechanical support structure, can be printed in one single process. In consequence, there are no alignment or mounting tolerances which need to be considered. The only remaining tolerance effect is the printing accuracy of the lithographic printing itself, which is well below 1 μm in all three dimensions. With a certain pre-knowledge and compensation of the shrinkage of the resist, the absolute geometry accuracy can even be smaller.

Mono-material

The printing of a combination of materials, e.g., different dispersion for chromatic corrections, is possible in general and has already been demonstrated. However, two materials require additional effort and writing time and may introduce alignment errors. Therefore, it is rational and economic to limit the material choice to only one polymer material, if possible. The standard photoresist employed for the following examples is IP-S (Nanoscribe GmbH, Germany), with a refractive index of $n = 1.5–1.54$ in the visible range. For coarse design estimates, we use a refractive index of $n = 1.5$.

Properties of Polymer Optics

The polymer-based printing materials typically exhibit a few disadvantages. For example, the range of materials and thus the available refractive indices as well as Abbe numbers is rather limited. Polymers are also more sensitive to temperature changes, are more difficult to coat, and offer less mechanical and chemical stability. Moreover stress-induced birefringence can be a problem.

Avoiding Ray Bending

In imaging optical design, it is good practice to avoid strong ray bending, a term which usually describes high angles between rays and the local surface normal at their point of incidence or exitance in air or glass. These angles are illustrated in Fig. 6.3 as γ_L and γ_G, respectively. High angles usually introduce large aberrations, which must be compensated elsewhere. Such systems are therefore sensitive to small changes and fabrication tolerances [60–62]. Furthermore, increasing ray bending leads to an increase in dielectric losses. This leads to transmission loss and enhanced stray light in the system. Therefore, a maximum local angle of incidence is typically included into the optimization merit function. In the example imaging optical systems shown below, we have limited the allowed maximum angle towards air to approximately 56.3° (Brewster's angle).

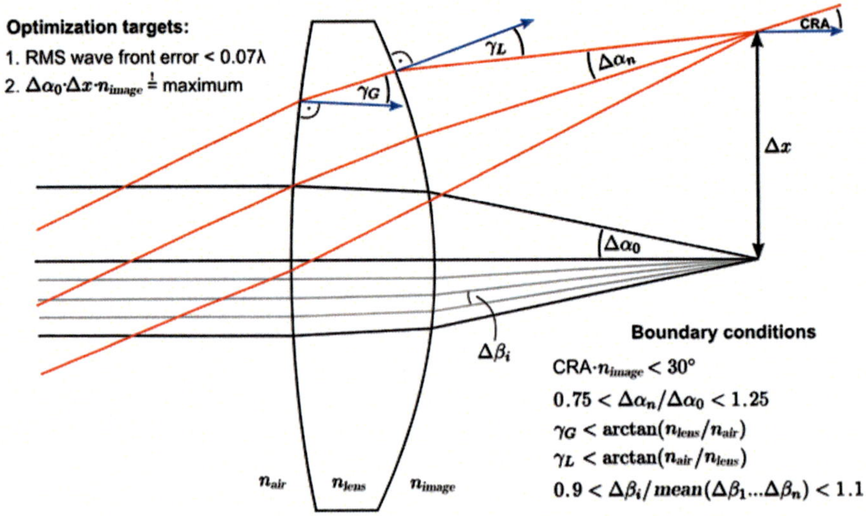

Fig. 6.3 Typical design targets and boundary conditions for the design of 3D printable highly aspherical lens systems

Avoiding Large Chief Ray Angles (CRA)

In most modern optical application, the microoptical system is used together with a CMOS image sensor or an imaging fiber bundle. Both elements do not allow large chief ray angles (CRA) in the image space. This angle is illustrated in Fig. 6.3. Usually the maximum CRA must be controlled below 30° for CMOS chips and below 20° for imaging fibers. In the following designs, we limit the CRA to 30° in air, or to 30°/n if the image space is immersed with a material of refractive index n.

Realization of Strong Aspheres and Freeform Surfaces

The printing process basically allows creating any 3D geometry, not only for the structural mount but also for the optical surfaces. In consequence, the limitation to classical optical surface shapes, such as rotational symmetric spheres, no longer exists. 3D printing naturally allows for aspheric or even freeform surfaces with large asphericity in the optical design. As a result of the added degrees of freedom, the performance of a design can be improved while keeping the same (or smaller) number of elements, and often also the total built volume can be significantly reduced.

In order to verify this hypothesis, we have compared the achievable performance of spherical to aspherical systems. As a performance figure of merit, we have compared the achievable space-bandwidth product of the designs, which connected to the product of the marginal ray angle $\Delta\alpha_0$ and the field size Δx, inside which the performance can be corrected to be diffraction limited. Figure 6.4 shows the resulting designs for 1–6 spherical surfaces in the upper row and 1–6 aspherical surfaces (with polynomial coefficients up to the tenth order) in the lower row. Below the merit function of the designs is illustrated.

From the design study and analysis, we find an almost linear dependence between the number of optical surfaces and the achievable figure of merit. From this behavior it may be concluded that all systems have been corrected to the same final level. The other main finding is that aspheric surfaces allow for a larger gradient of this dependence. This nicely verifies that aspheres provide a clear benefit for the achievable optical performance in terms of an increased space-bandwidth product as compared to an all-spherical design. However, it has to be noted that this dependence cannot be generalized to other design problems, as here the abovementioned boundary conditions on maximum incidence angles and CRA have been considered during the design.

The space-bandwidth product can also be converted into the number of resolvable image points. This number can be calculated from the maximum diffraction limited field size, divided by the diameter of the Airy disk, according to the Rayleigh criterion. The resulting quadratic dependence in Fig. 6.4 relative to the linear dependence of the space-bandwidth product results from the fact that the pixel number scales with the area (quadratic in field size).

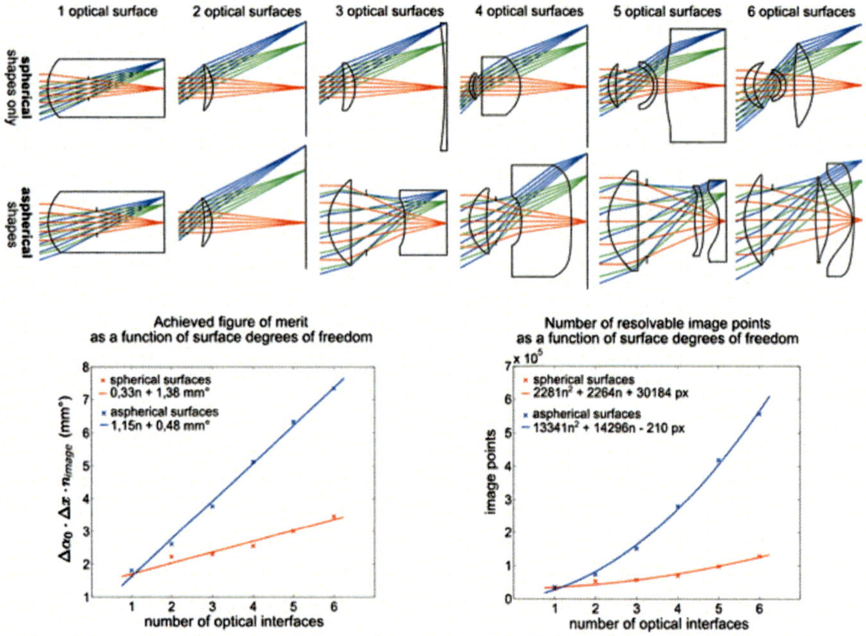

Fig. 6.4 Comparison of optimized spherical designs (upper row) versus aspherical designs (lower row) for an increasing number of surfaces. Below the corresponding space-bandwidth product is illustrated (left) and the respective number of resolvable image points (right)

Integration of Diffractive Structures

In addition to aspheric and freeform surfaces, multiphoton lithography, due to the small voxel size in lateral direction, allows for a direct printing of diffractive structures. These can add additional degrees of freedom and thus further improve the optical performance. In addition, the small size of the components does typically not require high line densities, which helps to avoid otherwise typical losses in diffraction efficiency. In general, diffractive elements are very useful in optical design, as they can not only compensate first-order chromatic aberrations, but also third-order monochromatic aberrations. These advantages must be compared to the well-known disadvantages of wavelength-dependent diffraction efficiency, large dispersion, and unavoidable residual light into unwanted diffraction orders.

Wave-Optical Effects

Diffractive elements are a prime example why a wave-optical analysis of the system is recommended as it allows for calculation of diffraction efficiency and chromatic effects. However, even without diffractive structures, the size of the printed elements is often small enough that diffraction effects from the apertures must be considered. As a criterion the so-called Fresnel number can be used. A significant diffraction effect must be expected when the Fresnel number approaches unity. It is known that in this case, diffraction will lead to an effective reduction of the focal length, or to diffraction-stimulated chromatic effects [63].

6.3 Examples of Printed Microoptical Imaging Systems

The following two examples of printed imaging systems shall illustrate the applications and limitations of 3D printing processes for optics in more detail.

Both systems have been optimized for 3D printing under the abovementioned design rules. In order to take full advantage of the high accuracy of the process, the design should be optimized such that it can be printed in one step, without any assembly or realignment. Moreover, openings in the structure and distances between the elements must be chosen such that the photoresist can be properly washed out. The latter condition is directly influencing the optical design.

In prior work it turned out that for uncoated surfaces, a surface number of 4, i.e., two aspheric lens elements, is a good compromise of optical quality and transmission loss. Fewer surfaces do not allow for a large enough space-bandwidth product, and more surfaces introduce too much material absorption, reflection losses, and stray light. Both designs have been furthermore optimized for IP-S photoresist, as mentioned above. The dispersion characteristics of IP-S were accurately measured over the visible range and included into the optical design software ZEMAX. The employed lithographic printing system (Nanoscribe GT Professional, Nanoscribe GmbH, Germany) at that time supported two different writing objectives and field diameters to be printed: a maximum diameter of 200 μm for the 63x objective and 500 μm for a 25× objective. In consequence the diameters of the optical systems were limited to 125 μm, respectively 450 μm. From the scaling laws of optical systems, it is known that smaller systems are easier to correct, since geometric and axial chromatic aberrations scale down with size. In addition, shrinkage effects of the polymer resist have less impact if the lens is smaller. Since the manufacturing of absorbing aperture stops (see later in this chapter) was not available at that time, both designs have their stop at the front surface. The image plane is located exactly at the substrate interface on which the print is fixed.

The two designs are microoptical versions of a tele-system (diameter 125 μm, field of view 20°) and a distortion corrected camera lens (diameter 450 μm, field of view 40°). Table 6.1 lists the most important parameters of both systems. In both systems aspheric surfaces are used, even if the asphericity is not strikingly visible in the lens drawing.

For the listed number of resolvable image points, the field of view was divided into five zones with equal area. In each zone the average radius of the Airy disk is used, if the image quality is diffraction limited, whereas the geometrical spot radius is used if geometrical aberrations are limiting.

Both designs are well corrected for distortion, which usually is additional effort or comes at the cost of other aberrations. This is one of the reasons why the camera

Table 6.1 Comparison of the two printed imaging systems, with four surfaces

Variant	Micro tele-lens	Camera lens
Aperture diameter	97 μm	390 μm
Maximum lens diameter	97 μm	390 μm
Length	204 μm	650 μm
Focal length	261 μm	544 μm
Full field of view	20°	40°
Numerical aperture (image space)	0.2	0.37
Object distance	∞	∞
Distortion	−0.1%	1%
Number of resolvable Image points	~2500	~40,000

lens could not be corrected to be diffraction limited over the full field of view at the employed f-number of 1.35. However still the number of resolvable pixels is much higher, and the spot size is smaller as compared to the tele-lens. Moreover, the absolute amount of light is larger due to the larger diameter. Figure 6.5 shows the optical layout of both designs.

After finishing the optical design, the surfaces are exported either via ZEMAX or via the mathematical surface description to a computer-aided design (CAD) model. In the CAD software (e.g., SolidWorks), the appropriate mechanical support structure is added, as illustrated in Fig. 6.6.

The height of the support structure is designed such that the image plane is exactly positioned on the substrate, which usually is a cover glass. If the lenses have convex shapes towards the bottom of the system, special care must be taken to print the vertex without them swimming in the resist.

Before the final printing process, CAD models must be triangulated and converted into a STL format. From that file the software "DeScribe " (Nanoscribe GmbH) creates a machine-readable format, which among other parameters contains the writing trajectories, as well as laser power, scan speed, and delay times.

The physical printing process is then performed using the "Dip-In" configuration with the 63×/1.4 NA objective (for the tele-lens) and the 25×/0.8 NA objective for the camera lens. The typical writing time is in the range of 2–3 h. After resist development and washout of the unused resist, the results are typically inspected with a digital microscope (Keyence VHX-1000), in order to assess the quality of the print. Figure 6.7 shows images of a print in comparison. However, this inspection only allows a qualitative first assessment.

For a more quantitative assessment of the imaging performance, a microscope arrangement can be employed. Here an object is imaged through the printed microoptics, and the created intermediate image is observed via a microscope system on the image side. As a standard object, the USAF-1951 test target is used, which is positioned at a hyperfocal distance. The imaging results are shown in Fig. 6.8, for both objectives. The visible part of the test chart is different, as the

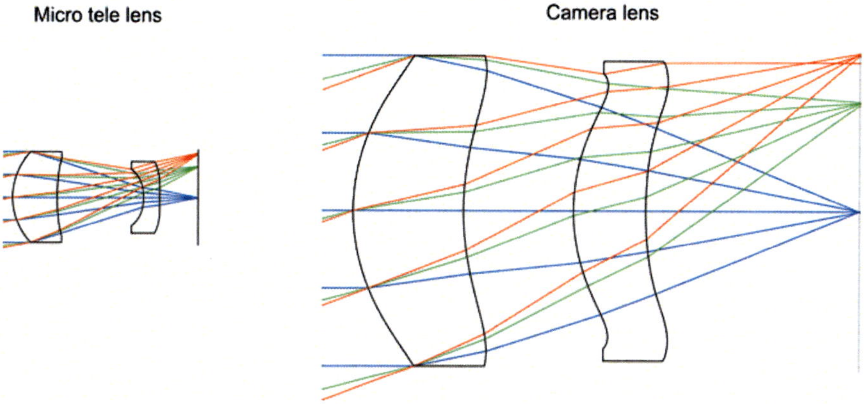

Fig. 6.5 Layout of the microoptical printed tele-lens and the camera lens

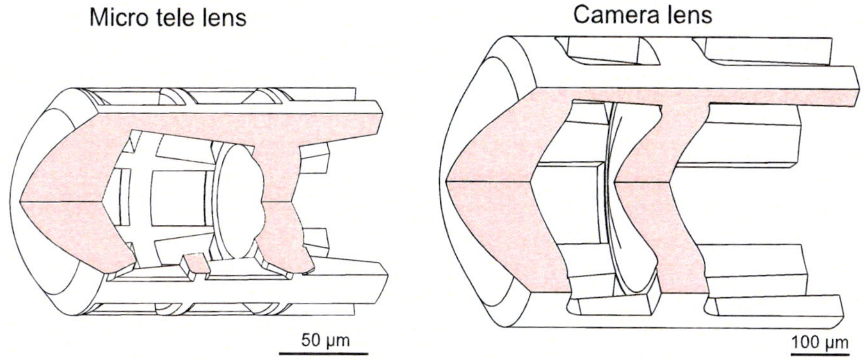

Fig. 6.6 CAD-model of the tele-lens and the camera lens including the support structure. For better visualization a quarter of the lens model has been removed

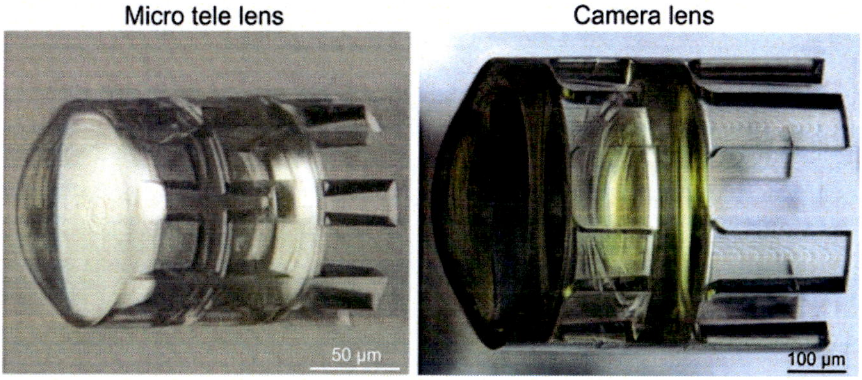

Fig. 6.7 Digital microscope image of printed version of the two designs

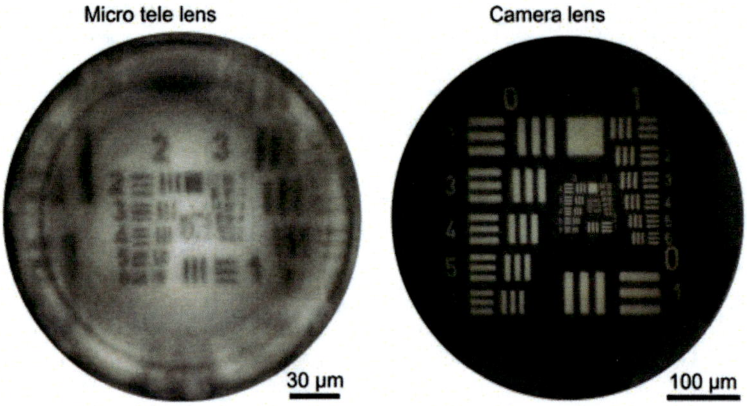

Fig. 6.8 Experimentally recorded image quality. For the camera lens, stray light was suppressed by external shielding

object is placed at different distances. However, the image quality already proves that the amount of details which can be transmitted is larger for the camera lens.

The contrast in both systems is different, since for the tele-lens, no measures for stray light suppression have been made; however the camera lens was surrounded by a black sheet of paper. This already brings us to another downside of mono-material systems. The fully transparent material of lens and mount will lead to a large amount of stray light, which even for perfect corrected systems can diminish the optical performance or even can make them useless. In consequence any optical system usually requires some way of stray light suppression, or at least a well-defined aperture. Otherwise light from undesired directions can enter the system or directly reach the image. Mono-material polymer optical system as shown above does not offer possibilities to integrate absorptive apertures, as all the material is transparent. Therefore, the optics will suffer from a loss of contrast, even if the illumination conditions are optimized. This is illustrated for another optical system as shown in Fig. 6.9. Here a fully transparent optical system, in this case a miniaturized Zeiss Hologon, is simulated via nonsequential ray tracing. The simulation drastically shows the difference between a system without any shielding, as compared to a system with absorptive walls and an absorptive aperture stop. The image simulation reveals that without stray light suppression, the contrast of the image is extremely poor which underlines the need for an integration of absorptive structures. Unfortunately, up to now no printable absorptive material is available. In addition, a second material would require an unwanted alignment step. Other possibilities are a post-process coating step [64] with an absorption layer, or the metal deposition out of a liquid suspension stimulated by the two-photon absorption [65].

Another option is the realization of refractive apertures or retro-reflective surfaces. Such elements will not absorb but rather redirect the light into uncritical directions. These structures and textures can be integrated into the printing process; however, they will work only within a limited range of incidence angles.

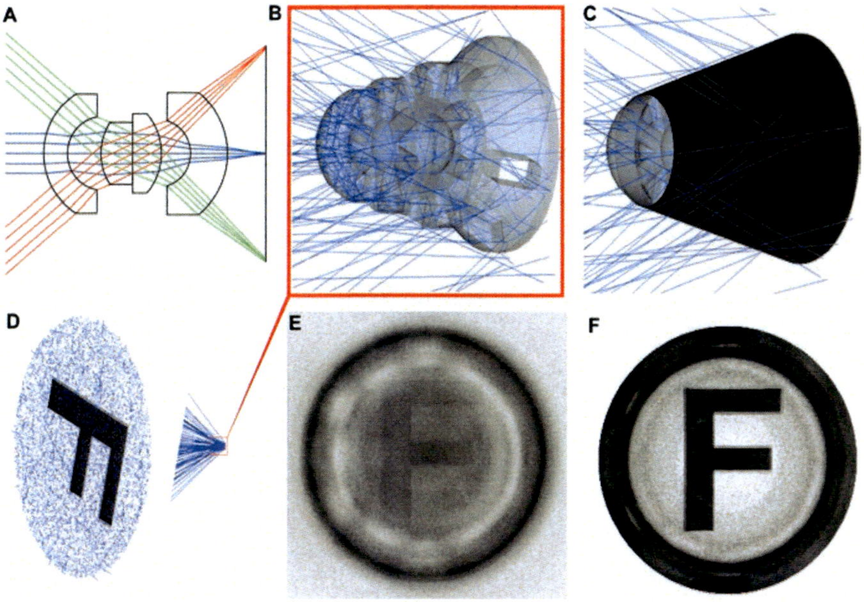

Fig. 6.9 Nonsequential ray tracing simulation of a miniaturized Zeiss Hologon. (**a**) Design layout. (**b**) Simulation for a fully transparent design. (**c**) Same simulation, but now with absorbing hull and absorbing aperture. (**d**) Employed source model for the letter "F." (**e**) Simulated image of this source for the transparent version. (**f**) Simulated image for the absorptive model

So far, the most promising method is the integration of microfluidic channels into the 3D model. These channels can later be filled with nontransparent ink. As the channels are printed with the same process, no additional alignment error is introduced[1]. The approach requires a good understanding of capillary forces, which are dominant at the microscale. Also, the choice of the liquid is important, since a large absorption is required. Metallic ink has proven to be a good candidate, as it also dries fast and can be sintered at low temperatures. The principle and corresponding manufacturing steps are illustrated in Fig. 6.10. Here a triplet lens is integrated into a special designed support structure, which is later filled with ink. The ink-filled channels form an absorptive hull, as well as an integrated aperture. Die filling process can be performed with several techniques. Either pressure is used to push the ink out of the capillary or electrical voltage. A passive technique where only the initial drop is created form a superfine inkjet printer (SIJ-S030) and the rest is self-filled via capillary forces has proven to work well. Figure 6.10 shows images of such a filling process in the lab.

The experimental results, as shown in Fig. 6.11, demonstrate the resulting improvement of contrast, this time for an endoscopic imaging system. Note that

[1]This method has been patented (EP 3162549 A1) and is currently under investigation.

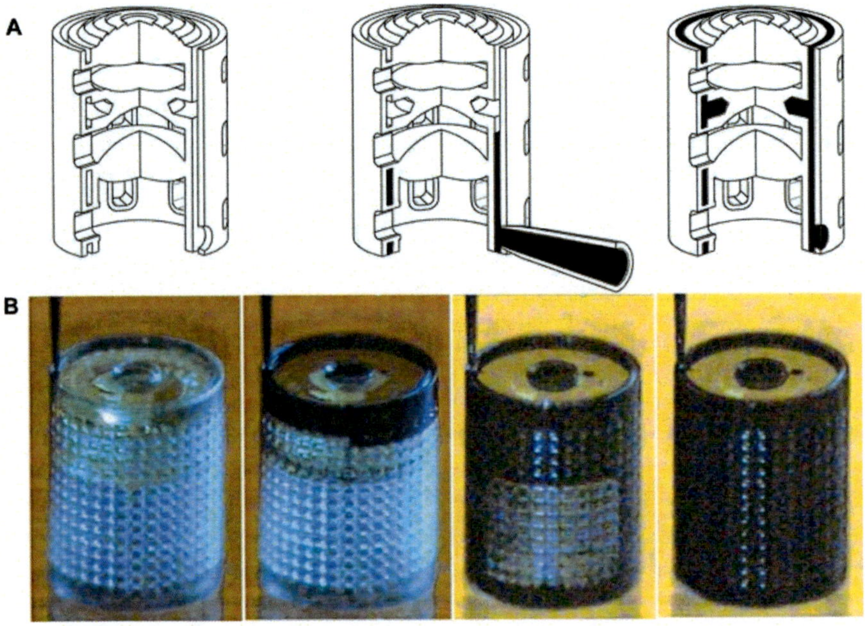

Fig. 6.10 Principle of generating absorptive structures. (**a**) Schematic process illustrated for a triplet with specially designed support structure. (**b**) Image sequence of the real filling process in the lab

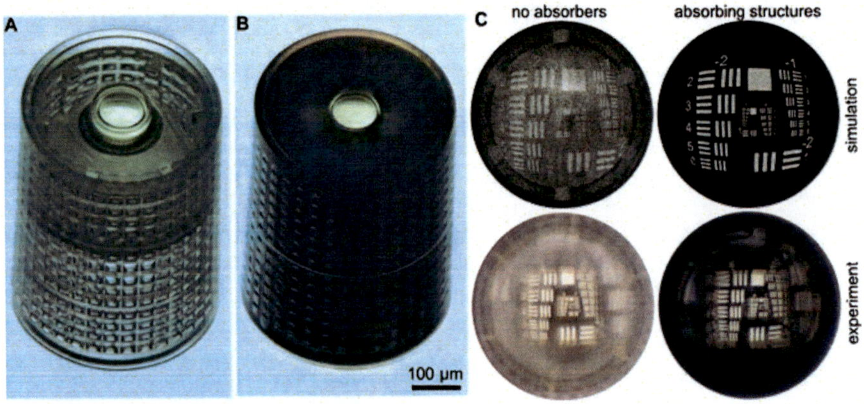

Fig. 6.11 Realized endoscopic imaging systems with absorptive elements. (**a**) Endoscopic system before filling with ink. (**b**) System after being filled with ink. (**c**) Comparison of the imaging results in theory and experiment

here the support hull is designed with small openings, large enough to let the ink dry, but too small to let the ink go through. The front aperture is realized by a liquid reservoir right on top of the structure. As a result of the absorptive ink, the image contrast is drastically improved in simulation, as well as in the experiment.

6.4 Printed Nonimaging Optics

The above chapter has proven that printed imaging systems are possible. However, since the geometry is not limited to refractive lenses, it is legitimate to also apply the technique for fabrication of nonimaging optical elements. Nonimaging optics describes systems which mainly collect radiation from a source volume into a target volume. In contrast to the above systems, they usually will not deliver an image from the object (source). Such systems are often employed for illumination but are also used for other purposes. Typical macroscopic applications of such systems can be found in automotive headlights, backlight displays, public lighting, flashlights, or solar collectors. On a smaller scale, corresponding elements can be found in sensors, fiber couplers, endoscopic systems, or photonic circuits.

One promising nonimaging application of direct 3D printing is the fabrication of concentration and collection systems on top of quantum dots for increased sensitivity. As the printing process allows for accurate alignment of the optics to the quantum dot, tailored freeform systems for beam shaping and collection are possible. In the following example, the printed optical system is designed to collect a large numerical aperture and collimate the light towards a secondary receiver system.

For this purpose, a two-element refractive collection system was designed with a numerical aperture of 0.7. The design was corrected for a field of view of a few micrometers, since the position of the quantum dot is only determined within certain tolerances. The quantum dot and the printed optics are contained within a cryo cell. Outside of the cryo cell, the light is captured through secondary optics with a numerical aperture 0.4. Figure 6.12 shows the design, the CAD model, and the manufactured optics.

Experimental tests at our cooperation partner have shown that the printed optics is operational at a temperature of a few degrees Kelvin and that the overall efficiency of the system can be improved significantly [66].

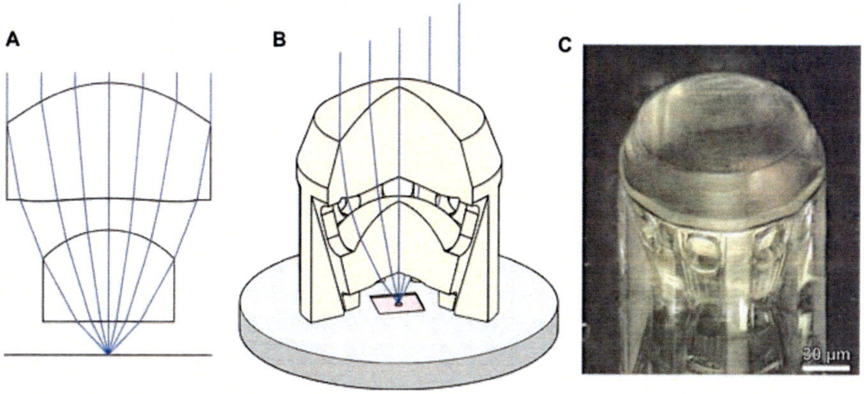

Fig. 6.12 Printed collection system on top of a quantum dot. (**a**) Optical design, (**b**) schematic setup, (**c**) microscope image of the printed system

Optical stimulation can, e.g., be achieved by a small laser focus. Also, for this purpose printed optical elements are of general use. For focusing narrowband light, diffractive elements can be employed which bring several advantages: the diffractive lens can be realized in a thin layer, which dramatically reduces printing volume and printing time. The thin optical volume also minimizes the material absorption and volume scattering effects, which are both known problems in some types of photoresists. Moreover, the multiphoton printing process intrinsically, due to the voxel shape, offers a higher lateral resolution as compared to the axial resolution, which is ideal for small feature sizes of the diffractive profile. General limitations of diffractive optics however are the large chromatic dispersion, the wavelength-dependent diffraction efficiency, and the limited efficiency at large deflection angles. In order to assess such limitations, a diffractive lens with a high numerical aperture of 0.8 was designed. The design of the diffractive phase profile itself is possible with any state-of-the-art optical design software in the geometrical limit, where physical optical effects, like diffraction efficiency, and multiple orders are not included. The ideal design was optimized for a wavelength of 543 nm and theoretically should deliver a diffraction limited spot. However, this is only true under the so-called thin element approximation (TEA) [67]. Even for a perfect kinoform with optimum step height, reduced diffraction efficiency is to be expected due to unavoidable self-shadowing effects for large angles.

The optimum profile step height h in air can be easily calculated via the refractive index n at the given wavelength $\lambda$to be $h = \lambda/(n-1)$. The efficiency loss of a real profile in comparison to the ideal TME can, e.g., be simulated using the "wave propagation method" (WPM) [68].

Figure 6.13 shows the design for a diameter of 180 μm and a focal length of 67.5 μm. From the theoretical phase function, the diffractive profile and the

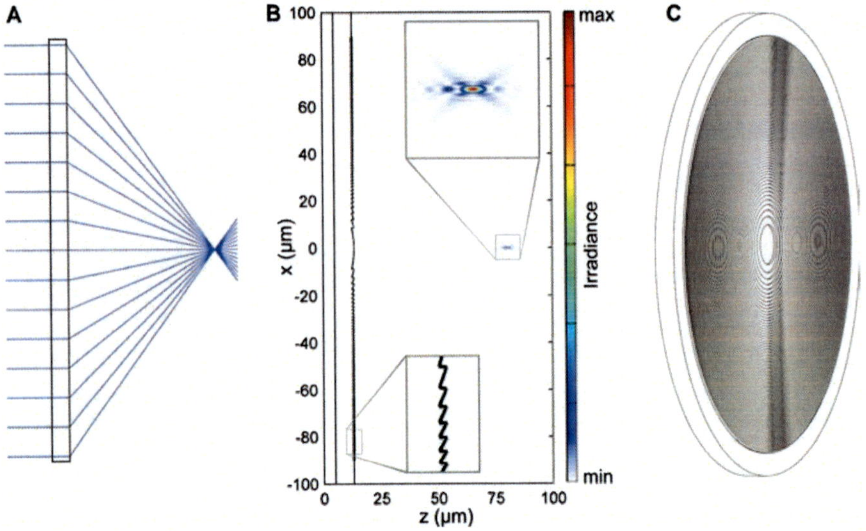

Fig. 6.13 Diffractive lens to be printed via multiphoton lithography. (**a**) Optical design for a wavelength = 543 nm and NA of 0.8. (**b**) Physical optical simulation of the profile via WPM propagation. (**c**) Corresponding CAD model

corresponding CAD model can be determined via modulo operation with the above step height. For this example, the resulting minimum period of the diffractive structure is about 700 nm at an ideal step height of 990 nm.

In a second step, the diffractive lens was printed, and the performance was assessed experimentally. The PSF was measured using a collimated laser beam under perpendicular incidence at the design wavelength. The spot in the focus was imaged with a microscope (100× objective with NA = 0.8). Figure 6.14 shows the result of the experiment in comparison with simulation. The resulting full widths at half maximum (FWHM) values are 303 nm (ZEMAX), 336 nm (WPM), und 452 nm (experimental, average of x- and y-section). In the diffraction limit, this corresponds to effective numerical apertures of 0.8 (ZEMAX), 0.75 (WPM), and 0.6 (experiment). It is assumed that the difference between simulation and experiment results from the unavoidable profile shape deviation, which however could be partly counter-compensated via profile optimization [69]. In summary the experiment proves that focusing into a subwavelength-sized spot is feasible with printed diffractive elements.

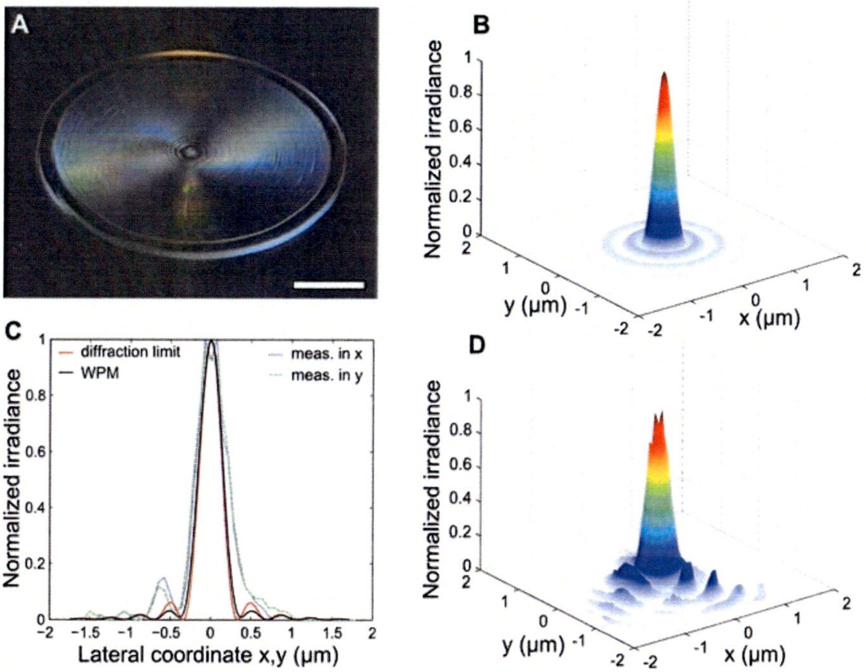

Fig. 6.14 Experimental results of a printed diffractive lens. (**a**) Microscopic image (scale bar: 50 µm). (**b**) WPM-simulated PSF. (**c**) Comparison of the diffraction-limited PSF versus the WPM and the experimental data. Resulting FWHM values: 303 nm (ZEMAX), 336 nm (WPM), and 452 nm (average of x- and y-profile). (**d**) Experimentally retrieved PSF

While the above examples were optimized for a point source or a point focus, we now are investigating a printed illumination system for an extended source[2]. Light-emitting diode (LED) light sources, in comparison to mono-mode lasers of fibers, exhibit a large-area angle product, or étendue. This results from the large angular emission characteristics in combination with an extended emitter area. The considerably large étendue limits the achievable level of collimation, especially if the collimator must be small in volume/diameter and close to the source. In consequence the collimation angle which can be realized with a printed microoptical component is limited, but nevertheless a reduction of the angular spectrum is possible.

We illustrate this with the following TIR collimation element, which was printed on top of an LED emitter surface and was able to reduce the angular radiation angle from 60° to 31.4° (Fig. 6.15). The emitter was a so-called Point Source LED (OSRAM F1372B), which is characterized by a very small emission area of only 80 μm in diameter at a wavelength of 650 nm (20 nm FWHM spectral width).

For the geometry of the TIR collimator, a combination of a dielectric total internal reflecting concentrator (DTIRC) and an aspherical surface was selected, as illustrated in Fig. 6.15. For the experimental verification, an LED of the TO-18 package was used, which allows for a simple handling, printing, and processing. As print material the standard photoresist IP-S was employed, which generally results in smooth printed surfaces. Figure 6.16 shows microscope images of the printed geometries. The angular emission characteristics of the light source were measured before and after the printing process, each time with the same driving current of 10 mA. As a test setup, the LED was placed under a Lambertian diffusor screen at 30 mm, far enough to neglect the spatial extent of the light source. The irradiance

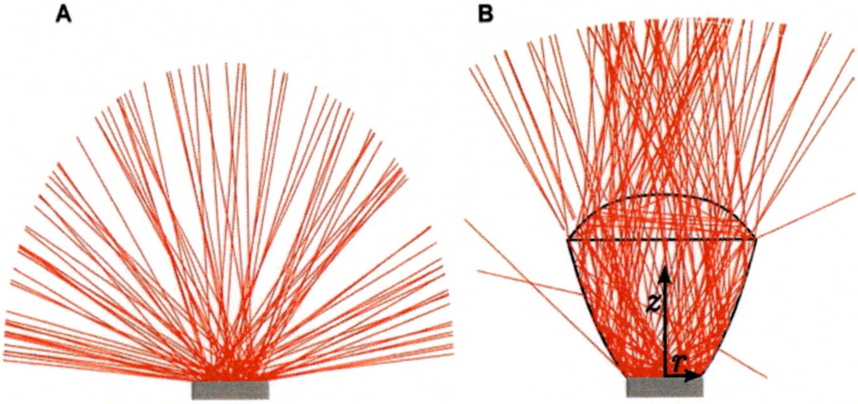

Fig. 6.15 (**a**) Intensity characteristics of the "Point Source LED" alone. (**b**) Reduced angular intensity width of the same source in combination with a printed TIR collimator. Reprinted with permission from reference [70]

[2]The corresponding results were published before in [70].

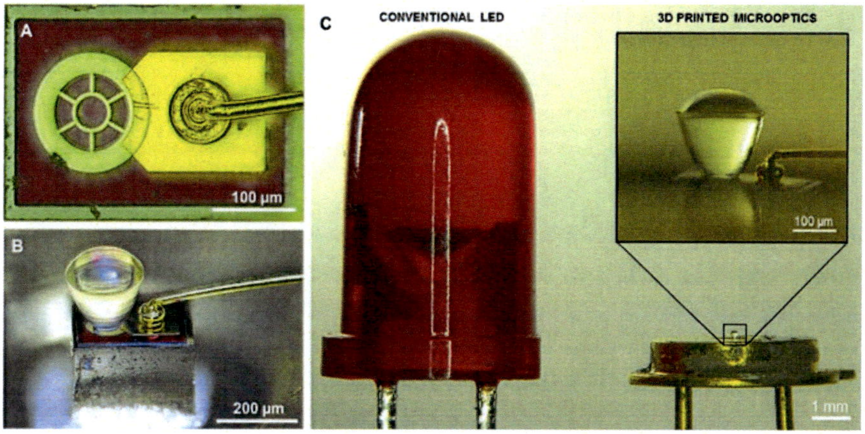

Fig. 6.16 Results of the 3D printing process. (**a**) "Point Source" LED chip before processing. (**b**) Chip after printing of a TIR collimator. (**c**) Comparison to a standard LED. Reprinted with permission from reference [70]

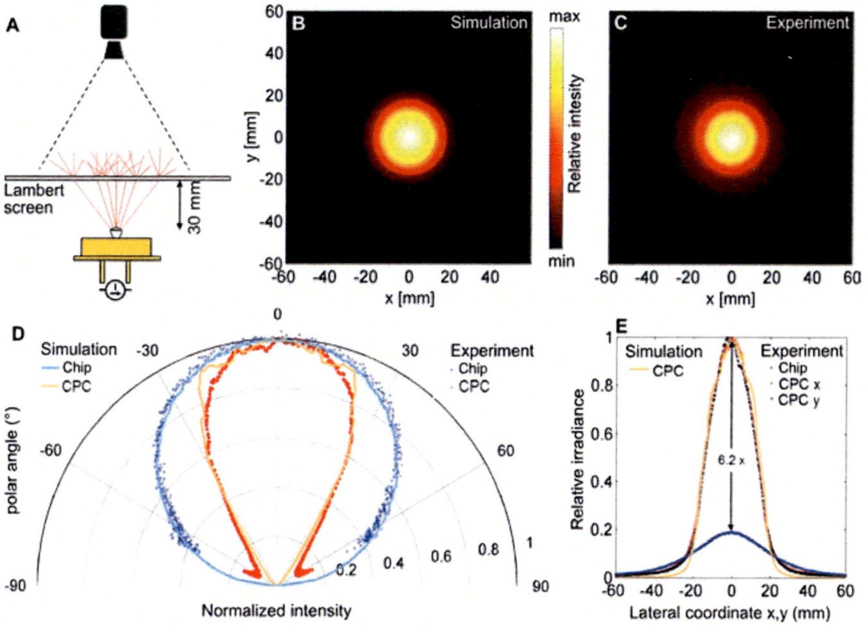

Fig. 6.17 Radiometric analysis and comparison to simulation. (**a**) Test setup for radiometric measurements. (**b**) Simulated irradiance on the diffusor in 30 mm distance. (**c**) Measured irradiance distribution. (**d**) Comparison of the normalized simulated and measured angular characteristics. (**e**) Corresponding relative irradiance profiles. Reprinted with permission from reference [70]

distribution on the diffusor was recorded with a standard color DSLR camera. The image was converted into gray levels by simple addition of the intensity of the three different color channels.

Again, a comparison between simulation and experimental results was performed. To do so, a nonsequential simulation model, containing the emission area and the TIR geometry, was set up in ZEMAX. Figure 6.17 shows the comparison of simulation results to the experimental results. Overall a very good match between both was achieved. The results predict that the printed collimation system can reduce the angular width from 60° to ~30°. The slight differences between simulation and experiment can most likely be explained by the remaining surface roughness on the TIR surface.

6.5 Summary

The fabrication method of femtosecond direct laser writing enables the 3D printing of complex optical components and systems. Due to its many degrees of freedom in terms of optical and structural design, specific rules were developed and introduced specifically for imaging systems. Two examples of doublet imaging objectives are presented and compared in design, fabrication, and imaging performance. A method which allows the direct assembly-free integration of absorbing structures is introduced and demonstrated experimentally. As three examples for nonimaging optics, coupling lenses for quantum dots, a diffractive lens for sub-μm focusing, and a collimation structure for LEDs are presented.

References

1. Wu E.-S., Strickler J. H., Harrell W. R., & Webb W. W. (1992). Two-photon lithography for microelectronic application. In *SPIE Proceedings* (p. 776). SPIE.
2. Maruo, S., Nakamura, O., & Kawata, S. (1997). Three-dimensional microfabrication with two-photon-absorbed photopolymerization. *Optics Letters, 22*, 132.
3. Maruo, S., & Kawata, S. (1998). Two-photon-absorbed near-infrared photopolymerization for three-dimensional microfabrication. *Journal of Microelectromechanical Systems, 7*, 411–415.
4. Sun, H.-B., Matsuo, S., & Misawa, H. (1999). Three-dimensional photonic crystal structures achieved with two-photon-absorption photopolymerization of resin. *Applied Physics Letters, 74*, 786–788.
5. Serbin, J., Egbert, A., Ostendorf, A., Chichkov, B. N., Houbertz, R., Domann, G., Schulz, J., Cronauer, C., Fröhlich, L., & Popall, M. (2003). Femtosecond laser-induced two-photon polymerization of inorganic–organic hybrid materials for applications in photonics. *Optics Letters, 28*, 301.
6. Straub, M., & Gu, M. (2002). Near-infrared photonic crystals with higher-order bandgaps generated by two-photon photopolymerization. *Optics Letters, 27*, 1824.
7. Korte, F., Serbin, J., Koch, J., Egbert, A., Fallnich, C., Ostendorf, A., & Chichkov, B. N. (2003). Towards nanostructuring with femtosecond laser pulses. *Applied Physics A, 77*, 229–235.

8. Straub, M., Nguyen, L. H., Fazlic, A., & Gu, M. (2004). Complex-shaped three-dimensional microstructures and photonic crystals generated in a polysiloxane polymer by two-photon microstereolithography. *Optical Materials, 27*, 359–364.
9. Serbin, J., Ovsianikov, A., & Chichkov, B. (2004). Fabrication of woodpile structures by two-photon polymerization and investigation of their optical properties. *Optics Express, 12*, 5221.
10. Deubel, M., Wegener, M., Linden, S., von Freymann, G., & John, S. (2006). 3D-2D-3D photonic crystal heterostructures fabricated by direct laser writing. *Optics Letters, 31*, 805.
11. Deubel, M., von Freymann, G., Wegener, M., Pereira, S., Busch, K., & Soukoulis, C. M. (2004). Direct laser writing of three-dimensional photonic-crystal templates for telecommunications. *Nature Materials, 3*, 444–447.
12. Guo, R., Li, Z., Jiang, Z., Yuan, D., Huang, W., & Xia, A. (2005). Log-pile photonic crystal fabricated by two-photon photopolymerization. *Journal of Optics A: Pure and Applied Optics, 7*, 396–399.
13. Wu, S., Serbin, J., & Gu, M. (2006). Two-photon polymerisation for three-dimensional microfabrication. *Journal of Photochemistry and Photobiology A: Chemistry, 181*, 1–11.
14. Gan, Z., Cao, Y., Evans, R. A., & Gu, M. (2013). Three-dimensional deep sub-diffraction optical beam lithography with 9 nm feature size. *Nature Communications, 4*, 2061.
15. Fischer, J., & Wegener, M. (2013). Three-dimensional optical laser lithography beyond the diffraction limit. *Laser & Photonics Reviews, 7*, 22–44.
16. Chidambaram, N., Kirchner, R., Fallica, R., Yu, L., Altana, M., & Schift, H. (2017). Selective surface smoothening of polymer microlenses by depth confined softening. *Advanced Materials Technologies, 2*, 1700018.
17. Sun H.-B., & Kawata S. (2004). Two-photon photopolymerization and 3d lithographic microfabrication. In *NMR • 3D Analysis • Photopolymerization* (Vol. 170, pp. 169–273). Berlin, Springer.
18. Yang, D., Jhaveri, S. J., & Ober, C. K. (2005). Three-dimensional microfabrication by two-photon lithography. *MRS Bulletin, 30*, 976–982.
19. Wippermann F. C., Radtke D., Zeitner U., Duparré J. W., Tünnermann A., Amberg M., Sinzinger S., Reinhardt C., Ovsianikov A., & Chichkov B. N.. (2006). Fabrication technologies for chirped refractive microlens arrays. In *SPIE Proceedings* (p. 628800). SPIE.
20. Ostendorf, A., & Chichkov, B. N. (2006). Two-photon polymerization: a new approach to micromachining. *Photonics Spectra, 40*, 72–80.
21. Guo, R., Xiao, S., Zhai, X., Li, J., Xia, A., & Huang, W. (2006). Micro lens fabrication by means of femtosecond two photon photopolymerization. *Optics Express, 14*, 810.
22. Chen, Q.-D., Wu, D., Niu, L.-G., Wang, J., Lin, X.-F., Xia, H., & Sun, H.-B. (2007). Phase lenses and mirrors created by laser micronanofabrication via two-photon photopolymerization. *Applied Physics Letters, 91*, 171105.
23. Nishiyama, H., Nishii, J., Mizoshiri, M., & Hirata, Y. (2009). Microlens arrays of high-refractive-index glass fabricated by femtosecond laser lithography. *Applied Surface Science, 255*, 9750–9753.
24. Wu, D., Wu, S.-Z., Niu, L.-G., Chen, Q.-D., Wang, R., Song, J.-F., Fang, H.-H., & Sun, H.-B. (2010). High numerical aperture microlens arrays of close packing. *Applied Physics Letters, 97*, 31109.
25. Malinauskas, M., Gilbergs, H., Žukauskas, A., Purlys, V., Paipulas, D., & Gadonas, R. (2010). A femtosecond laser-induced two-photon photopolymerization technique for structuring microlenses. *Journal of Optics, 12*, 35204.
26. Žukauskas, A., Malinauskas, M., Reinhardt, C., Chichkov, B. N., & Gadonas, R. (2012). Closely packed hexagonal conical microlens array fabricated by direct laser photopolymerization. *Applied Optics, 51*, 4995–5003.
27. Hu, Y., Chen, Y., Ma, J., Li, J., Huang, W., & Chu, J. (2013). High-efficiency fabrication of aspheric microlens arrays by holographic femtosecond laser-induced photopolymerization. *Applied Physics Letters, 103*, 141112.

28. Malinauskas, M., Žukauskas, A., Purlys, V., Belazaras, K., Momot, A., Paipulas, D., Gadonas, R., Piskarskas, A., Gilbergs, H., Gaidukevičiūtė, A., Sakellari, I., Farsari, M., & Juodkazis, S. (2010). Femtosecond laser polymerization of hybrid/integrated micro-optical elements and their characterization. *Journal of Optics, 12*, 124010.
29. Brasselet, E., Malinauskas, M., Žukauskas, A., & Juodkazis, S. (2010). Photopolymerized microscopic vortex beam generators. Precise delivery of optical orbital angular momentum. *Applied Physics Letters, 97*, 211108.
30. Malinauskas, M., Žukauskas, A., Belazaras, K., Tikuišis, K., Purlys, V., Gadonas, R., & Piskarskas, A. (2012). Laser fabrication of various polymer microoptical components. *European Physical Journal Applied Physics, 58*, 20501.
31. Osipov, V., Doskolovich, L. L., Bezus, E. A., Cheng, W., Gaidukeviciute, A., & Chichkov, B. (2012). Fabrication of three-focal diffractive lenses by two-photon polymerization technique. *Applied Physics A: Materials Science & Processing, 107*, 525–529.
32. Thiel M, Radke A., Fries B., Eicke D., Niesler F., Baretzky C., Bückmann T., & Wegener M. (2013). High-speed 3D direct laser writing of micro-optical elements. in CLEO: Science and Innovations. Part of Conference on Lasers and Electro-Optics, 9–14 May 2013. San Jose, CA, OSA technical digest (online): OSA The Optical Society, ATu2N.4.
33. Malinauskas, M., Farsari, M., Piskarskas, A., & Juodkazis, S. (2013). Ultrafast laser nanostructuring of photopolymers. A decade of advances. *Physics Reports, 533*, 1–31.
34. Schumann, M., Bückmann, T., Gruhler, N., Wegener, M., & Pernice, W. (2014). Hybrid 2D–3D optical devices for integrated optics by direct laser writing. *Light: Science and Applications, 3*, e175.
35. Lightman, S., Hurvitz, G., Gvishi, R., & Arie, A. (2017). Tailoring lens functionality by 3{D} laser printing. *Applied Optics, 56*, 9038–9043.
36. Lightman, S., Hurvitz, G., Gvishi, R., & Arie, A. (2017). Miniature wide-spectrum mode sorter for vortex beams produced by 3{D} laser printing. *Optica, 4*, 605.
37. Williams, H. E., Freppon, D. J., Kuebler, S. M., Rumpf, R. C., & Melino, M. A. (2011). Fabrication of three-dimensional micro-photonic structures on the tip of optical fibers using SU-8. *Optics Express, 19*, 22910–22922.
38. Žukauskas, A., Malinauskas, M., & Brasselet, E. (2013). Monolithic generators of pseudo-nondiffracting optical vortex beams at the microscale. *Applied Physics Letters, 103*, 181122.
39. Žukauskas, A. (2014). Improvement of the fabrication accuracy of fiber tip microoptical components via mode field expansion. *Journal of Laser Micro/Nanoengineering, 9*, 68–72.
40. Gissibl, T., Thiele, S., Herkommer, A., & Giessen, H. (2016). Two-photon direct laser writing of ultracompact multi-lens objectives. *Nature Photonics, 10*, 554–560.
41. Weiss I., & Marom D. M. (2016). Direct 3D nanoprinting on fiber tip of collimating lens and OAM mode converter in one compound element. In *2016 Optical Fiber Communications Conference and Exhibition (OFC)* (p. Th3E.2), 20–24 March 2016. IEEE.
42. Cojoc, G., Liberale, C., Candeloro, P., Gentile, F., Das, G., de Angelis, F., & Di Fabrizio, E. (2010). Optical micro-structures fabricated on top of optical fibers by means of two-photon photopolymerization. *Microelectronic Engineering, 87*, 876–879.
43. Bianchi, S., Rajamanickam, V. P., Ferrara, L., Di Fabrizio, E., Liberale, C., & Di Leonardo, R. (2013). Focusing and imaging with increased numerical apertures through multimode fibers with micro-fabricated optics. *Optics Letters, 38*, 4935–4938.
44. Gissibl, T., Thiele, S., Herkommer, A., & Giessen, H. (2016). Sub-micrometre accurate free-form optics by three-dimensional printing on single-mode fibres. *Nature Communications, 7*, 11763.
45. Weiss I., & Marom D. M. (2015). Direct 3D nano-printing on optical fiber tip. In *2015 International Conference on Optical MEMS and Nanophotonics, 2–5 August 2015* (pp. 1–2). Jerusalem: OMN2015, Jersusalem: proceedings. IEEE.
46. Malinauskas, M., Žukauskas, A., Purlys, V., Gaidukevičiu¯tė, A., Balevičius, Z., Piskarskas, A., Fotakis, C., Pissadakis, S., Gray, D., Gadonas, R., Vamvakaki, M., & Farsari, M. (2012). 3D microoptical elements formed in a photostructurable germanium silicate by direct laser writing. *Optics and Lasers in Engineering, 50*, 1785–1788.

47. Gissibl, T., Schmid, M., & Giessen, H. (2016). Spatial beam intensity shaping using phase masks on single-mode optical fibers fabricated by femtosecond direct laser writing. *Optica, 3*, 448.
48. Kowalczyk, M., Haberko, J., & Wasylczyk, P. (2014). Microstructured gradient-index antireflective coating fabricated on a fiber tip with direct laser writing. *Optics Express, 22*, 12545–12550.
49. Li, J., Fejes, P., Lorenser, D., Quirk, B. C., Noble, P. B., Kirk, R. W., Orth, A., Wood, F. M., Gibson, B. C., Sampson, D. D., & McLaughlin, R. A. (2018). Two-photon polymerisation 3D printed freeform micro-optics for optical coherence tomography fibre probes. *Scientific Reports, 8*, 14789.
50. Lightman, S., Gvishi, R., Hurvitz, G., & Arie, A. (2015). Shaping of light beams by 3D direct laser writing on facets of nonlinear crystals. *Optics Letters, 40*, 4460–4463.
51. Schleunitz A., Klein J. J., Houbertz R., & Krupp A. (2017). Novel hybrid polymers for fabrication of advanced micro-optics. In *SPIE Newsroom*. SPIE.
52. Sun, Y.-L., Dong, W.-F., Niu, L.-G., Jiang, T., Liu, D.-X., Zhang, L., Wang, Y.-S., Chen, Q.-D., Kim, D.-P., & Sun, H.-B. (2014). Protein-based soft micro-optics fabricated by femtosecond laser direct writing. *Light: Science and Applications, 3*, e129.
53. Kotz, F., Plewa, K., Bauer, W., Schneider, N., Keller, N., Nargang, T., Helmer, D., Sachsenheimer, K., Schäfer, M., Worgull, M., Greiner, C., Richter, C., & Rapp, B. E. (2016). Liquid glass. a facile soft replication method for structuring glass. *Advanced Materials (Deerfield Beach, Fla.), 28*, 4646–4650.
54. Cadarso, V. J., Perera-Núñez, J., Jacot-Descombes, L., Pfeiffer, K., Ostrzinski, U., Voigt, A., Llobera, A., Grützer, G., & Brugger, J. (2011). Microlenses with defined contour shapes. *Optics Express, 19*, 18665–18670.
55. Cheng, Y., Sugioka, K., Midorikawa, K., Masuda, M., Toyoda, K., Kawachi, M., & Shihoyama, K. (2003). Three-dimensional micro-optical components embedded in photosensitive glass by a femtosecond laser. *Optics Letters, 28*, 1144.
56. Gale M. T., & Knop K. (1983). The fabrication of fine lens arrays by laser beam writing. In *SPIE Proceedings* (pp. 347–353). SPIE.
57. Göppert-Mayer, M. (1931). Über Elementarakte mit zwei Quantensprüngen. *Ann. Phys., 401*, 273–294.
58. Spence, D. E., Kean, P. N., & Sibbett, W. (1991). 60-fsec pulse generation from a self-mode-locked Ti. Sapphire laser. *Optics Letters, 16*, 42.
59. Sibbett, W., Lagatsky, A. A., & Brown, C. T. A. (2012). The development and application of femtosecond laser systems. *Optics Express, 20*, 6989–7001.
60. Jeffs M. (2002). Reduced manufacturing sensitivity in multi-element lens systems. In *SPIE Proceedings* (p. 104). SPIE.
61. Isshiki M., Sinclair D. C., & Kaneko S. (2006). *Lens design. Global optimization of both performance and tolerance sensitivity* (Vol. 6342, p. 63420N). International Society for Optics and Photonics.
62. Bates R.. (2010). Performance and tolerance sensitivity optimization of highly aspheric miniature camera lenses. In *SPIE Proceedings* (p. 779302). SPIE.
63. Ruffieux, P., Scharf, T., Herzig, H. P., Völkel, R., & Weible, K. J. (2006). On the chromatic aberration of microlenses. *Optics Express, 14*, 4687.
64. Rill, M. S., Kriegler, C. E., Thiel, M., von Freymann, G., Linden, S., & Wegener, M. (2009). Negative-index bianisotropic photonic metamaterial fabricated by direct laser writing and silver shadow evaporation. *Optics Letters, 34*, 19.
65. Ma, Z.-C., Zhang, Y.-L., Han, B., Chen, Q.-D., & Sun, H.-B. (2018). Femtosecond-laser direct writing of metallic micro/nanostructures. From fabrication strategies to future applications. *Small Methods, 2*, 1700413.
66. Fischbach, S., Schlehahn, A., Thoma, A., Srocka, N., Gissibl, T., Ristok, S., Thiele, S., Kaganskiy, A., Strittmatter, A., Heindel, T., Rodt, S., Herkommer, A., Giessen, H., & Reitzenstein, S. (2017). Single quantum dot with microlens and 3{D}-printed micro-objective as integrated bright single-photon source. *ACS Photonics, 4*, 1327–1332.

67. Schmidt, S., Tiess, T., Schröter, S., Hambach, R., Jäger, M., Bartelt, H., Tünnermann, A., & Gross, H. (2016). Wave-optical modeling beyond the thin-element-approximation. *Optics Express, 24*, 30188–30200.
68. Schmidt, S., Thiele, S., Herkommer, A., Tünnermann, A., & Gross, H. (2017). Rotationally symmetric formulation of the wave propagation method-application to the straylight analysis of diffractive lenses. *Optics Letters, 42*, 1612–1615.
69. Korolkov, V. P., Nasyrov, R. K., & Shimansky, R. V. (2006). Zone-boundary optimization for direct laser writing of continuous-relief diffractive optical elements. *Applied Optics, 45*, 53.
70. Thiele, S., Gissibl, T., Giessen, H., & Herkommer, A. M. (2016). Ultra-compact on-chip LED collimation optics by 3D femtosecond direct laser writing. *Optics Letters, 41*, 3029–3032.

Chapter 7
Hybrid Polymers for Conventional and Additive Manufacturing of Microoptical Elements

Martin Herder, Jan Jasper Klein, Marko Vogler, Maria-Melanie Russew, Arne Schleunitz, and Gabi Grützner

Abstract Hybrid polymers constitute a class of optical materials combining properties of inorganic glass and organic polymers. The flexible synthesis and processing allows for specific tailoring of their properties as required for the fabrication of high-performance and reliable microoptical elements. While wafer-scale fabrication of microoptics using hybrid polymers is widely used in an industrial environment, they gain ever-increasing attention in additive manufacturing and 3D printing technologies. This chapter introduces the chemical concepts behind hybrid polymers, discusses their synthesis and processing, and gives a record on their application for the fabrication of microoptical and photonic elements using established wafer-scale as well as emerging additive manufacturing processes, in particular inkjet printing and two-photon polymerization direct laser writing.

Keywords Optical polymers · Microoptics · Hybrid polymers · ORMOCERs Photolithography · UV molding · Nanoimprint lithography · Additive manufacturing · Inkjet printing · 3D printing · Two-photon polymerization

7.1 Introduction

Microoptical elements constitute a key enabling technology for consumer electronics, telecommunication, sensing, imaging, lighting, as well as virtual and augmented reality applications [1–5]. Their sophisticated design, miniaturization, and integration are crucial for the realization of new devices with high functional complexity and energy efficiency. Current wafer-scale production technology allows for mass fabrication of microoptical elements such as microlens arrays, gratings, diffractive

M. Herder (✉) · J. J. Klein · M. Vogler · M.-M. Russew · A. Schleunitz · G. Grützner
micro resist technology GmbH, Berlin, Germany
e-mail: m.herder@microresist.de; j.klein@microresist.de

A. Heinrich (ed.), *3D Printing of Optical Components*, Springer Series in Optical Sciences 233, https://doi.org/10.1007/978-3-030-58960-8_7

optical elements (DOEs), or waveguides, as well as their integration into complex optoelectronic devices [6].

Modern 3D printing methods bring unprecedented opportunities to the fabrication of microoptics [7]: straightforward fabrication of freeform and true 3D microoptical elements as well as rapid prototyping with turnover times from the optical design to the fabricated part in the range of minutes to few hours will allow for completely new concepts in design and optimization. Furthermore, fabrication of integrated devices is simplified by direct printing onto sensors, emitters, or optical fibers.

Importantly, the rise of 3D printing as new fabrication technology goes hand in hand with the development of innovative optical materials with superior performance and processability. While a large number of conventional optical materials such as inorganic glass or organic polymers exist, their applicability in 3D printing requires adaptation or is very challenging as in the case of optical glass [8–10]. Furthermore, from a commercial viewpoint, materials must fulfill stringent quality requirements regarding not only optical performance but also environmental and long-term stability, processability, as well as cost-effectiveness.

In this respect, inorganic-organic hybrid polymers constitute a material platform showing unique properties and superior performance [11–13]. They are established commercial materials for the wafer-scale fabrication of microoptics and hold great promise for applications realized with additive manufacturing. In this review, we will point out the advantages of hybrid polymers in the context of optical materials. An introduction into synthetic routes and properties of hybrid polymers is given, and finally their application in the fabrication of microoptical components by wafer-scale and additive manufacturing technologies will be discussed.

7.2 Optical Materials and Fabrication Processes

7.2.1 Glass and Polymers for Optical Applications

The most important material parameter for the fabrication of optical elements is the ability to reflect, refract, or diffract light while being transparent at the wavelength of interest. In addition, refractive index, dispersion, and birefringence are crucial for the design and performance of optical elements. The transparency of any material is determined by several factors: (1) For transparency in the near-UV (NUV) and visible range, the electronic band gap of the material, i.e., the energy required for the excitation of electrons, must be higher than the light energy. (2) At infrared (IR) wavelengths, transparency strongly depends on the frequency and type of atomic and molecular vibrations. (3) The size of individual structural features of the material, i.e., grain boundaries and defects, must be smaller than the wavelength of light to avoid loss by scattering.

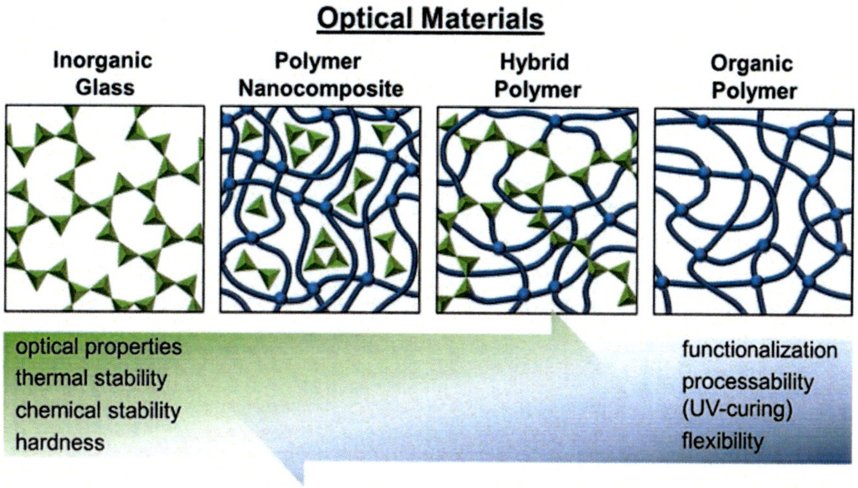

Fig. 7.1 Architecture and properties of inorganic and organic units combined in optical materials

In this respect, historically inorganic glass is the material of choice for the fabrication of optics. Its valence electrons have no energy level available that can be reached by absorption of a visible photon, its transparency ranges to near-IR (NIR) wavelengths, and its structure is highly amorphous showing no long-range order (Fig. 7.1 left). Other inorganic ceramic and semiconducting materials applied in optics, e.g., chalcogenide glasses, silicon, silicon carbide, GaAs, or InP, generally have lower band gaps than silica glass. Thus, they are not necessarily transparent in the visible range, but show other distinct properties, e.g., increased refractive index or transparency in the IR range [3]. On the other hand, an ever-increasing number of optical quality polymers exist, which possess highly amorphous (Fig. 7.1 right) or semicrystalline structures and consist of molecular building blocks with high band gap. Typical representatives are silicones (e.g., PDMS), poly(methyl methacrylate) (PMMA), polystyrene (PS), polycarbonate (PC), acrylonitrile butadiene styrene (ABS), or cyclic olefin (co)polymer (COC/COP).

Notably, inorganic glass and organic polymers are complementary in their properties (Fig. 7.1): most glasses show maximum transparency over a wide range of wavelengths from the NUV over visible to NIR wavelengths. In addition, they are available with a wide range of refractive indices and dispersion while generally showing low birefringence [14]. In comparison, transparency of polymers is reduced, in particular in the UV range, due to absorption of specific functional groups in the backbone and increased scattering effects. The range of refractive indices and dispersion properties is much smaller than that of glass [15]. While most optical polymers are naturally isotropic, their processing often induces marked birefringence due to orientation of the polymer chains [16].

Regarding mechanical properties, differences are even more pronounced. Glasses are generally hard and brittle materials with superior chemical inertness and stability against thermal and light damage. In contrast, the mechanical properties of polymers can be tuned over an extremely broad range from soft and flexible to hard and tough by changing their molecular architecture. In addition, organic chemistry allows for application-specific tailoring of bulk and surface properties such as gas permeability and hydrophilicity. However, due to the organic nature of the building blocks, the inertness and stability of polymers is lower compared to inorganic glass.

Importantly, these material properties dictate the choice of processes for the fabrication of optical components. Glass components in the macro- to microscale are fabricated using subtractive methods, i.e., cutting, grinding, turning, and polishing processes. Molding techniques—offering a highly parallel and cost-effective way for mass production—are limited to specific glass types possessing low T_g [14]. In contrast, utilizing polymers allows for a large range of processing conditions and thus much larger freedom in the design of optical components. They are mass fabricated from thermoplastic polymers, i.e., polymers that can be melted by heating, using hot embossing or injection molding [17]. Alternatively, UV curing, i.e., light-induced polymerization and cross-linking of resins containing acrylate, epoxide, or thiol-ene functions, is advantageously used for structuring polymer layers by spatial UV exposure through a photomask (UV lithography) or pressing a stamp into the material prior to UV flood exposure and separation (UV imprinting). In particular, with thermal or UV nanoimprint lithography (NIL), optical components with micro- and nanoscale resolution can be realized using wafer-scale and roll-to-roll fabrication [18, 19]. During these processes, either the structured polymer itself becomes integral part of the optical component, or it is used as resist for transferring the structure into the underlying glass or semiconductor substrate by an etching step.

In general, due to their more facile and versatile processing, polymers are the material of choice for most 3D printing processes [20]. On the one hand, thermoplastics are employed in fused deposition modeling (FDM) and selective laser sintering (SLS), whereby these non-cross-linked polymers are heated above their glass transition temperature T_g becoming liquid and being extruded or molten into the desired architecture. On the other hand, light-based 3D printing methods such as UV-assisted inkjet printing, stereolithography (SLA), or two-photon polymerization direct laser writing (2PP-DLW) rely on photocurable thermosets. Using these, maximum resolution, being of tremendous importance in the context of microoptics, is achieved due to localized curing with the high spatial and temporal resolution inherent to the application of light. In contrast to polymers, direct 3D structuring or printing of glass materials requires much harsher conditions or can only be achieved using sophisticated chemistry [8–10].

7.2.2 *Hybrid Materials*

Regarding the complementary properties of glass and polymers, it is tempting to combine both into a unique material and profit from the advantages of both worlds. This is achieved with so-called hybrid materials [21, 22], which can be divided into class I hybrids, e.g., polymer nanocomposites, and class II hybrids, e.g., hybrid polymers (Fig. 7.1 middle). In polymer nanocomposites clusters or particles of an inorganic material are dispersed in a polymeric matrix, and only weak interactions, such as van der Waals interactions or hydrogen bonds, exist between the two domains. In contrast, in hybrid polymers the inorganic and organic domains are grafted together by strong covalent interactions. Thus, the inorganic domain forms clusters or an extended network interpenetrating with the organic polymer. In order to be suited for applications in optics, the length scale of the individual domains should be smaller than the wavelength of light, i.e., usually in the range of few tens of nanometers, to avoid scattering effects.

Hybrid materials do not only combine the properties of the individual components, but new properties may emerge due to the nanometer length scale and resulting large interface between the inorganic and organic domain. Thus, they show an extraordinarily high mechanical, thermal, and chemical resistance as compared to bare organic polymers. At the same time, their mechanical properties are tunable showing greater flexibility and toughness compared to inorganic glass. The combination of sol-gel synthesis (*vide infra*) and polymer chemistry in hybrid materials open up new and versatile ways of material fabrication and processing. Most importantly, optical properties of hybrid materials such as refractive index and dispersion can be tuned by varying the composition of inorganic and organic domains exceeding the range offered by conventional polymers [23].

Consequently, it turns out that for applications in microoptics and photonics with high demands for optical quality, stability, and processing, high-performing hybrid materials are the optimal choice. However, for application in a commercial environment, further considerations have to be made: synthesis of a material should be scalable and cost-effective, and it should possess a shelf life of several months to years. In this respect, hybrid polymers clearly outperform polymer nanocomposites. The latter often suffer from aging and aggregation phenomena during synthesis and storage. Furthermore, stringent quality control during every stage of the production process is essential, in particular for applications in optics, to gain full control over material properties and guarantee reproducible specifications and reliable performance.

In this realm, hybrid polymers based on the ORMOCER® (ORganically MOdified CERamics) technology [24] being commercialized by *micro resist technology GmbH* under the names OrmoComp®, OrmoClear®, OrmoClear®FX, and OrmoStamp® [25] enjoy a tremendous popularity for the fabrication of microoptical components and are used worldwide in industry and academia.

7.2.3 Production and Processing of Hybrid Polymers

The ORMOCER technology initially developed by the *Fraunhofer Institute for Silicate Research* is based on the sol-gel processing of organically modified alkoxides as hybrid molecular precursors. These alkoxides are mainly based on silicon, but also other metal alkoxides (e.g., Al, Ti, Zr) may be added. The resulting material consists of domains of inorganic oxidic glass (silica network), organically modified polysiloxanes, and cross-linked organic polymers. As the covalent silicon–carbon bond is stable against hydrolysis, a permanent and strong connection between the inorganic and organic domains of the hybrid polymer is obtained. Starting from the mid-1980s, the basic chemistry and processing of ORMOCER-based hybrid polymers have been developed [26–35] leading to today's numerous applications not only in the field of optics but also as coatings, dental fillings, membranes, microelectronics, and energy conversion [21, 30].

The fabrication of hybrid polymers proceeds in three main steps (Fig. 7.2):

1. Sol-gel synthesis of the inorganic domain resulting in a solvent-free base resin.
2. Formulation by addition of comonomers, photoinitiator, and functional additives.
3. UV curing of the organic domain during processing. In the following brief details on each of the three steps are given with emphasis on how distinct choice of chemical entities or parameters influences the properties of the final hybrid polymer.

7.2.3.1 Sol-Gel Synthesis of the Base Resin

Sol-gel synthesis [36, 37] relies on the subsequent hydrolysis and condensation of silicon alkoxides $R'_xSi(OR)_{4-x}$ with R′ being an optional organic residue and OR a simple alcohol. In a first step, Si–OR bonds are hydrolyzed by the attack of water, and in a second step, Si–O–Si bonds are formed by condensation of silanol groups (Fig. 7.2). The condensation step is irreversible, thus with ongoing reaction time,

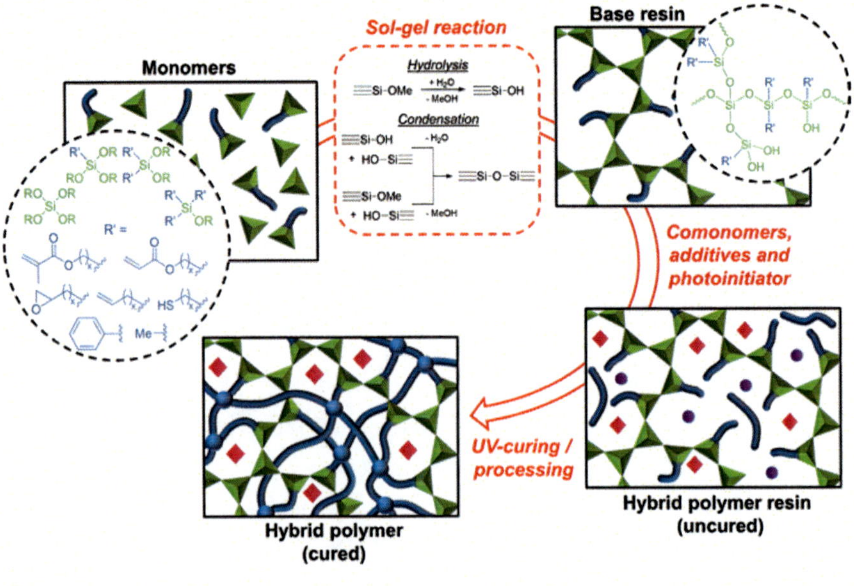

Fig. 7.2 Synthesis and processing of hybrid polymers in three steps. Exemplary structures of silane precursors as well as possible structural motifs of the inorganic domain are shown in the insets

small, soluble particles evolve (the sol), which start to interconnect and aggregate to larger clusters finally forming an extended three-dimensional network inducing gelation of the material. Notably, during the sol-gel synthesis, the reaction is precisely controlled in order to stop it just before gelation. Thus, liquid base resins with tunable viscosities are obtained, which are easily processable by means of micro- and nanofabrication in later steps (*vide infra*). Importantly, the type of silane precursor and reaction conditions (solvent, catalysts, temperature, pH) determine the relative rates of hydrolysis and condensation. This allows for control over the Si–OH content of the resin and the size of inorganic particles, which are typically in the range of 1–3 nm [35, 38] ensuring high optical quality of the material and allowing for its structuring with nanometer resolution.

A huge variety of silane precursors $R'_xSi(OR)_{4-x}$ for the sol-gel reaction is available. Some examples are depicted in Fig. 7.2. Importantly, the number of organic residues R′ on the precursor dictates the structure of the resulting inorganic network ranging from extended 3D structures over silsesquioxane clusters and linear chains to short oligomers. Residues R′ either can be inert (e.g., alkyl or phenyl groups) or possess polymerizable functions such as acrylate, methacrylate, epoxy, vinyl, thiol, or amine groups. Size and number of organic groups again influence the network density of the inorganic domain by steric interactions, while the amount and type of polymerizable groups determine the structure of the organic network after curing. Their main task is to counteract the brittle nature of the inorganic glass-like domain and induce toughness or a certain degree of flexibility to the material after UV curing.

Usually at the end of the sol-gel reaction, base resins are dried, all solvents are removed, and the material is fine-filtered to ensure reproducible quality and tight specifications.

7.2.3.2 Formulation

After sol-gel synthesis of the base resin, its properties and processability are fine-tuned by using different additives: organic comonomers as well as chain transfer agents fine-tune the structure of the organic network and optimize the curing chemistry. Furthermore, comonomers are also used to tune viscosity of the uncured resin as well as mechanical and surface properties of the cured hybrid polymer. Sol-gel additives, i.e., unreacted organoalkoxysilanes, optimize processing, in particular as adhesion promoter for coatings, thin films, and wafer-scale fabrication processes.

The choice of the photoinitiator is of particular importance for light-induced curing of the organic network. Depending on the polymerization chemistry utilized (epoxy-, vinyl- or acrylate-based systems), a large variety of photoacid generators or radical photoinitiators is available [39]. Critical parameters are optimal photosensitivity at the curing wavelength, high reactivity, low sensitivity towards oxygen inhibition, as well as nontoxicity of photoproducts. In addition, for the fabrication of optical components, full bleaching of the initiator as well as minimization of discoloration and luminescence caused by photoproducts is essential and requires careful optimization of the initiation system. In this respect, phosphine oxides (e.g., TPO) and α-hydroxy ketones show superior performance.

Further additives may include polymerization stabilizers to ensure high shelf life of the uncured hybrid polymer resins, thermo- and UV stabilizers for improving long-term stability of final products, dyes for modification of optical properties, as well as surfactants to modify surface tension and coating behavior of the resins.

7.2.3.3 Processing and UV Curing

Uncured hybrid polymers being solvent-free viscous liquids can easily be processed using standard wafer-scale fabrication techniques, such as UV lithography, UV molding, nanoimprint lithography, and direct laser writing. Exemplary processing for UV lithography and UV imprinting, typically employed for the fabrication of microoptical components, is shown in Fig. 7.3 [11]. Generally, hybrid polymers behave as a typical negative-tone resist. After spin coating or dispensing of the resin, UV curing is induced under a mask in proximity exposure or while pressing a structured stamp into the material. Post-exposure bake, development and hardbake steps follow. Cross-linking of the organic domain is completed during UV curing, while the optional thermal annealing steps increase adhesion to the substrate, stabilize optical properties of the material, and further increase its thermal and environmental stability due to relief of residual internal stress.

Critical parameters during processing are the irradiation setup, O_2 sensitivity, and shrinkage of the hybrid polymer during curing. Hybrid polymers are designed

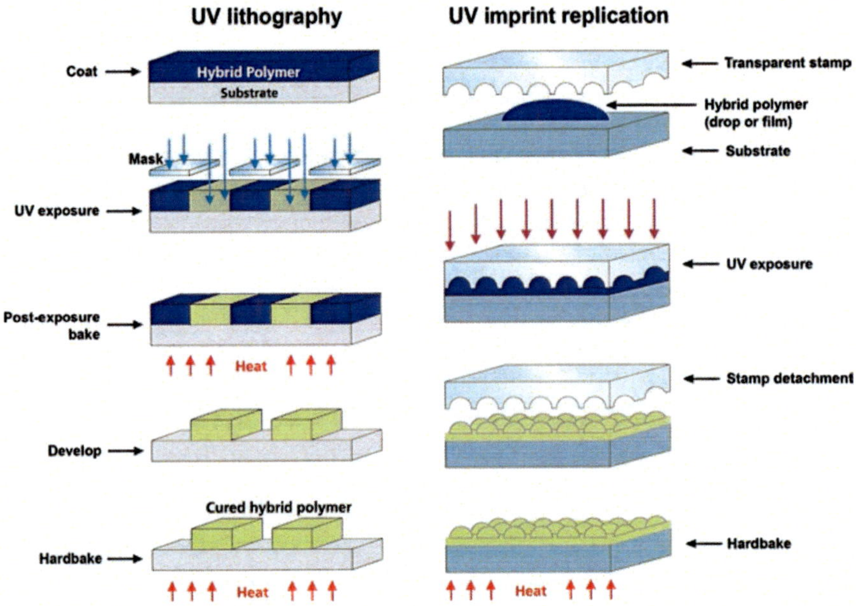

Fig. 7.3 Process schemes for the wafer-scale direct structuring of hybrid polymers using UV lithography or pattern replication using UV imprinting

to have a broad processing window in terms of wavelengths and intensity used, e.g., broadband UV, i-line (365 nm), or h-line (405 nm) irradiation. However, specific UV curing conditions prevailing, for example, during 3D printing, may require an adaptation of the photoinitiator content or type (*vide infra*). Notably, UV curing of commercial hybrid polymers [25] is based on free radical polymerization of acrylate and methacrylate groups, thus showing very fast curing speeds in the range of seconds to few minutes with low-intensity UV light. This is in contrast to epoxy-based optical polymers, which need significantly longer processing times including thermal steps. Efficient bleaching of the photoinitiator and high conversion of the polymerizable groups leads to absence of yellowing and increased long-term stability of the optical properties.

Inhibition effects by the presence of oxygen during UV curing are typical for free radical polymerizations [40] and strongly depend on the chemistry of the base resin and additives. Notably, formulations such as OrmoComp and OrmoClearFX readily cure under ambient atmosphere and oxygen permeable stamps (e.g., PDMS stamps).

Volume shrinkage of the polymer during curing is of particular importance for the creation of structures with high fidelity and resolution. In general, compared to conventional acrylate polymers, often showing more than 20% volume shrinkage, it is drastically reduced in hybrid polymers due to the preformed inorganic network [29]. Commercial hybrid polymers exhibit shrinkage from 2% up to 7% [25], while in academic work, it has been further reduced to practically zero by incorporation of ZrO_2 into the inorganic domain [41].

7.2.4 Properties of Hybrid Polymers

7.2.4.1 Optical Transparency

Hybrid polymers for optical applications show full transparency (>99%) for visible wavelengths >400 nm up to the NIR range. Additionally, in the NUV range (350–400 nm), only very little absorption is observed in transmission spectra (Fig. 7.4). Importantly, the outstanding transparency is hardly affected in long-term applications. Post-curing yellowing due to degradation of the polymer backbone by thermal and climate stress, as observed in most purely organic polymers, is minimized by the hybrid structure (*vide infra*). Prerequisites for optimal transparency and low yellowing are the selection of an efficiently bleaching photoinitiator, optimized curing conditions, and careful quality control during synthesis and formulation of the resins concerning selection of raw materials and monitoring of impurities.

With their high optical transparency, low haze and tunable refractive index hybrid polymers are also suited for the application as waveguide materials. Therefore, specifically designed hybrid polymers can be operated with typical datacom wavelengths in the NIR range. Low optical loss at these wavelengths is achieved by two factors:

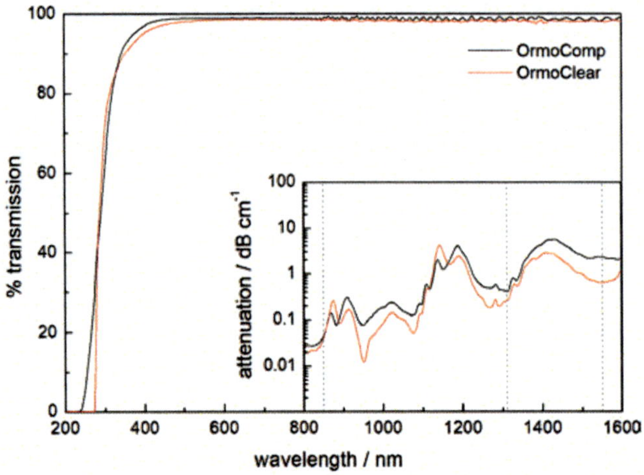

Fig. 7.4 Optical transmission of cured thin films ($d = 20\ \mu m$) of hybrid polymers on quartz substrates. Inset: optical attenuation of uncured hybrid polymers in the visible and NIR range

1. Reduction of the CH content in the organic domain by the choice of suited monomers [35]
2. Reduction of the Si–OH content in the inorganic domain by specific, nonaqueous sol-gel conditions [31]

The effect is shown in the inset of Fig. 7.4: the attenuation of uncured OrmoClear, with its low Si–OH-containing base resin, is significantly lower compared to OrmoComp, with its base resin obtained from aqueous sol-gel chemistry. After curing of OrmoClear, its optical attenuation is further reduced to values of 0.2–0.3 dB cm^{-1} at 1310 nm and 0.5–0.6 dB cm^{-1} at 1550 nm [33].

7.2.4.2 Autofluorescence

Autofluorescence of polymers is either intrinsic due to the backbone structure of the polymer or is induced by impurities and degradation phenomena evolving during curing or long-term operation [42]. Notably, the hybrid structure and corresponding stability impose very low levels of autofluorescence to hybrid polymers, in particular OrmoComp and OrmoClearFX [43–46]. With the background signal being comparable to that of Borofloat glass and with its transparency in the NUV and Vis range, these materials are therefore highly suited for the fabrication of cell substrates, optofluidic chips, and imaging probes based on fluorescence detection.

7.2.4.3 Refractive Index and Dispersion

The refractive index of hybrid polymers can be tuned by:

1. Alternation of precursors and conditions of the sol-gel process
2. Chemistry of the organic domain
3. Utilization of other metal alkoxides

In particular the variation of the sol-gel-derived base resin by co-reacting alkoxysilane precursors with $Zr(OR)_4$ or $Ti(OR)_4$ is broadly applied to increase the RI of the resulting hybrid polymers. For commercially available materials and prototypes, currently a range of refractive indices between 1.45 and 1.65 at 589 nm can be realized [25, 30].

For the fabrication of microoptical components, the combination of high refractive index and low dispersion, i.e., high Abbe number, often is desired in order to make structures as compact as possible while keeping acceptable levels of chromatic aberration [14]. However, there is a strict trade-off between the two parameters for organic polymers [15]. As inorganic glass covers significantly wider regions in the Abbe diagram, the combination of inorganic and organic domains [23] in hybrid polymers broadens the achievable refractive index/Abbe number combinations to some extent. Standard hybrid polymers possess Abbe numbers from 30 to 51, depending on the type of material [25].

Compared to conventional organic polymers, hybrid polymers show pronounced negative thermooptical behavior with $dn/dT \sim -2{\cdot}10^{-4}$ K^{-1} and an outstandingly low birefringence in the range of $n_{TE} - n_{TM} < 1$ to $7{\cdot}10^{-4}$. Together with low optical loss (*vide supra*), this makes them good candidates for photonic applications such as thermooptical waveguides or sensors based on microring resonators and Bragg gratings [47–49].

7.2.4.4 Mechanical Properties

By varying the network densities of both the inorganic and organic domains, the mechanical properties such as Young's modulus and coefficient of thermal expansion (CTE) can be tailored over a wide range [29]. Variation of network density is achieved by changing relative amounts of mono-, di-, and trialkoxy silane precursors, the amount of organic polymerizable groups, and the linker length in functionalized organosilanes. With increasing inorganic content and network density, the CTE of hybrid polymers decreases, with typical values being in the range 100–150 ppm K^{-1} for commercial materials [25]. Hybrid polymers possessing Young's modulus from around 100 MPa up to several GPa can be synthesized [29, 32]. Typical values for commercial hybrid polymers for optical applications range between 0.6 and 1.2 GPa [25]. The increased mechanical strength of hybrid polymers in comparison to conventional organic polymers is also manifested in significantly improved scratch and abrasion resistance [30, 50].

7.2.4.5 Thermal and Climate Stability

Hybrid polymers show remarkable stability against physical stress, such as temperature, humidity, and light. Thus, thermogravimetric analysis of hybrid polymers OrmoComp and OrmoClear (Fig. 7.5a) shows beginning decomposition only at temperatures above 360 °C and 300 °C, respectively, which is higher than conventional cross-linkable acrylates. Consequently, these hybrid polymers tolerate long-term heating to 270 °C and short-term heating to 300 °C [25]. Thus, soldering and dicing during subsequent processing of the device can be performed without deterioration of optical properties. Notably, during standardized climate stability tests, in particular temperature cycling and aging under elevated temperature and humidity, hybrid polymers do not show any significant change of their optical characteristics (Fig. 7.5b). Furthermore, under high-intensity irradiation with visible light, hybrid polymers show significantly longer lifetimes than conventional organic UV-curable polymers [33].

7.2.4.6 Biocompatibility

Remarkably high biocompatibility of OrmoComp directly after fabrication of test structures and without any additional surface treatment has been proven in several studies using a number of different cell lines [51–54]. Cell viability on the hybrid polymer is as good as on control substrates. Importantly, UV curing of the substrate should be sufficiently long to reduce the presence of unreacted cytotoxic photoinitiator and acrylates [55]. Furthermore, performing a hardbake after UV curing increases cell viability. In some cases a protein coating [56, 57] or specific functionalization [45] of the OrmoComp surface may be applied to increase cell adherence

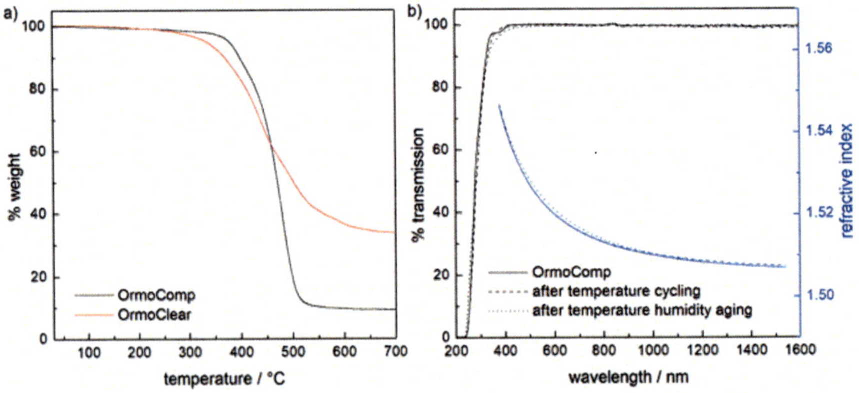

Fig. 7.5 (**a**) Thermogravimetry of hybrid polymers OrmoComp and OrmoClear. (**b**) Transmission and refractive index dispersion of thin films of OrmoComp (d = 20 μm) before and after temperature cycling (100 cycles between 45 and 85 °C) and temperature humidity aging (85% RH, 85 °C, 1000 h) tests

and viability. Consequently, in addition to their applications as optical materials, hybrid polymers are advantageously used for the fabrication of micro- and optofluidic devices [43, 58–67], medical diagnostics [44, 66], as well as scaffolds for tissue engineering (*vide infra*).

7.3 Fabrication of Microoptical Elements by UV Lithography and Replication Using Hybrid Polymers

High-precision and large-scale production of microoptical and photonic components in 2D using wafer-based processing highly profits from the past technological developments in the semiconductor industries. Thus, optical components are manufactured with micro- and nanometer resolution using classic UV lithography and subsequent etching of the structure into silicon or glass [6]. When using UV-curable polymers as optical materials, most microoptical components are fabricated by cost-effective UV imprinting and replication techniques, besides direct structuring via UV lithography or direct laser writing. These standard techniques are advantageously applied to structure hybrid polymers for the fabrication of microoptical components [11, 12]. Notably, due to the wafer-based processing and restrictions during replication steps, fabrication is limited to 2D and 2.5D structures.

7.3.1 UV Lithography and Direct Laser Writing

Employing hybrid polymers as negative-tone resists and their direct patterning using UV lithography and direct laser writing in principle yields binary structures, i.e., a film of the hybrid polymer is fully cured at certain positions and washed away at others during the development step. Consequently, these techniques are mainly utilized for binary photonic structures such as waveguides using OrmoCore and OrmoClad—hybrid polymers specifically designed for waveguide applications [31, 67–70]. Stacks of waveguides were realized by layer-wise repetition of the coating and structuring process [71, 72].

Also nonbinary structures can be realized using combined nanoimprint and UV lithography: by applying a mask pre-structured with an aperture lens on a thick layer of hybrid polymer, an array of high-aspect-ratio conic microlenses was fabricated [34, 73]. By profiting from diffraction effects during proximity exposure, concave microchannels acting as mirror elements were fabricated into OrmoComp, enhancing fluorescent single-cell imaging [62].

Besides these optical and photonic applications, direct structuring of hybrid polymers by UV lithography and direct laser writing is used for fabrication of microfluidic chips [43, 58, 61, 74]. Hybrid polymers may also be used as sacrificial photoresist in UV lithography for pattern transfer in etching processes with high etch selectivity [75].

7.3.2 UV Imprint and Replication

More complex 2.5D microoptical structures can be fabricated with hybrid polymers using replication from a master stamp by NIL. In particular, structures with high resolution down to few nanometers, combining elements on completely different length scales from nano to micro, or designed as freeform optical elements can be realized, given a properly structured master original is available. The latter may be fabricated by conventional UV lithography, direct laser writing, electron-beam lithography, or focused ion beam milling with high resolution and large freedom in design. However, as master fabrication is very laborious and cost-intensive, it is advantageous to replicate it into a working stamp, which in turn is used to perform multiple imprints into the optical polymer.

In this respect, hybrid polymers offer superior properties not only for the fabrication of microoptical components itself but also for the fabrication of working stamps for their replication. A dedicated commercial material—OrmoStamp—is widely used as hard stamp material suited for thermal and UV-NIL with low oxygen permeability, superior resolution and structure fidelity, as well as high durability during repetitive imprinting processes [76–80].

Microlens arrays constitute one of the most important microoptical elements and are indispensable for imaging and illumination applications in today's scientific and medical instruments and most importantly consumer products [81]. The masters for microlens arrays can be obtained by photolithographic structuring of a photoresist and subsequent thermal reflow to obtain spherical or cylindrical lenses [82, 83]. These structures can be etched into the substrate or transferred by molding into a working stamp. By imprinting into the optical polymer such as OrmoComp, microlens arrays are fabricated directly on CMOS detectors [83] or VCSELs [84] and are implemented into stacks of functional elements within wafer-scale production processes. Low shrinkage and high thermal stability of hybrid polymers (*vide supra*) are advantageous for high structural precision and stability during further handling of functional dies including soldering and dicing at elevated temperatures. Recently microlens arrays were realized by UV imprinting of OrmoComp on flexible fluorinated polymer substrates being suited for roll-to-roll fabrication processes (Fig. 7.6a). Structured foils could be bent to low radii without deterioration of the optical properties of the lenses [85].

NIL fabrication is similarly suited to replicate structures containing features with very different length scales. Thus, macroscopic millimeter-sized lenses containing a nanostructured surface as intrinsic moth-eye antireflective layer could be replicated with excellent structural fidelity into the hybrid polymer OrmoClear (Fig. 7.6b) [11]. The master for the replication process was obtained by combining thermoformed lenses with a nanostructured foil [86]. Similar hierarchically structured optical elements such as a microlens array with an antireflective layer [87] and a diffusor element with a diffractive nanostructure on top [88] were fabricated by replication into OrmoComp.

Freeform refractive lens arrays utilized as multi-aperture objective for ultra-slim cameras were fabricated using OrmoComp as the optical material (Fig. 7.6c) [89, 90]. As photoresist reflow cannot be applied for master fabrication due to the freeform structure of the lens array, a step-and-repeat process starting from an ultra-precision machined single lens array master was applied to replicate the structure on wafer-scale. For superior quality of the optical surface, the stamp tool was moved in z-direction during curing to compensate for shrinkage of the material.

A number of photonic components such as waveguides for datacom applications [91–96] and microring resonators for sensing [47, 97] have been fabricated into low-optical-loss hybrid polymers OrmoClear, OrmoCore, and OrmoClad using UV-NIL. Figure 7.6d shows a microring resonator functioning with high Q-factor

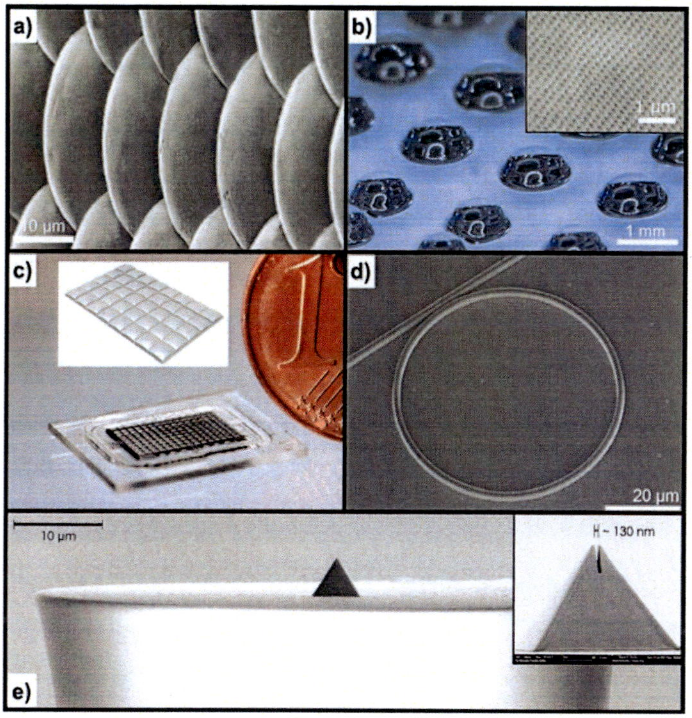

Fig. 7.6 Examples of replicated microoptical structures by UV imprinting of hybrid polymers: (**a**) SEM micrograph of OrmoComp microlens array on flexible fluorinated ethylene propylene (FEP) substrate [85]. (**b**) OrmoClear macro lenses with nanostructured surface [11]. (**c**) Multi-aperture objective for ultrathin camera modules containing two refractive freeform microlens arrays fabricated using OrmoComp. Inset: 3D model of one quarter of the lens array [90]. (**d**) SEM micrograph of a microring resonator with designed gap width of 250 nm fabricated using OrmoCore [47]. (**e**) SEM micrograph of a campanile probe at the facet of an optical fiber imprinted into OrmoComp [102]

which was fabricated on wafer scale with nanometer resolution. Further examples of NIL replicated photonic structures consist of photonic fences [98], Bragg gratings [48, 49], and metasurfaces [99].

Of particular interest is the generation of photonic nanostructures on the facet of an optical fiber to modify and improve its beam shaping and collecting properties. Thus, a 3D beam splitter, Fresnel lens, and campanile probe were realized by direct nanoimprinting on the facet of a fiber [100–102]. The process includes several molding steps using the hybrid polymer OrmoComp as mold and imprint material. Alignment of the mold and the fiber is guided by red light being coupled into the fiber, while final curing of the structure on the tip is achieved by sending blue light through it. By realizing a nanometer-sized gap on the campanile probe, sub-diffraction imaging was demonstrated (Fig. 7.6e).

7.4 Hybrid Polymers in Additive Manufacturing

In contrast to structuring by UV lithography and UV imprinting, which are basically limited to 2.5D structures, additive manufacturing allows for fabrication with full freedom in the three-dimensional space, paving the way for new designs and functions of microoptical components. In addition to wafer-based processing, other substrates, including pre-patterned and 3D-shaped components, can be directly functionalized with microoptical elements. The technological development in fabrication goes hand in hand with development of improved and new materials, which are specifically adapted to the requirements of the numerous 3D printing processes.

Due to the variability of the synthesis of hybrid polymers, they can be easily adapted to multiple 3D printing processes for the fabrication of optics. Thus, extrusion-based methods, such as nozzle extrusion [103] and inkjet printing (see Sect. 7.4.1), as well as vat polymerization methods, such as stereolithography [104–106] and in particular 2PP-DLW (see Sect. 7.4.2), have been employed for the fabrication of microoptical components using hybrid polymers. Notably, the material OrmoComp is very popular in additive manufacturing due to its fast curing kinetics, oxygen insensitivity, acceptable shrinkage, high biocompatibility, as well as excellent optical and mechanical properties.

7.4.1 *Inkjet Printing and Dispensing*

Inkjet printing and dispensing may be advantageously applied for the fabrication of microlenses and microlens arrays with enhanced performance due to higher structural variability. The processes are easy to implement and can be applied on wafer scale directly on functional dies. Due to the additive nature of the processes, the printing material can be deposited in a highly controlled manner, thus giving precise control over the position, size, and shape of microlenses. Thus, in particular for the fabrication of individualized lenses, they are cost-effective and fast alternatives to laborious master fabrication using, e.g., diamond machining or UV lithography and thermal reflow.

Inkjet printing of arrays of microlenses using hybrid polymers was demonstrated on glass substrates [107–110]. An inkjet printing compatible formulation of OrmoComp (InkOrmo) was developed by dilution with suited solvents. This ink was deposited in form of individual droplets onto the surface, prebaked to remove the solvent, and finally UV-cured to give spherical microlenses. It was shown that their size and shape, depending on the free energy balance between the droplet formulation, substrate, and surrounding atmosphere, could be tuned by adjusting the surface's hydrophobicity and the number of droplets per microlens.

To widen the structural variety of inkjet-printed microlenses, a process was developed using a surface pre-patterned with pedestals [111, 112]. The size of the pedestal as well as the amount of ink deposited on top defines the shape of the microlens, allowing for precise control over its focal length and numerical aperture.

An improved optical performance was obtained using hybrid polymer materials for both the patterned surface and the microlenses [113]. For this purpose, nanoimprint lithography for the generation of pedestals was combined with inkjet printing of the microlenses (Fig. 7.7a). With three different pedestal sizes, precise and predictable numerical apertures from 0.45 to 0.9 as well as focal lengths between 10 and 100 μm were obtained by simply adjusting the number of droplets on individual pedestals (Fig. 7.7b). Notably, microlenses with varying focal lengths can easily be combined into multifocal microlens arrays (Fig. 7.7c). Single arrays with up to nine different microlenses were demonstrated.

Recently, sophisticated hybrid microoptical components were fabricated on a wafer scale using a combination of UV-NIL, UV lithography, and inkjet printing [114, 115]. By dispensing a hybrid polymer ink on a micro- and nanostructured template, a combination of refractive and diffractive elements in a single microlens was achieved (Fig. 7.8a). First, a nanostructure consisting of a laminar grating with lines and spaces of 500 nm width and 500 nm depth was replicated using an organic NIL resist. Then, a layer of SU-8 with a thickness of 150 μm was deposited on top of the nanoimprinted grating and photolithographically structured. This gave circular confining structures with a diameter of 1000 μm. The resulting resist master was replicated twice to obtain an exact copy using different soft and hard material combinations as mold and substrate (materials A and B; for details see [115]). Thus, replicas, e.g., in hard OrmoComp on silicon wafer or more flexible OrmoClearFX on PC foil, were generated. The hybrid polymer ink was dispensed into the cavities of the replica and soft-baked to release the solvent. Several repetitions of the dispensing and soft-baking steps might be necessary to achieve the final lens shape. The total amount of dispensed hybrid polymer defines the focal length of the refractive part of the microlens. Finally, UV curing, hard-baking, and demolding yield the hybrid refractive-diffractive microlens (Fig. 7.8b). The structural fidelity during the replication processes is very high with shrinkage of the step height below 5% over the whole process chain. Optical testing using a collimated expanded laser beam gave the expected image of three partial beams due to diffraction of the laminar

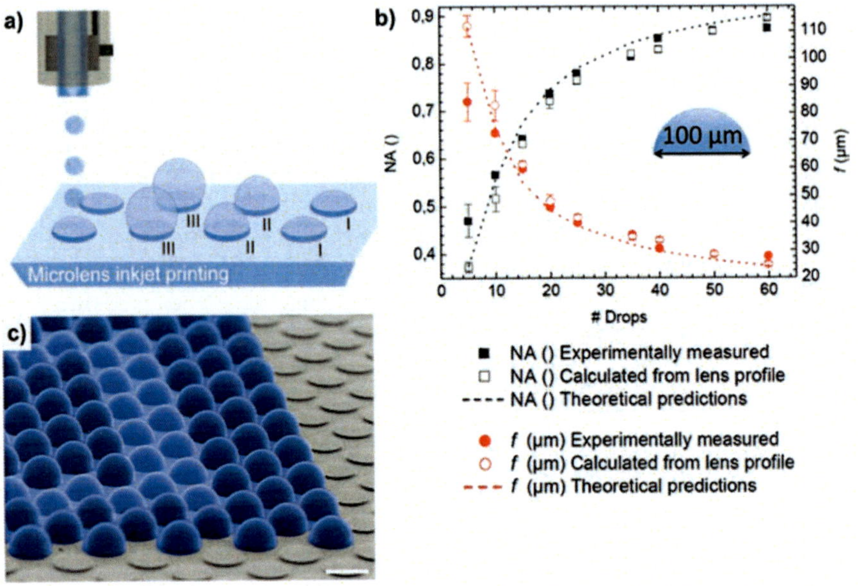

Fig. 7.7 Inkjet printing of microlenses on preformed pedestals: (**a**) Illustration of the printing process. (**b**) Dependence of numerical aperture and focal length of the lens on amount of dispensed ink in case of 100 µm diameter pedestals. (**c**) Artificially colored SEM micrograph of an array of microlenses with two different focal lengths (scale bar = 100 µm) [113]

grating well-focused by the refractive part of the microlens (Fig. 7.8c). This work demonstrates the power of combining conventional UV lithography and replication techniques for the generation of functional elements, i.e., a diffractive pattern, with inkjet printing as an additive manufacturing process defining the size and shape of the entire microoptical component.

While the abovementioned inkjet printing processes rely on solvent-based hybrid polymer formulations with low viscosities (<25 mPa s), it would be desirable to utilize solvent-free inks, thereby avoiding time-consuming heating and solvent evaporation steps. In one approach [116] the low-viscosity requirement was met by dilution of OrmoComp resin with an acrylate-based reactive diluent at volume contents up to 70%. Thus, large-area microlens arrays could be inkjet printed with minimum deviations of the individual lens geometries and a largely reduced overall processing time compared to solvent-based inks.

By developing a new picoliter dispensing technique, which is capable of dispensing liquids with a wide range of viscosities (200–10,000 mPa s), the fabrication of microlens arrays and single individualized microlenses was demonstrated using undiluted OrmoComp and OrmoClearFX [117]. A small stamp is used to transfer single polymer droplets in picoliter volumes from a reservoir onto a surface or into a preformed cavity. The shape of the resulting microlens is defined by the amount of transferred material, the diameter of the stamp, viscosity of the polymer resin, and hydrophobicity of the substrate. Being able to dispense viscous, undiluted hybrid

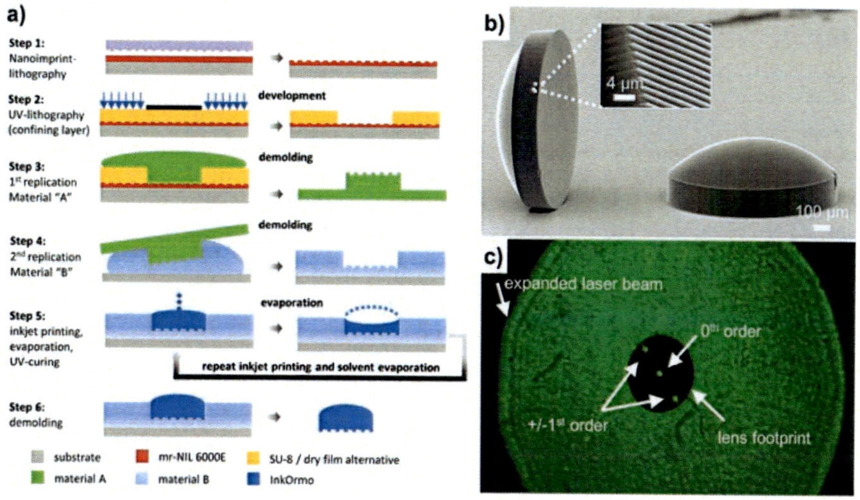

Fig. 7.8 Refractive-diffractive microlens by combined UV-NIL replication and inkjet printing of hybrid polymers [114, 115]. (**a**) Process chain. (**b**) SEM micrograph of two InkOrmo refractive-diffractive microlenses with a laminar grating as diffractive pattern. (**c**) Screen photograph of the 0th and +/− 1st diffraction orders generated by the diffraction grating and focused by the refractive part of the microlens

polymer resins allows for fast process times and unaltered material properties and thus performance of the fabricated microoptical component.

7.4.2 Two-Photon Polymerization Direct Laser Writing (2PP-DLW)

In recent years 2PP-DLW-based 3D printing developed into an exciting technology for the fabrication of optical components with micro- and nanometer resolution way beyond the diffraction limit of light [118]. 2PP-DLW allows for highest resolutions among today's 3D printing methods, complete architectural freedom in the design of components, fabrication of monolithic materials with smooth surfaces, and the direct printing onto various (functionalized) substrates. A large variety of UV-curable optical materials is utilized for 2PP-DLW [119], among them epoxy-based organic polymers such as SU-8, or more recently mr-DWL, as well as acrylate-based materials such as IP resins (resin formulations optimized to 2PP-DLW proprietary to Nanoscribe GmbH). While these all-organic polymers are optimized for distinct 2PP-DLW processes, their applicability for the fabrication of optical components is limited due to their optical and mechanical properties or stability. Thus, for the fabrication of optical components, using 2PP-DLW direct structuring of hybrid polymers enjoys a tremendous popularity. In particular, the commercially available

OrmoComp and the academia-originated zirconia containing material SZ2080 are widely used and implemented into various 2PP-DLW platforms [120–123].

7.4.2.1 Material Adaptation

Though standard hybrid polymers such as OrmoComp are well suited for 2PP-DLW, specific adaptation of the material to the distinct irradiation conditions can significantly improve the performance [13, 124]. Particularly, the shape accuracy and surface roughness of the printed structures, shrinkage during polymerization, and the usable dynamic range of laser power and scanning speed are important parameters.

It was reported that in a slight modification of the organic curing chemistry, i.e., introducing styrene moieties by the sol-gel process, the resolution of a hybrid polymer resin was significantly improved to the sub-100 nm range [125]. Furthermore, by co-condensation of an organosilane with 20 mol% of zirconium alkoxides $Zr(OR)_4$, a hybrid polymer labeled as SZ2080 was created enjoying much attention in 2PP-DLW 3D printing [41]. Due to its zirconia content, SZ2080 shows negligible shrinkage during polymerization allowing for the fabrication of structures with highest fidelity [126]. Furthermore, damage of the material by high-intensity laser light is significantly reduced broadening the processing window [127, 128]. However, compared to conventional solvent-free hybrid polymers, SZ2080 is a solvent-based formulation; thus spin coating and pre-baking steps to evaporate the solvent are mandatory.

The performance of standard OrmoComp in 2PP-DLW could be further improved by variations in type and content of the photoinitiator as well as content of the stabilizer, generally utilized to prevent premature polymerization in the dark [124]. For nine different formulations, the dynamic range, i.e., the range between the threshold laser power under which no polymerization occurs and overexposure leading to deteriorated shapes and laser damage, was compared (Fig. 7.9a). It turned out that using a photoinitiator with higher two-photon absorption (TPA) cross-section (formulations 1–3) led to lowering of the polymerization threshold but also to reduced resolution and structure fidelity due to the high amount of initiator radicals produced. Increasing the concentration of the standard phosphine oxide photoinitiator (examples 4–6) led to lower threshold values, higher dynamic range, and better structure fidelity (Fig. 7.9b, c). Finally, increasing the concentration of the stabilizer (examples 7–9) led to further improvement of structure fidelity though an increased polymerization threshold was observed. With optimized formulation 9, a complex prism array was realized by 2PP-DLW (Fig. 7.9d), which served as template for the fabrication of a mold and multiple replicas thereof using injection molding [129].

This and other examples [13, 125, 130, 131] show that the performance of a resin in 2PP-DLW is a complex interplay between exposure dose, TPA cross-section of the initiator, radical initiation, and inhibition efficiency as well as polymerization chemistry. Recently it was found that 2PP-DLW with SZ2080 without any added photoinitiator gives structures with improved resolution, surface quality, and stability against laser damage [128]. On the other side, current developments into sophisticated photoinitiators with high TPA cross-section [132] as well as chemical

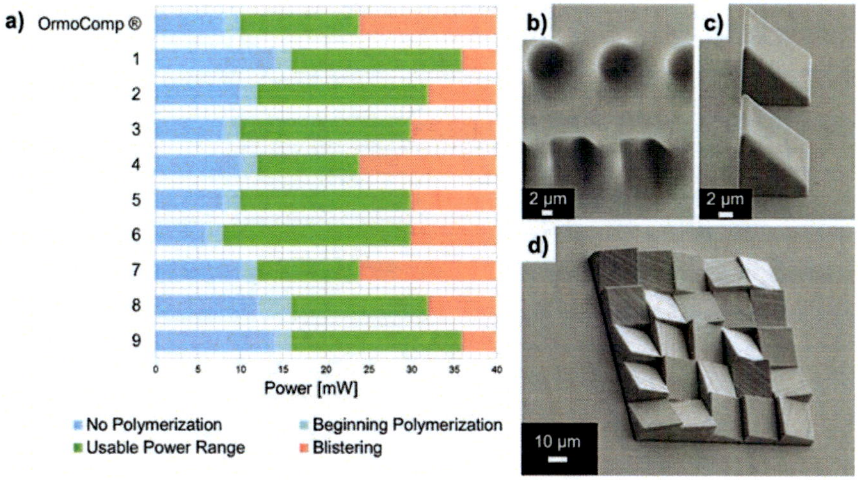

Fig. 7.9 3D printing by 2PP-DLW in OrmoComp: (**a**) dynamic range of the usable laser power for standard OrmoComp and nine adapted formulations with tuned photoinitiator and inhibitor content (for details see [124]). (**b**, **c**) SEM micrographs of printed test structures showing the effect of photoinitiator content (0.5% vs. 1.25%) on structure fidelity. (**d**) SEM micrograph of the fabricated prism array with individual inclination and tilt angles of each prism [124]

and instrumental inhibition strategies [133–135] are promising for boosting fabrication speed and attainable resolution of 2PP-DLW.

7.4.2.2 Fabrication of Optical Components with 2PP-DLW

Using 2PP-DLW with hybrid polymers, a broad range of optical and photonic components reaching from the nano- to the macroscale have been realized [118, 136]: spherical, axicon, and freeform lenses [66, 137–145], phase plates [146, 147], Fresnel lenses [13, 140], gratings [148], DOEs [140, 149], waveguides [13, 140, 150], photonic fences [98], microcavities [151, 152], photonic crystals [41, 153–155], and metasurfaces [156]. These examples demonstrate that with 2PP-DLW, almost no limitation in shape or size of the architecture exists. An exemplary selection of structures with differing length scales is given in Fig. 7.10.

Particularly remarkable is the fabrication of optics with high resolution and smooth surfaces directly on functional elements: OrmoComp was used to print spherical collimation lenses of different sizes directly onto the facet of an optical fiber (Fig. 7.10a) [141]. Though a layer-wise printing process was utilized, the surface of the lens is smooth, and excellent light output was achieved with the material. The same groups demonstrated printing of a variety of fully freeform structures and compound lenses onto optical fibers and sensor chips using OrmoComp and IP-S resins [141, 157, 158]. Recently, a microlens cascade fabricated from IP-S and OrmoComp using 2PP-DLW was integrated into a functional microendoscopic probe for high-resolution deep tissue imaging [66].

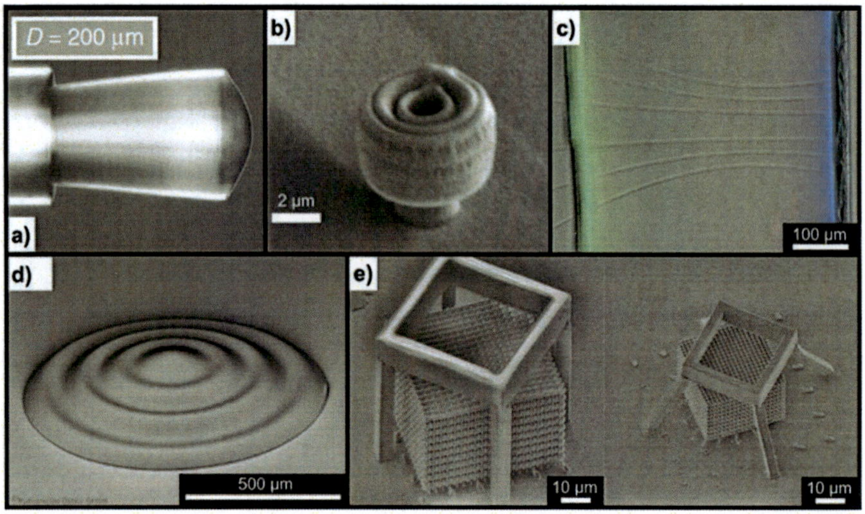

Fig. 7.10 Microoptical components fabricated by 2PP-DLW using hybrid polymers: (**a**) spherical collimation lens made of OrmoComp on facet of single-mode optical fiber [141]. (**b**) Freeform light-directing nanophotonic lens printed form OrmoComp on top of GaAs nanowire [144]. (**c**) Optical microscope image of six waveguide cores embedded in a cladding matrix of an adapted ORMOCER-based hybrid polymer [13]. (**d**) Millimeter-sized freeform lens fabricated by contour structuring of OrmoComp [145]. (**e**) Photonic crystal structure made of SZ2080 before (left) and after (right) pyrolysis [159]

A remarkable resolution was achieved using standard OrmoComp for fabrication of a nanophotonic lens on emitting GaAs nanowire (Fig. 7.10b) [144]. An evolutionary algorithm was utilized to design the lens structure with nanometer-sized features. By interference effects the isotropic photoluminescence of the GaAs nanowire was turned into highly directional emission over a broad spectral range. The designed structure could be fabricated with high fidelity by 2PP-DLW. The excellent transparency and low autofluorescence of OrmoComp allowed the experimental demonstration of the dramatic angular redistribution of emitted light.

For beam shaping and light distribution applications, instead of sculpturing the surface of a material with a homogeneous refractive index, an alternative approach would be to have a uniform structure and vary the refractive index within. Such a GRIN (gradient refractive index) element was realized using SZ2080 [160]. Within the broad dynamic range of this polymer, the double-bond conversion and similarly the refractive index increases with the laser exposure dose. Thus, by varying scanning speed during 2PP-DLW fabrication, the refractive index was dynamically modulated with a maximum change of $\Delta n = 0.01$ within a single structure.

The dependence of refractive index on exposure dose was further exploited to demonstrate straightforward coupling of optoelectronic components by 2PP-DLW fabrication of waveguides in a cladding matrix. For this purpose, a uniform film of a hybrid polymer resin is coated onto the optoelectronic components, the whole film is cured by UV flash illumination, and finally waveguides are written within the cured material by 2PP-DLW, locally promoting further polymerization and an increase in refractive index. The principle was demonstrated by fabrication of waveguides between two glass slides in a film of an OrmoComp formulation adapted to 2PP-DLW having a large processing window (Fig. 7.10c) [13]. As the voxel diameter of the writing laser is smaller than the waveguide profile, different sizes (single vs. multimode) and shapes (circular, square, and elliptical) were realized by hatching of the waveguide's cross-section [140]. Importantly, such hybrid polymer waveguides integrated into functional chips are sufficiently stable to survive further lamination, bonding, and dicing steps [161]. Similar to refractive index, the mechanical properties of hybrid polymers change with increasing curing time. Thus, domains with differing elasticity and stiffness may be imparted to a single structure by varying exposure dose [13]. However, close to the polymerization threshold, structures become weak, show much higher shrinkage, and tend to collapse [126, 128, 155].

A common bottleneck in 2PP-DLW fabrication is the limitation in size of the fabricated objects as well as production time. However, recent improvements of instrument manufacturers and the application of sophisticated laser focusing and scanning strategies allowed for significant improvements of both parameters [145]. As an example, the macroscopic ($d = 1$ mm) freeform lens shown in Fig. 7.10e was rapidly manufactured from OrmoComp using contouring exposure, i.e., the outer shape of the lens was polymerized with the laser, and only after the development step, the inner volume was cured using UV flood illumination. Importantly, for high-throughput 2PP-DLW printing, the materials have to show high reactivity to keep up with the fast laser trajectories. More information on recent developments in 2PP-DLW hardware and processes can be found in Chap. 5 of this book.

The unique combination of inorganic and organic domains in hybrid polymers offers an additional processing step after fabrication of the 3D-printed structure: by pyrolysis and sintering at temperatures above 1000 °C, the organic domain decomposes leaving behind an amorphous inorganic glass or ceramic structure [162, 163]. Therefore, the structure undergoes considerable isotropic shrinkage lowering the minimum structural feature size. Applied to a photonic crystal fabricated with SZ2080, pyrolysis at 1000 °C for 2 h gave shrinkage of ca 40%, while the material stayed amorphous with good structural fidelity (Fig. 7.10d) [128, 159]. Pyrolysis at higher temperatures gave transition into crystalline phases accompanied by deterioration of the structure. Much milder thermal annealing of a hybrid polymer photonic crystal at 300 °C resulted in only partly decomposition of the organic groups further densifying the material. Thus, precise control over shrinkage was achieved resulting in a shift of the photonic band gap from the NIR into the visible range [154].

7.4.2.3 Fabrication of Cell Scaffolds and Microfluidic Devices by 2PP-DLW

Besides optical components, 2PP-DLW of hybrid polymers is widely used for the fabrication of microfluidic structures as well as scaffolds for cell adhesion and tissue engineering. Again OrmoComp plays a remarkable role as a material with excellent structuring capability being at the same time highly biocompatible, generally without any additional coating step (see Sect. 7.2.4). Additionally, its negligible autofluorescence makes it superior over other 3D printing materials concerning imaging of fluorescently labeled biological structures.

Consequently, 2PP-DLW-derived scaffolds made of OrmoComp are utilized for studying cell adhesion, growth, migration, and differentiation. To this purpose a large variety of patterned surfaces and complex 3D architectures have been realized [54–56, 164–167]. As an example a tubular microtower for the cultivation of neuronal cells is shown in Fig. 7.11a. Precisely controlled surface properties such as wetting behavior, mechanical stiffness of the structure, as well as structural confinements strongly influence cell behavior.

Interestingly, 3D scaffolds made of two or three different materials with distinct binding properties were realized using consecutive printing steps [57, 168, 169]. OrmoComp, showing high tendency to bind proteins, was combined with a photoactivatable and a protein-repellent acrylate resist. Thus, the spatial functionalization of the scaffold with different proteins and subsequent controlled and type-specific adhesion of cells was realized (Fig. 7.11b) [57].

While micrometer-scaled fluidic devices generally are fabricated using molding, imprinting, or one-photon 3D printing, specific fluidic elements requiring nanometer resolution can be created using 2PP-DLW. As such sophisticated elements actively generating microfluidic currents [170–172] as well as arrays of microneedles with different shapes and high aspect ratios used for drug delivery (Fig. 7.11c) [51, 52, 63, 173] were realized using OrmoComp. These applications profit from exceptionally high inertness, biocompatibility, and mechanical strength of the material.

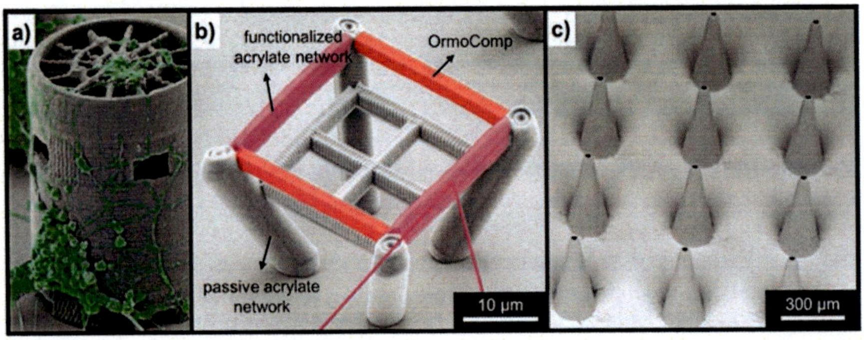

Fig. 7.11 SEM micrographs of 2PP-DL-printed structures for tissue engineering and medical applications. (**a**) Tubular microtower ($h = 150$ μm) fabricated form OrmoComp as 3D platform for the cultivation of neuronal cells [166]. (**b**) Multimaterial 3D cell scaffold selectively functionalized with two distinct adhesive proteins (highlighted in red and pink color) for the guidance of cell attachment [57]. (**c**) Array of microneedles fabricated with OrmoComp. Printing of microneedles directly on top of microfluidic channels fabricated in PMMA was demonstrated for the realization of point-of-care systems [63]

7.5 Conclusions

Hybrid polymers were demonstrated to be a superior material platform for the manufacturing of reliable and high-performance microoptical components. With their unique structure, derived from the combination of inorganic glass-like with organic polymeric domains in one material, hybrid polymers unify excellent optical properties, environmental stability, and versatility in processing. Importantly, by adaptation of chemical synthesis, formulation, and processing conditions, the properties of hybrid polymers can be tuned over a broad range and tailored to specific applications.

By wafer-scale processing of hybrid polymers using UV lithography and nanoimprint lithography, sophisticated microoptical components are routinely manufactured with high precision, high throughput, and low costs. Synthesis of hybrid polymers and their processing employing these technologies are established in industrial environments serving markets, for example, in consumer electronics, telecommunication, and lighting and sensing applications.

Additive manufacturing and 3D printing of microoptics brings new perspectives for the design of optical elements as well as their fabrication, such as exploitation of all three dimensions and rapid prototyping. The ever-increasing number of additive manufacturing processes and their technological advance require specific adaptations of materials to be printed in order to gain best performance in terms of quality and throughput. Besides their commercial applications in wafer-level optics, hybrid polymers are perfectly suited for additive manufacturing of optical and other components, in particular using inkjet printing and 2PP-DLW. Therefore, we foresee the utility of hybrid polymers in the direct fabrication of individualized microoptical components with high demands on optical and mechanical properties as well as environmental and long-term stability.

A current challenge in the development of hybrid polymers as optical materials consists of extending the range of achievable refractive indices and dispersion without deteriorating the optical performance and reliability. Specific challenges related to additive manufacturing are the adaptation of hybrid polymer resins to low-viscosity requirements, as set by, e.g., inkjet printing, as well as the precise tailoring of optical density, curing speed, shrinkage, and oxygen sensitivity, as required by laser-assisted 3D printing methods. Fine-tuning of these parameters by resin synthesis and formulation utilizing sophisticated chemical concepts enables large improvements in writing speed and resolution [174–176]. In addition, the simultaneous additive processing of multiple materials with different mechanical and optical properties [177] and implementation of special material functions becoming active after fabrication [178–180] hold great promise for future developments.

We actively pursue the development of hybrid polymers towards these directions in order to improve established applications and help to create exciting innovations in microoptics and related fields.

References

1. Salt, M., & Rossi, M. (2006). Replicated micro-optics for multimedia products. *Proceedings of SPIE, 6196*, 61960F.
2. Duparré, J., & Völkel, R. (2006). Novel optics/micro-optics for miniature imaging systems. *Proceedings of SPIE, 6196*, 619607a.
3. Zappe, H. (2012). Micro-optics: A micro-tutorial. *Advanced Optical Technologies, 1*, 117–126.
4. Mönch, W. (2015). Micro-optics in lighting applications. *Advanced Optical Technologies, 4*, 79–85.
5. Kress, B. C. (2019). Digital optical elements and technologies (EDO19): Applications to AR/VR/MR. *Proceedings of SPIE, 11062*, 1106222.
6. Völkel, R. (2012). Wafer-scale micro-optics fabrication. *Advanced Optical Technologies, 1*, 135–150.
7. Camposeo, A., Persano, L., Farsari, M., & Pisignano, D. (2019). Additive manufacturing: Applications and directions in photonics and optoelectronics. *Advanced Optical Materials, 7*, 1800419.
8. Kotz, F., Arnold, K., Bauer, W., Schild, D., Keller, N., Sachsenheimer, K., Nargang, T. M., Richter, C., Helmer, D., & Rapp, B. E. (2017). Three-dimensional printing of transparent fused silica glass. *Nature, 544*, 337–339.
9. Nguyen, D. T., Meyers, C., Yee, T. D., Dudukovic, N. A., Destino, J. F., Zhu, C., Duoss, E. B., Baumann, T. F., Suratwala, T., Smay, J. E., & Dylla-Spears, R. (2017). 3D-printed transparent glass. *Advanced Materials, 29*, 1701181.
10. Gal-Or, E., Gershoni, Y., Scotti, G., Nilsson, S. M. E., Saarinen, J., Jokinen, V., Strachan, C. J., af Gennäs, G. B., Yli-Kauhaluoma, J., & Kotiaho, T. (2019). Chemical analysis using 3D printed glass microfluidics. *Analytical Methods, 11*, 1802–1810.
11. Grützner, G., Klein, J., Vogler, M., & Schleunitz, A. (2014). UV-curable hybrid polymers for optical applications: Technical challenges, industrial solutions, and future developments. *Proceedings of SPIE, 8974*, 897406.
12. Schleunitz, A., Klein, J. J., Houbertz, R., Vogler, M., & Grützner, G. (2015). Towards high precision manufacturing of 3D optical components using UV-curable hybrid polymers. *Proceedings of SPIE, 9368*, 93680E.
13. Schleunitz, A., Klein, J. J., Krupp, A., Stender, B., Houbertz, R., & Grützner, G. (2017). Evaluation of hybrid polymers for high-precision manufacturing of 3D optical interconnects by two-photon absorption lithography. *Proceedings of SPIE, 10109*, 1010905.
14. Hartmann, P., Jedamzik, R., Reichel, S., & Schreder, B. (2010). Optical glass and glass ceramic historical aspects and recent developments: A Schott view. *Applied Optics, 49*, D157–D176.
15. Brinkmann, M., Hayden, J., Letz, M., Reichel, S., Click, C., Mannstadt, W., Schreder, B., Wolff, S., Ritter, S., Davis, M., Bauer, T., Ren, H., Fan, Y.-H., Wu, S.-T., Bonrad, K., Krätzig, E., Buse, K., & Paquin, R. (2007). Optical materials and their properties. In F. Träger (Ed.), *Springer hand-book of lasers and optics* (pp. 249–372). New York, NY: Springer.
16. Tagaya, A., & Koike, Y. (2012). Compensation and control of the birefringence of polymers for photonics. *Polymer Journal, 44*, 306–314.
17. Schaub, M. P. (2012). A tutorial on plastic consumer optics. *Advanced Optical Technologies, 1*, 21–29.
18. Schift, H. (2008). Nanoimprint lithography: An old story in modern times? A review. *Journal of Vacuum Science and Technology B, 26*, 458–480.
19. Schift, H. (2014) *NaPANIL library of processes* (3rd ed.), published by the NaPANIL-consortium represented by J. Ahopelto, ISBN: 978-3-00-038372-4. Retrieved from http://www.psi.ch/lmn/helmut-schift
20. Ligon, S. C., Liska, R., Stampfl, J., Gurr, M., & Mülhaupt, R. (2017). Polymers for 3D printing and customized additive manufacturing. *Chemical Reviews, 117*, 10212–10290.

21. Sanchez, C., Julián, B., Belleville, P., & Popall, M. (2005). Applications of hybrid organic–inorganic nanocomposites. *Journal of Materials Chemistry, 15*, 3559–3592.
22. Lebeau, B., & Innocenzi, P. (2011). Hybrid materials for optics and photonics. *Chemical Society Reviews, 40*, 886–906.
23. Werdehausen, D., Burger, S., Staude, I., Pertsch, T., & Decker, M. (2019). Nanocomposites—A Route to better and smaller optical Elements? In *Proceedings of OSA Optical Design and Fabrication* (p. OT2A.2). Washington, DC: Optical Society of America.
24. Trademark of the Fraunhofer-Gesellschaft zur Förderung der angewandten Forschung e.V. München.
25. Hybrid polymers manufactured by micro resist technology GmbH. Retrieved from https://www.microresist.com
26. Schmidt, H., Scholze, H., & Kaiser, A. (1984). Principles of hydrolysis and condensation reaction of alkoxysilanes. *Journal of Non-Crystalline Solids, 63*, 1–11.
27. Schmidt, H., & Seiferling, B. (1986). Chemistry and applications of inorganic-organic polymers (organically modified silicates). *MRS Proceedings, 73*, 739.
28. Schmidt, H., & Wolter, H. (1990). Organically modified ceramics and their applications. *Journal of Non-Crystalline Solids, 121*, 428–435.
29. Haas, K.-H., & Wolter, H. (1999). Synthesis, properties and applications of inorganic–organic copolymers (ORMOCER®s). *Current Opinion in Solid State & Materials Science, 4*, 571–580.
30. Haas, K.-H. (2000). Hybrid inorganic–organic polymers based on organically modified sialkoxides. *Advanced Engineering Materials, 2*, 571–582.
31. Buestrich, R., Kahlenberg, F., Popall, M., Dannberg, P., Müller-Fiedler, R., & Rösch, O. (2001). ORMOCER®s for optical interconnection technology. *Journal of Sol-Gel Science and Technology, 20*, 181–186.
32. Haas, K.-H., & Rose, K. (2003). Hybrid inorganic/organic polymers with nanoscale building blocks: Precursors, processing, properties and applications. *Reviews on Advanced Materials Science, 5*, 47–52.
33. Houbertz, R., Domann, G., Cronauer, C., Schmitt, A., Martin, H., Park, J.-U., Fröhlich, L., Buestrich, R., Popall, M., Streppel, U., Dannberg, P., Wächter, C., & Bräuer, A. (2003). Inorganic–organic hybrid materials for application in optical devices. *Thin Solid Films, 442*, 194–200.
34. Houbertz, R., Fröhlich, L., Popall, M., Streppel, U., Dannberg, P., Bräuer, A., Serbin, J., & Chichkov, B. N. (2003). Inorganic–organic hybrid polymers for information technology: From planar technology to 3D nanostructures. *Advanced Engineering Materials, 5*, 551–555.
35. Kahlenberg, F., & Popall, M. (2004). ORMOCER®s (organic-inorganic hybrid polymers) for telecom applications: Structure/property correlations. *MRS Online Proceeding Library Archive, 847*, EE14.4.
36. Schubert, U. (2015). Chemistry and fundamentals of the sol–gel process. In D. Levy & M. Zayat (Eds.), *The sol-gel handbook* (pp. 1–28). New York: Wiley.
37. Schubert, U. (2018). Sol-Gel-Chemie. *Chemie in unserer Zeit, 52*, 18–25.
38. Fessel, S., Schneider, A. M., Steenhusen, S., Houbertz, R., & Behrens, P. (2012). Towards an atomistic model for ORMOCER®-I: Application of forcefield methods. *Journal of Sol-Gel Science and Technology, 63*, 356–365.
39. Green, W. A. (2010). *Industrial photoinitiators: A technical guide*. Boca Raton: CRC Press.
40. Ligon, S. C., Husár, B., Wutzel, H., Holman, R., & Liska, R. (2014). Strategies to reduce oxygen inhibition in photoinduced polymerization. *Chemical Reviews, 114*, 557–589.
41. Ovsianikov, A., Viertl, J., Chichkov, B., Oubaha, M., MacCraith, B., Sakellari, I., Giakoumaki, A., Gray, D., Vamvakaki, M., Farsari, M., & Fotakis, C. (2008). Ultra-low shrinkage hybrid photosensitive material for two-photon polymerization microfabrication. *ACS Nano, 2*, 2257–2262.
42. Piruska, A., Nikcevic, I., Lee, S. H., Ahn, C., Heineman, W. R., Limbach, P. A., & Seliskar, C. J. (2005). The autofluorescence of plastic materials and chips measured under laser irradiation. *Lab on a Chip, 5*, 1348–1354.

43. Sikanen, T., Aura, S., Heikkilä, L., Kotiaho, T., Franssila, S., & Kostiainen, R. (2010). Hybrid ceramic polymers: New, nonbiofouling, and optically transparent materials for microfluidics. *Analytical Chemistry, 82*, 3874–3882.
44. Tadayon, M. A., Pavlova, I., Martyniuk, K. M., Mohanty, A., Roberts, S. P., Barbosa, F., Denny, C. A., & Lipson, M. (2018). Microphotonic needle for minimally invasive endoscopic imaging with sub-cellular resolution. *Scientific Reports, 8*, 10756.
45. Kidwell, D. A., Lee, W.-K., Perkins, K., Gilpin, K. M., O'Shaughnessy, T. J., Robinson, J. T., Sheehan, P. E., & Mulvaney, S. P. (2019). Chemistries for making additive nanolithography in OrmoComp permissive for cell adhesion and growth. *ACS Applied Materials & Interfaces, 11*, 19793–19798.
46. Schmid, M., Ludescher, D., & Giessen, H. (2019). Optical properties of photoresists for femtosecond 3D printing: Refractive index, extinction, luminescence-dose dependence, aging, heat treatment and comparison between 1-photon and 2-photon exposure. *Optical Materials Express, 9*, 4564–4577.
47. Kirchner, R., Finn, A., Landgraf, R., Nueske, L., Teng, L., Vogler, M., & Fischer, W. (2014). Direct UV-imprinting of hybrid-polymer photonic microring resonators and their characterization. *Journal of Lightwave Technology, 32*, 1674–1681.
48. Foerthner, M., Rumler, M., Stumpf, F., Fader, R., Rommel, M., Frey, L., Girschikofsky, M., Belle, S., Hellmann, R., & Klein, J. J. (2016). Hybrid polymers processed by substrate conformal imprint lithography for the fabrication of planar Bragg gratings. *Applied Physics A: Materials Science & Processing, 122*, 240.
49. Förthner, M., Girschikofsky, M., Rumler, M., Stumpf, F., Rommel, M., Hellmann, R., & Frey, L. (2018). One-step nanoimprinted Bragg grating sensor based on hybrid polymers. *Sensors and Actuators A: Physical, 283*, 298–304.
50. Schottner, G., Rose, K., & Posset, U. (2003). Scratch and abrasion resistant coatings on plastic lenses—State of the art, current developments and perspectives. *Journal of Sol-Gel Science and Technology, 27*, 71–79.
51. Doraiswamy, A., Jin, C., Narayan, R. J., Mageswaran, P., Mente, P., Modi, R., Auyeung, R., Chrisey, D. B., Ovsianikov, A., & Chichkov, B. (2006). Two photon induced polymerization of organic-inorganic hybrid biomaterials for microstructured medical devices. *Acta Biomaterialia, 2*, 267–275.
52. Doraiswamy, A., Ovsianikov, A., Gittard, S. D., Monteiro-Riviere, N. A., Crombez, R., Montalvo, E., Shen, W., Chichkov, B. N., & Narayan, R. J. (2010). Fabrication of microneedles using two photon polymerization for transdermal delivery of nanomaterials. *Journal of Nanoscience and Nanotechnology, 10*, 6305–6312.
53. Yoon, S.-H., Kim, Y. K., Han, E. D., Seo, Y.-H., Kim, B. H., & Mofrad, M. R. K. (2012). Passive control of cell locomotion using micropatterns: The effect of micropattern geometry on the migratory behavior of adherent cells. *Lab on a Chip, 12*, 2391–2402.
54. Ovsianikov, A., Schlie, S., Ngezahayo, A., Haverich, A., & Chichkov, B. N. (2007). Two-photon polymerization technique for microfabrication of CAD-designed 3D scaffolds from commercially available photosensitive materials. *Journal of Tissue Engineering and Regenerative Medicine, 1*, 443–449.
55. Trautmann, A., Rüth, M., Lemke, H.-D., Walther, T., & Hellmann, R. (2018). Two-photon polymerization based large scaffolds for adhesion and proliferation studies of human primary fibroblasts. *Optics and Laser Technology, 106*, 474–480.
56. Turunen, S., Käpylä, E., Lähteenmäki, M., Ylä-Outinen, L., Narkilahti, S., & Kellomäki, M. (2014). Direct laser writing of microstructures for the growth guidance of human pluripotent stem cell derived neuronal cells. *Optics and Lasers in Engineering, 55*, 197–204.
57. Richter, B., Hahn, V., Bertels, S., Claus, T. K., Wegener, M., Delaittre, G., Barner-Kowollik, C., & Bastmeyer, M. (2017). Guiding cell attachment in 3D microscaffolds selectively functionalized with two distinct adhesion proteins. *Advanced Materials, 29*, 1604342.
58. Aura, S., Sikanen, T., Kotiaho, T., & Franssila, S. (2008). Novel hybrid material for microfluidic devices. *Sensors and Actuators B: Chemical, 132*, 397–403.

59. Fernandez-Cuesta, I., Laura Palmarelli, A., Liang, X., Zhang, J., Dhuey, S., Olynick, D., & Cabrini, S. (2011). Fabrication of fluidic devices with 30 nm nanochannels by direct imprinting. *Journal of Vacuum Science and Technology B, 29*, 06F801.
60. Sikanen, T., Aura, S., Franssila, S., Kotiaho, T., & Kostiainen, R. (2012). Microchip capillary electrophoresis-electrospray ionization–mass spectrometry of intact proteins using uncoated Ormocomp microchips. *Analytica Chimica Acta, 711*, 69–76.
61. Singh, A., Scotti, G., Sikanen, T., Jokinen, V., & Franssila, S. (2014). Laser direct writing of thick hybrid polymers for microfluidic chips. *Micromachines, 5*, 472–485.
62. Bonabi, A., Cito, S., Tammela, P., Jokinen, V., & Sikanen, T. (2017). Fabrication of concave micromirrors for single cell imaging via controlled over-exposure of organically modified ceramics in single step lithography. *Biomicrofluidics, 11*, 034118.
63. Trautmann, A., Roth, G.-L., Nujiqi, B., Walther, T., & Hellmann, R. (2019). Towards a versatile point-of-care system combining femtosecond laser generated microfluidic channels and direct laser written microneedle arrays. *Microsystems & Nanoengineering, 5*, 6.
64. Fernandez-Cuesta, I., West, M. M., Montinaro, E., Schwartzberg, A., & Cabrini, S. (2019). A nanochannel through a plasmonic antenna gap: An integrated device for single particle counting. *Lab on a Chip, 19*, 2394–2403.
65. Esmek, F. M., Bayat, P., Pérez-Willard, F., Volkenandt, T., Blick, R. H., & Fernandez-Cuesta, I. (2019). Sculpturing wafer-scale nanofluidic devices for DNA single molecule analysis. *Nanoscale, 11*, 13620–13631.
66. Tadayon, M. A., Chaitanya, S., Martyniuk, K. M., McGowan, J. C., Roberts, S. P., Denny, C. A., & Lipson, M. (2019). 3D microphotonic probe for high resolution deep tissue imaging. *Optics Express, 27*, 22352–22362.
67. Mohamed, R., Razali, N., Ehsan, A. A., & Shaari, S. (2005). Characterisation and process optimisation of photosensitive acrylates for photonics applications. *Science and Technology of Advanced Materials, 6*, 375–382.
68. Rezzonico, D., Guarino, A., Herzog, C., Jazbinsek, M., & Günter, P. (2006). High-finesse laterally coupled organic-inorganic hybrid polymer microring resonators for VLSI photonics. *IEEE Photonics Technology Letters, 18*, 865–867.
69. Wang, S., Borden, B., Li, Y., & Goel, P. (2006). Laser direct writing of inorganic-organic hybrid polymeric channel waveguide for optical integrated circuits. *Proceedings of SPIE, 6389*, 63890B.
70. Madani, A., & Azarinia, H. R. (2017). Design and fabrication of all-polymeric photonic waveguides in optical integrated circuits. *Optik, 138*, 33–39.
71. Streppel, U., Dannberg, P., Wächter, C., Bräuer, A., Nicole, P., Fröhlich, L., Houbertz, R., & Popall, M. (2001). Development of a new fabrication method for stacked optical waveguides using inorganic-organic copolymers. In *Proc. IEEE Conference on Polymers and Adhesives in Microelectronics and Photonics* (pp. 329–335).
72. Streppel, U., Dannberg, P., Wächter, C., Bräuer, A., Fröhlich, L., Houbertz, R., & Popall, M. (2003). New wafer-scale fabrication method for stacked optical waveguide interconnects and 3D micro-optic structures using photoresponsive (inorganic–organic hybrid) polymers. *Optical Materials, 21*, 475–483.
73. Streppel, U., Dannberg, P., Wächter, C., Bräuer, A., & Kowarschik, R. (2003). Formation of micro-optical structures by self-writing processes in photosensitive polymers. *Applied Optics, 42*, 3570–3579.
74. Järvinen, P., Bonabi, A., Jokinen, V., & Sikanen, T. (2020). Simultaneous culturing of cell monolayers and spheroids on a single microfluidic device for bridging the gap between 2D and 3D cell assays in drug research. *Advanced Functional Materials, 30*, 2000479.
75. Tsougeni, K., Bourkoula, A., Petrou, P., Tserepi, A., Kakabakos, S. E., & Gogolides, E. (2014). Photolithography and plasma processing of polymeric lab on chip for wetting and fouling control and cell patterning. *Microelectronic Engineering, 124*, 47–52.
76. Klukowska, A., Vogler, M., Kolander, A., Reuther, F., Grützner, G., Mühlberger, M., Bergmair, I., & Schöftner, R. (2008). Alternative approach to transparent stamps for UV-based nanoimprint lithography—Techniques and materials. *Proceedings of SPIE, 6792*, 67920J.

77. Schift, H., Spreu, C., Saidani, M., Bednarzik, M., Gobrecht, J., Klukowska, A., Reuther, F., Grützner, G., & Solak, H. (2009). Transparent hybrid polymer stamp copies with sub-50-nm resolution for thermal and UV-nanoimprint lithography. *Journal of Vacuum Science and Technology B, 27*, 2846–2849.
78. Schleunitz, A., Spreu, C., Mäkelä, T., Haatainen, T., Klukowska, A., & Schift, H. (2011). Hybrid working stamps for high speed roll-to-roll nanoreplication with molded sol–gel relief on a metal backbone. *Microelectronic Engineering, 88*, 2113–2116.
79. Schleunitz, A., Vogler, M., Fernandez-Cuesta, I., Schift, H., & Grützner, G. (2013). Innovative and tailor-made resist and working stamp materials for advancing NIL-based production technology. *Journal of Photopolymer Science and Technology, 26*, 119–124.
80. Kulmala, T. S., Rawlings, C. D., Spieser, M., Glinsner, T., Schleunitz, A., Bullerjahn, F., & Holzner, F. (2018). Single-nanometer accurate 3D nanoimprint lithography with master templates fabricated by NanoFrazor lithography. *Proceedings of SPIE, 10584*, 1058412.
81. Stevens, R., & Miyashita, T. (2010). Review of standards for microlenses and microlens arrays. *The Imaging Science Journal, 58*, 202–212.
82. Dannberg, P., Mann, G., Wagner, L., & Bräuer, A. (2000). Polymer UV-molding for micro-optical systems and O/E-integration. *Proceedings of SPIE, 4179*, 137–145.
83. Dannberg, P., Wippermann, F., Brückner, A., Matthes, A., Schreiber, P., & Bräuer, A. (2014). Wafer-level hybrid integration of complex micro-optical modules. *Micromachines, 5*, 325–340.
84. Gimkiewicz, C., Moser, M., Obi, S., Urban, C., Pedersen, J. S., Thiele, H., Zschokke, C., & Gale, M. T. (2004). Waferscale replication and testing of micro-optical components for VCSELs. *Proceedings of SPIE, 5453*, 13–26.
85. Jucius, D., Lazauskas, A., Grigaliūnas, V., Abakevičiene, B., Smetona, S., & Tamulevičius, S. (2018). UV-NIL replication of microlens arrays on flexible fluoropolymer substrates. *Microsystem Technologies, 24*, 1115–1125.
86. Senn, T., Kutz, O., Weniger, C., Li, J., Schoengen, M., Löchel, H., Wolf, J., Göttert, P., & Löchel, B. (2011). Integration of moth-eye structures into a poly(dimethylsiloxane) stamp for the replication of functionalized microlenses using UV-nanoimprint lithography. *Journal of Vacuum Science and Technology B, 29*, 061601.
87. Xie, S., Wan, X., Yang, B., Zhang, W., Wei, X., & Zhuang, S. (2019). Design and fabrication of wafer-level microlens array with moth-eye antireflective nanostructures. *Nanomaterials, 9*, 747.
88. Moharana, A. R., Außerhuber, H. M., Mitteramskogler, T., Haslinger, M. J., & Mühlberger, M. M. (2020). Multilayer nanoimprinting to create hierarchical stamp masters for nanoimprinting of optical micro- and nanostructures. *Coatings, 10*, 301.
89. Dunkel, J., Wippermann, F., Brückner, A., Reimann, A., & Bräuer, A. (2013). Fabrication of an array-like freeform molding tool for UV-replication using a step and repeat process. *Proceedings of SPIE, 8763*, 876330.
90. Brückner, A., Oberdörster, A., Dunkel, J., Reimann, A., Müller, M., & Wippermann, F. (2014). Ultra-thin wafer-level camera with 720p resolution using micro-optics. *Proceedings of SPIE, 9193*, 91930W.
91. Kim, W.-S., Lee, J.-H., Shin, S.-Y., Bae, B.-S., & Kim, Y.-C. (2004). Fabrication of ridge waveguides by UV embossing and stamping of sol-gel hybrid materials. *IEEE Photonics Technology Letters, 16*, 1888–1890.
92. Boersma, A., Wiegersma, S., Offrein, B. J., Duis, J., Delis, J., Ortsiefer, M., van Steenberge, G., Karpinen, M., van Blaaderen, A., & Corbett, B. (2013). Polymer-based optical interconnects using nanoimprint lithography. *Proceedings of SPIE, 8630*, 86300Y.
93. Hiltunen, M., Hiltunen, J., Stenberg, P., Aikio, S., Kurki, L., Vahimaa, P., & Karioja, P. (2014). Polymeric slot waveguide interferometer for sensor applications. *Optics Express, 22*, 7229–7237.
94. Hermannsson, P. G., Sørensen, K. T., Vannahme, C., Smith, C. L. C., Klein, J. J., Russew, M.-M., Grützner, G., & Kristensen, A. (2015). All-polymer photonic crystal slab sensor. *Optics Express, 23*, 16529–16539.

95. Khan, M. U., Justice, J., Petäjä, J., Korhonen, T., Boersma, A., Wiegersma, S., Karppinen, M., & Corbett, B. (2015). Multi-level single mode 2D polymer waveguide optical interconnects using nano-imprint lithography. *Optics Express, 23*, 14630–14639.
96. Zhou, C., Hedayati, M. K., & Kristensen, A. (2018). Multifunctional waveguide interferometer sensor: Simultaneous detection of refraction and absorption with size-exclusion function. *Optics Express, 26*, 24372–24383.
97. Morarescu, R., Pal, P. K., Beneitez, N. T., Missinne, J., Steenberge, G. V., Bienstman, P., & Morthier, G. (2016). Fabrication and characterization of high-optical-quality-factor hybrid polymer microring resonators operating at very near infrared wavelengths. *IEEE Photonics Journal, 8*, 6600409.
98. Cadarso, V. J., Kiefer, T., Auzelyte, V., Atasoy, H., Grützner, G., & Brugger, J. (2014). Direct imprinting of organic–inorganic hybrid materials into high aspect ratio sub-100 nm structures. *Microsystem Technologies, 20*, 1961–1966.
99. Zhu, X., Vannahme, C., Højlund-Nielsen, E., Mortensen, N. A., & Kristensen, A. (2015). Plasmonic colour laser printing. *Nature Nanotechnology, 11*, 325.
100. Calafiore, G., Koshelev, A., Allen, F. I., Dhuey, S., Sassolini, S., Wong, E., Lum, P., Munechika, K., & Cabrini, S. (2016). Nanoimprint of a 3D structure on an optical fiber for light wavefront manipulation. *Nanotechnology, 27*, 375301.
101. Koshelev, A., Calafiore, G., Piña-Hernandez, C., Allen, F. I., Dhuey, S., Sassolini, S., Wong, E., Lum, P., Munechika, K., & Cabrini, S. (2016). High refractive index Fresnel lens on a fiber fabricated by nanoimprint lithography for immersion applications. *Optics Letters, 41*, 3423–3426.
102. Calafiore, G., Koshelev, A., Darlington, T. P., Borys, N. J., Melli, M., Polyakov, A., Cantarella, G., Allen, F. I., Lum, P., Wong, E., Sassolini, S., Weber-Bargioni, A., Schuck, P. J., Cabrini, S., & Munechika, K. (2017). Campanile near-field probes fabricated by nanoimprint lithography on the facet of an optical fiber. *Scientific Reports, 7*, 1651.
103. Lorang, D. J., Tanaka, D., Spadaccini, C. M., Rose, K. A., Cherepy, N. J., & Lewis, J. A. (2011). Photocurable liquid core–fugitive shell printing of optical waveguides. *Advanced Materials, 23*, 5055–5058.
104. Neumeister, A., Himmelhuber, R., Materlik, C., Temme, T., Pape, F., Gatzen, H.-H., & Ostendorf, A. (2008). Properties of three-dimensional precision objects fabricated by using laser based micro stereo lithography. *Journal of Laser Micro/Nanoengineering, 3*, 67–72.
105. Stampfl, J., Baudis, S., Heller, C., Liska, R., Neumeister, A., Kling, R., Ostendorf, A., & Spitzbart, M. (2008). Photopolymers with tunable mechanical properties processed by laser-based high-resolution stereolithography. *Journal of Micromechanics and Microengineering, 18*, 125014.
106. Overmeyer, L., Neumeister, A., & Kling, R. (2011). Direct precision manufacturing of three-dimensional components using organically modified ceramics. *CIRP Annals, 60*, 267–270.
107. Kim, J. Y., Brauer, N. B., Fakhfouri, V., Boiko, D. L., Charbon, E., Grützner, G., & Brugger, J. (2011). Hybrid polymer microlens arrays with high numerical apertures fabricated using simple ink-jet printing technique. *Optical Materials Express, 1*, 259–269.
108. Voigt, A., Ostrzinski, U., Pfeiffer, K., Kim, J. Y., Fakhfouri, V., Brugger, J., & Grützner, G. (2011). New inks for the direct drop-on-demand fabrication of polymer lenses. *Microelectronic Engineering, 88*, 2174–2179.
109. Kim, J. Y., Pfeiffer, K., Voigt, A., Grützner, G., & Brugger, J. (2012). Directly fabricated multi-scale microlens arrays on a hydrophobic flat surface by a simple ink-jet printing technique. *Journal of Materials Chemistry, 22*, 3053–3058.
110. Grützner, G., Fink, M., Pfeiffer, K., Brugger, J., Fakhfouri, V., & Kim, J. Y. (2016). *Micro optical articles, process for their production and uses*. Patent EP2159040B1.
111. Jacot-Descombes, L., Gullo, M. R., Cadarso, V. J., & Brugger, J. (2012). Fabrication of epoxy spherical microstructures by controlled drop-on-demand inkjet printing. *Journal of Micromechanics and Microengineering, 22*, 074012.

112. Kim, J. Y., Martin-Olmos, C., Baek, N. S., & Brugger, J. (2013). Simple and easily controllable parabolic-shaped microlenses printed on polymeric mesas. *Journal of Materials Chemistry C, 1*, 2152–2157.
113. Jacot-Descombes, L., Cadarso, V. J., Schleunitz, A., Grützner, S., Klein, J. J., Brugger, J., Schift, H., & Grützner, G. (2015). Organic-inorganic-hybrid-polymer microlens arrays with tailored optical characteristics and multi-focal properties. *Optics Express, 23*, 25365–25376.
114. Wolf, J., Grützner, S., Ferstl, M., Lawal, J., Schleunitz, A., & Grützner, G. (2018). Fabrication of polymeric micro-optical components with integrated nano-topography for advanced photonic applications. In *Proceedings of the 7th GMM-Workshop: Mikro-Nano-Integration* (pp. 74–77), Dortmund: VDE Verlag.
115. Wolf, J., Grützner, S., Ferstl, M., Schleunitz, A., & Grützner, G. (2019). Assessment of additive manufacturing processes for monolithic diffractive-refractive micro-components. In *Proc. IEEE 20th International Conference on Solid-State Sensors, Actuators and Microsystems Eurosensors XXXIII* (pp. 466–469). Berlin.
116. Beckert, E., Kemper, F., Schreiber, P., Reif, M., Dannberg, P., & Sauva, S. (2019). Inkjet printing of microlens arrays on large, lithographic structured substrates. *Proceedings of SPIE, 10930*, 109300C.
117. Wachholz, P., Wolf, J., Marx, S., Weber, D., Klein, J., & Schröder, H. (2020). Novel technology for dispensing liquid polymers of a wide viscosity range on a picoliter scale for photonic applications. *Proceedings of SPIE, 11286*, 112860L.
118. Ovsianikov, A., Passinger, S., Houbertz, R., & Chichkov, B. N. (2007). Three dimensional material processing with femtosecond lasers. In C. Phipps (Ed.), *Laser ablation and its applications* (pp. 121–157). Boston, MA: Springer US.
119. Selimis, A., Mironov, V., & Farsari, M. (2015). Direct laser writing: Principles and materials for scaffold 3D printing. *Microelectronic Engineering, 132*, 83–89.
120. Multiphoton Optics GmbH. Retrieved from https://multiphoton.net/
121. Nanoscribe GmbH. Retrieved from https://www.nanoscribe.com/
122. Microlight3D. Retrieved from http://www.microlight.fr
123. Femtika. Retrieved from https://www.femtika.lt/
124. Harnisch, E., Russew, M., Klein, J., König, N., Crailsheim, H., & Schmitt, R. (2015). Optimization of hybrid polymer materials for 2PP and fabrication of individually designed hybrid microoptical elements thereof. *Optical Materials Express, 5*, 456–461.
125. Burmeister, F., Steenhusen, S., Houbertz, R., Zeitner, U. D., Nolte, S., & Tünnermann, A. (2012). Materials and technologies for fabrication of three-dimensional microstructures with sub-100 nm feature sizes by two-photon polymerization. *Journal of Laser Applications, 24*, 042014.
126. Ovsianikov, A., Shizhou, X., Farsari, M., Vamvakaki, M., Fotakis, C., & Chichkov, B. N. (2009). Shrinkage of microstructures produced by two-photon polymerization of Zr-based hybrid photosensitive materials. *Optics Express, 17*, 2143–2148.
127. Žukauskas, A., Batavičiūtė, G., Ščiuka, M., Jukna, T., Melninkaitis, A., & Malinauskas, M. (2014). Characterization of photopolymers used in laser 3D micro/nanolithography by means of laser-induced damage threshold (LIDT). *Optical Materials Express, 4*, 1601–1616.
128. Jonušauskas, L., Gailevičius, D., Mikoliūnaitė, L., Sakalauskas, D., Šakirzanovas, S., Juodkazis, S., & Malinauskas, M. (2017). Optically clear and resilient free-form μ-optics 3D-printed via ultrafast laser lithography. *Materials, 10*, 12.
129. Harnisch, E. M., & Schmitt, R. (2017). Two-photon polymerization as a structuring technology in production: Future or fiction? *Proceedings of SPIE, 10115*, 101150Q.
130. Steenhusen, S., Stichel, T., Houbertz, R., & Sextl, G. (2010). Multi-photon polymerization of inorganic-organic hybrid polymers using visible or IR ultrafast laser pulses for optical or optoelectronic devices. *Proceedings of SPIE, 7591*, 759114.
131. Perevoznik, D., Perevoznik, D., Nazir, R., Nazir, R., Kiyan, R., Kiyan, R., Kiyan, R., Kurselis, K., Koszarna, B., Gryko, D. T., Gryko, D. T., Chichkov, B. N., Chichkov, B. N., & Chichkov, B. N. (2019). High-speed two-photon polymerization 3D printing with a microchip laser at its fundamental wavelength. *Optics Express, 27*, 25119–25125.

132. Zhou, R., Malval, J.-P., Jin, M., Spangenberg, A., Pan, H., Wan, D., Morlet-Savary, F., & Knopf, S. (2019). A two-photon active chevron-shaped type I photoinitiator designed for 3D stereo-lithography. *Chemical Communications, 55*, 6233–6236.
133. Gan, Z., Cao, Y., Evans, R. A., & Gu, M. (2013). Three-dimensional deep sub-diffraction optical beam lithography with 9 nm feature size. *Nature Communications, 4*, 2061.
134. Barner-Kowollik, C., Bastmeyer, M., Blasco, E., Delaittre, G., Müller, P., Richter, B., & Wegener, M. (2017). 3D laser micro- and nanoprinting: Challenges for chemistry. *Angewandte Chemie, International Edition, 56*, 15828–15845.
135. Wollhofen, R., Buchegger, B., Eder, C., Jacak, J., Kreutzer, J., & Klar, T. A. (2017). Functional photoresists for sub-diffraction stimulated emission depletion lithography. *Optical Materials Express, 7*, 2538–2559.
136. Selimis, A., & Farsari, M. (2016). Hybrid materials for multiphoton polymerization. In J. Stampfl, R. Liska & A. Ovsianikov (Eds.), *Multiphoton lithography* (pp. 167–181). Weinheim:Wiley-VCH.
137. Malinauskas, M., Žukauskas, A., Purlys, V., Belazaras, K., Momot, A., Paipulas, D., Gadonas, R., Piskarskas, A., Gilbergs, H., Gaidukevičiūtė, A., Sakellari, I., Farsari, M., & Juodkazis, S. (2010). Femtosecond laser polymerization of hybrid/integrated micro-optical elements and their characterization. *Journal of Optics, 12*, 124010.
138. Žukauskas, A., Malinauskas, M., Reinhardt, C., Chichkov, B. N., & Gadonas, R. (2012). Closely packed hexagonal conical microlens array fabricated by direct laser photopolymerization. *Applied Optics, 51*, 4995–5003.
139. Sanchez-Padilla, B., Žukauskas, A., Aleksanyan, A., Balčytis, A., Malinauskas, M., Juodkazis, S., & Brasselet, E. (2016). Wrinkled axicons: Shaping light from cusps. *Optics Express, 24*, 24075–24082.
140. Steenhusen, S., Burmeister, F., Eckstein, H.-C., & Houbertz, R. (2015). Two-photon polymerization of hybrid polymers for applications in micro-optics. *Proceedings of SPIE, 9353*, 93530K.
141. Gissibl, T., Thiele, S., Herkommer, A., & Giessen, H. (2016). Sub-micrometer accurate free-form optics by three-dimensional printing on single-mode fibres. *Nature Communications, 7*, 11763.
142. Steenhusen, S., Hasselmann, S., & Domann, G. (2017). Strategies for rapid and reliable fabrication of microoptical structures using two-photon polymerization. *Proceedings of SPIE, 10115*, 101150S.
143. Petrov, A. K., Bessonov, V. O., Abrashitova, K. A., Kokareva, N. G., Safronov, K. R., Barannikov, A. A., Ershov, P. A., Klimova, N. B., Lyatun, I. I., Yunkin, V. A., Polikarpov, M., Snigireva, I., Fedyanin, A. A., & Snigirev, A. (2017). Polymer X-ray refractive nano-lenses fabricated by additive technology. *Optics Express, 25*, 14173–14181.
144. Johlin, E., Mann, S. A., Kasture, S., Koenderink, A. F., & Garnett, E. C. (2018). Broadband highly directive 3D nanophotonic lenses. *Nature Communications, 9*, 4742.
145. Stender, B., Hilbert, F., Dupuis, Y., Krupp, A., Mantei, W., & Houbertz, R. (2019). Manufacturing strategies for scalable high-precision 3D printing of structures from the micro to the macro range. *Advanced Optical Technologies, 8*, 225–231.
146. Brasselet, E., Malinauskas, M., Žukauskas, A., & Juodkazis, S. (2010). Photopolymerized microscopic vortex beam generators: Precise delivery of optical orbital angular momentum. *Applied Physics Letters, 97*, 211108.
147. Žukauskas, A., Malinauskas, M., & Brasselet, E. (2013). Monolithic generators of pseudo-nondiffracting optical vortex beams at the microscale. *Applied Physics Letters, 103*, 181122.
148. Woggon, T., Kleiner, T., Punke, M., & Lemmer, U. (2009). Nanostructuring of organic-inorganic hybrid materials for distributed feedback laser resonators by two-photon polymerization. *Optics Express, 17*, 2500–2507.
149. Jia, B., Serbin, J., Kim, H., Lee, B., Li, J., & Gu, M. (2007). Use of two-photon polymerization for continuous gray-level encoding of diffractive optical elements. *Applied Physics Letters, 90*, 073503.

150. Houbertz, R., Wolter, H., Schmidt, V., Kuna, L., Satzinger, V., Wüchter, C., & Langer, G. (2007). Optical waveguides embedded in PCBs—A real world application of 3D structures written by TPA. *MRS Online Proceeding Library Archive, 1054*, 1054-FF01-04.
151. Grossmann, T., Schleede, S., Hauser, M., Beck, T., Thiel, M., von Freymann, G., Mappes, T., & Kalt, H. (2011). Direct laser writing for active and passive high-Q polymer microdisks on silicon. *Optics Express, 19*, 11451–11456.
152. Siegle, T., Schierle, S., Kraemmer, S., Richter, B., Wondimu, S. F., Schuch, P., Koos, C., & Kalt, H. (2017). Photonic molecules with a tunable inter-cavity gap. *Light: Science & Applications, 6*, e16224.
153. Serbin, J., Egbert, A., Ostendorf, A., Chichkov, B. N., Houbertz, R., Domann, G., Schulz, J., Cronauer, C., Fröhlich, L., & Popall, M. (2003). Femtosecond laser-induced two-photon polymerization of inorganic–organic hybrid materials for applications in photonics. *Optics Letters, 28*, 301–303.
154. Li, J., Jia, B., & Gu, M. (2008). Engineering stop gaps of inorganic-organic polymeric 3D woodpile photonic crystals with post-thermal treatment. *Optics Express, 16*, 20073–20080.
155. Sun, Q., Juodkazis, S., Murazawa, N., Mizeikis, V., & Misawa, H. (2010). Femtosecond laser photopolymerization of photonic and free-movable microstructures in sol-gel hybrid resist. *Proceedings of SPIE, 7591*, 75910K.
156. Sakellari, I., Yin, X., Nesterov, M. L., Terzaki, K., Xomalis, A., & Farsari, M. (2017). 3D chiral plasmonic metamaterials fabricated by direct laser writing: The twisted omega particle. *Advanced Optical Materials, 5*, 1700200.
157. Gissibl, T., Thiele, S., Herkommer, A., & Giessen, H. (2016). Two-photon direct laser writing of ultracompact multi-lens objectives. *Nature Photonics, 10*, 554–560.
158. Thiele, S., Arzenbacher, K., Gissibl, T., Giessen, H., & Herkommer, A. M. (2017). 3D-printed eagle eye: Compound microlens system for foveated imaging. *Science Advances, 3*, e1602655.
159. Gailevičius, D., Padolskytė, V., Mikoliūnaitė, L., Šakirzanovas, S., Juodkazis, S., & Malinauskas, M. (2019). Additive manufacturing of 3D glass-ceramics down to nanoscale resolution. *Nanoscale Horizons, 4*, 647–651.
160. Žukauskas, A., Matulaitienė, I., Paipulas, D., Niaura, G., Malinauskas, M., & Gadonas, R. (2015). Tuning the refractive index in 3D direct laser writing lithography: Towards GRIN microoptics. *Laser & Photonics Reviews, 9*, 706–712.
161. Houbertz, R., Satzinger, V., Schmid, V., Leeb, W., & Langer, G. (2008). Optoelectronic printed circuit board: 3D structures written by two-photon absorption. *Proceedings of SPIE, 7053*, 70530B.
162. Eckel, Z. C., Zhou, C., Martin, J. H., Jacobsen, A. J., Carter, W. B., & Schaedler, T. (2016). Additive manufacturing of polymer-derived ceramics. *Science, 351*, 58–62.
163. Zanchetta, E., Cattaldo, M., Franchin, G., Schwentenwein, M., Homa, J., Brusatin, G., & Colombo, P. (2016). Stereolithography of SiOC ceramic microcomponents. *Advanced Materials, 28*, 370–376.
164. Marino, A., Ciofani, G., Filippeschi, C., Pellegrino, M., Pellegrini, M., Orsini, P., Pasqualetti, M., Mattoli, V., & Mazzolai, B. (2013). Two-photon polymerization of sub-micrometric patterned surfaces: "Investigation of cell-substrate interactions and improved differentiation of neuron-like cells". *ACS Applied Materials & Interfaces, 5*, 13012–13021.
165. Marino, A., Filippeschi, C., Genchi, G. G., Mattoli, V., Mazzolai, B., & Ciofani, G. (2014). The Osteoprint: A bioinspired two-photon polymerized 3-D structure for the enhancement of bone-like cell differentiation. *Acta Biomaterialia, 10*, 4304–4313.
166. Turunen, S., Joki, T., Hiltunen, M. L., Ihalainen, T. O., Narkilahti, S., & Kellomäki, M. (2017). Direct laser writing of tubular microtowers for 3D culture of human pluripotent stem cell-derived neuronal cells. *ACS Applied Materials & Interfaces, 9*, 25717–25730.
167. Buchroithner, B., Hartmann, D., Mayr, S., Oh, Y. J., Sivun, D., Karner, A., Buchegger, B., Griesser, T., Hinterdorfer, P., Klar, T. A., & Jacak, J. (2020). 3D multiphoton lithography using bio-compatible polymers with specific mechanical properties. *Nanoscale Advances, 2*, 2422–2428.

168. Klein, F., Richter, B., Striebel, T., Franz, C. M., von Freymann, G., Wegener, M., & Bastmeyer, M. (2011). Two-component polymer scaffolds for controlled three-dimensional cell culture. *Advanced Materials, 23*, 1341–1345.
169. Rekštyte, S., Kaziulionyte, E., Balciunas, E., Kaškelyte, D., & Malinauskas, M. (2014). Direct laser fabrication of composite material 3D microstructured scaffolds. *Journal of Laser Micro/Nanoengineering, 9*, 25–30.
170. Qiu, F., Zhang, L., Peyer, K. E., Casarosa, M., Franco-Obregón, A., Choi, H., & Nelson, B. J. (2013). Noncytotoxic artificial bacterial flagella fabricated from biocompatible ORMOCOMP and iron coating. *Journal of Materials Chemistry B, 2*, 357–362.
171. Gao, Z., Li, H., Chen, X., & Zhang, H. P. (2015). Using confined bacteria as building blocks to generate fluid flow. *Lab on a Chip, 15*, 4555–4562.
172. Bertin, N., Spelman, T. A., Combriat, T., Hue, H., Stéphan, O., Lauga, E., & Marmottant, P. (2017). Bubble-based acoustic micropropulsors: Active surfaces and mixers. *Lab on a Chip, 17*, 1515–1528.
173. Plamadeala, C., Gosain, S. R., Hischen, F., Buchroithner, B., Puthukodan, S., Jacak, J., Bocchino, A., Whelan, D., O'Mahony, C., Baumgartner, W., & Heitz, J. (2020). Bio-inspired microneedle design for efficient drug/vaccine coating. *Biomedical Microdevices, 22*, 8.
174. Stocker, M. P., Li, L., Gattass, R. R., & Fourkas, J. T. (2011). Multiphoton photoresists giving nanoscale resolution that is inversely dependent on exposure time. *Nature Chemistry, 3*, 223–227.
175. Tumbleston, J. R., Shirvanyants, D., Ermoshkin, N., Janusziewicz, R., Johnson, A. R., Kelly, D., Chen, K., Pinschmidt, R., Rolland, J. P., Ermoshkin, A., Samulski, E. T., & DeSimone, J. M. (2015). Continuous liquid interface production of 3D objects. *Science, 347*, 1349–1352.
176. de Beer, M. P., van der Laan, H. L., Cole, M. A., Whelan, R. J., Burns, M. A., & Scott, T. F. (2019). Rapid, continuous additive manufacturing by volumetric polymerization inhibition patterning. *Science Advances, 5*, eaau8723.
177. Dolinski, N. D., Page, Z. A., Callaway, E. B., Eisenreich, F., Garcia, R. V., Chavez, R., Bothman, D. P., Hecht, S., Zok, F. W., & Hawker, C. J. (2018). Solution mask liquid lithography (SMaLL) for one-step, multimaterial 3D printing. *Advanced Materials, 30*, 1800364.
178. Gernhardt, M., Blasco, E., Hippler, M., Blinco, J., Bastmeyer, M., Wegener, M., Frisch, H., & Barner-Kowollik, C. (2019). Tailoring the mechanical properties of 3D microstructures using visible light post-manufacturing. *Advanced Materials, 31*, 1901269.
179. del Barrio, J., & Sánchez-Somolinos, C. (2019). Light to shape the future: From photolithography to 4D printing. *Advanced Optical Materials, 7*, 1900598.
180. Hippler, M., Blasco, E., Qu, J., Tanaka, M., Barner-Kowollik, C., Wegener, M., & Bastmeyer, M. (2019). Controlling the shape of 3D microstructures by temperature and light. *Nature Communications, 10*, 232.